PROCESSING, FABRICATION, AND MANUFACTURING OF COMPOSITE MATERIALS
— 1992 —

presented at
THE WINTER ANNUAL MEETING OF
THE AMERICAN SOCIETY OF MECHANICAL ENGINEERS
ANAHEIM, CALIFORNIA
NOVEMBER 8–13, 1992

sponsored by
THE MATERIALS DIVISION, ASME

edited by
T. S. SRIVATSAN
UNIVERSITY OF AKRON

E. J. LAVERNIA
UNIVERSITY OF CALIFORNIA, IRVINE

THE AMERICAN SOCIETY OF MECHANICAL ENGINEERS
345 East 47th Street • United Engineering Center • New York, N.Y. 10017

Cover: Scanning Electron Micrograph of Discontinuously Reinforced Al-4Si/TiB_2 Metal Matrix Composite Processed by Spray Atomization and Co-Deposition

Statement from By-Laws: The Society shall not be responsible for statements or opinions advanced in papers . . . or printed in its publications (7.1.3)

ISBN No. 0-7918-1089-5

Library of Congress
Catalog Number 92-56544

Copyright © 1992 by
THE AMERICAN SOCIETY OF MECHANICAL ENGINEERS
All Rights Reserved
Printed in U.S.A.

FOREWORD

Interest in hybrid materials or composites as engineering materials is rapidly increasing. As current materials reach their functional limits, designers are examining several of these hybrid materials, ranging from metal-matrix composites to ceramic-matrix composites and including intermetallic-matrix and polymeric-matrix composites, as attractive alternatives which can provide the additional strength, stiffness and high temperature capabilities required for advanced engineering applications. Whereas, few commercial products are currently manufactured, in entirety, using composite materials, the trend is bound to rapidly change as more technologies mature and emerge.

Over the years, basic and applied research and development efforts in composite materials technology has increased along with the growing interest in using these advanced materials in the transportation industry where there exists a critical need for weight-critical and stiffness-critical structures. This need has resulted in the development and emergence of novel and innovative methods of processing these materials, starting from the raw product. In order to enhance the use of composite materials in several structural and non-structural applications in the commercial industry, new and innovative methods of manufacturing the final product have to be developed, tested and implemented.

This volume includes the papers presented at the symposium, the second in the series, entitled: "Processing and Manufacturing of Composite Materials II" held in November 1992 during the Winter Annual Meeting of The American Society of Mechanical Engineers, in Anaheim, California. The four session symposium was sponsored by the Materials Division of The American Society of Mechanical Engineers (ASME). The organizers of this symposium are also the editors of this symposium proceedings.

The symposium brought together researchers from academia, industry and research laboratories. Several of the authors are acknowledged experts in their respective fields and have made notable contributions to the techniques they describe. We wish to express our sincere thanks and appreciation to the authors for their participation in the symposium and for the cooperation extended in the difficult task of preparing the manuscripts. Each manuscript was reviewed by a technical reviewer and a technical editor. Mandatory and optional revisions were returned to the authors for incorporation in their manuscripts. We gratefully acknowledge the contribution of the authors, their institutes and funding agencies to the success of this symposium.

Our thanks to the reviewers for their diligence and promptness. The executive committee of the Materials Division receives our thanks for providing the opportunity and encouragement to organize this symposium. A special "Thank You" is extended to Dr. Vijay K. Stokes of the Materials Division for his unstinted support, patience, timely, valuable and meticulous advice which helped in a smooth transition from conception to completion of the project. We express profound thanks and appreciation to Ms. Barbara Signorelli and her crew for their assistance in the timely completion of this volume.

We hope this symposium volume will be of interest to several different types of readers: engineers, scientists, researchers in academia and industry, potential users of composite materials, and most importantly to students of all ages in manufacturing, mechanical engineering and materials engineering.

T. S. Srivatsan
University of Akron

E. J. Lavernia
University of California, Irvine

CONTENTS

POLYMER-MATRIX COMPOSITES

State of the Art of the Formation of High Performance Fibers for Composites
  A. S. Abhiraman, P. Desai, and B. Wade 1

An Experimental Investigation Into Pitting of Hole Surfaces When Drilling Graphite/Epoxy Materials
  K. Colligan and M. Ramulu 11

Axial Response of a Multilayer Strand Composed of Viscoelastic Filaments
  T. A. Conway and G. A. Costello 27

Delamination-Free Drilling of Composite Laminates
  Sanjeev Jain and Daniel C. H. Yang 45

Elastic Properties of Two-Dimensional Composites Containing Polygonal Holes
  I. Jasiuk, J. Chen, and M. F. Thorpe 61

Fibre Diameter/Mechanical Behavior Correlation in Carbon Fibre Processing
  S. Ozbek and D. H. Isaac 75

Microstructure and Texture Development During Carbon Fibre Processing
  S. Thitipoomdeja, S. Ozbek, and D. H. Isaac 87

Fatigue Life Characterisation of AS4/PEEK Composites
  K. S. Saib, W. J. Evans, and D. H. Isaac 103

Thermoplastic Powder Composite Manufacturing Using a Wet Slurry Method
  Karthik Ramani, Michail Tryfonidis, Chris Hoyle, and John Gentry 115

METAL-MATRIX COMPOSITE MATERIALS

Ductile Fracture Process of Discontinuous Fiber Reinforced Composites
  S. B. Biner 131

Influence of Spray Atomization and Deposition on the Microstructure and Mechanical Behavior of Aluminum-Copper Based Metal Matrix Composites
  Manoj Gupta, T. S. Srivatsan, Farghalli A. Mohamed, and Enrique J. Lavernia 149

A Comparison of Creep Behavior of 6061 Al and SiC-6061 Al
  Kyung-Tae Park, Enrique J. Lavernia, and Farghalli A. Mohamed 177

Evaluation of Analytical and Numerical Models for the Elastic-Plastic Response of Particulate Composites
  David G. Taggart, Jialiang Qin, and Mark D. Adley 187

Elevated Temperature Behavior of Fine-Grained $Mg-9Li-B_4C$ Composites
  J. Wolfenstine, G. Gonzalez-Doncel, and O. D. Sherby 201

INTERMETALLIC-MATRIX AND CERAMIC-MATRIX COMPOSITE MATERIALS

Effect of Non-Homogeneous Compaction on Densification of Ceramic Matrix Composites
  L. R. Dharani and Wei Hong 209

High Temperature Rupture Mechanisms in a Particulate Reinforced Intermetallic Matrix Composite
  H. K. Kim, X. Liang, Enrique J. Lavernia, and J. C. Earthman 221

Role of Back Stresses in the Creep of Molydisilicide Composites
  K. Sadananda and C. R. Feng 231

On the Notch-Sensitivity and Toughness of a Ceramic Composite
   Keith T. Kedward and Peter W. R. Beaumont........................... 247

The Influence of Processing on the High Temperature Mechanical Properties of a Whisker-Reinforced Alumina Composite
   Kenong Xia, John R. Porter, and Terence G. Langdon 253

Author Index 265

STATE OF THE ART OF THE FORMATION OF HIGH PERFORMANCE FIBERS FOR COMPOSITES

A. S. Abhiraman
School of Chemical Engineering

P. Desai and B. Wade
School of Textile and Fiber Engineering

Polymer Education and Research Center
Georgia Tech
Atlanta, Georgia

ABSTRACT

A brief analysis of the prospects for advances in the formation of high performance fibers is presented. Synthesis and processing of new materials as well as possible refinements in existing fibers have been considered. The focus is primarily on process-related issues in the formation of organic and inorganic fibers for structural composites. The analysis reveals the need for a major synthesis-led effort that is cognizant of the need for integration of chemical evolution with plausible process routes.

INTRODUCTION

Formation of a broad range of new high performance fibers via integrated efforts in synthesis and processing is among the most significant recent advances in materials science and engineering. Materials with a wide spectrum of physical properties are being developed to meet a variety of highly specialized needs. These materials offer the potential for developing fibrous structures with unique combinations of mechanical properties, such as high stiffness and strength, high temperature performance, and optical and electrical properties. Of particular interest in structural applications are composites of fibers of organic and inorganic polymers possessing various combinations of high specific strength, modulus and work of rupture, and high temperature stability in oxidizing and non-oxidizing environments. We include here a broad range of materials, from relatively flexible composites, such as tires, to rigid composites in low and high temperature applications, such as those in automotive and aircraft structural and engine applications. Fibers of polyethylene terephthalate (PET), Nylon 66, ultrahigh molecular weight polyethylene (UHMWPE), rigid and semi-rigid rod-like polymers such as poly(paraphenylene terephthalamide) and poly(benzobisthiazole), carbon, silicon carbide, silicon nitride, and various mono- and mixed-metal oxides are well known examples of the fibers in such composites.

Extensive research is being carried out in numerous laboratories around the world on synthesis and processing of new fiber forming materials and also on mechanisms by which the structural fibers in current commercial production can be improved. Table I summarizes the mechanical properties of many of the fibers that are currently either in commercial production or at various stages of development. The data in Table I were compiled from the information given in references 1 through 12.

We present in the following a brief summary of the critical process-related issues pertaining to the formation of "new" high performance fibers and the mechanisms through which they are being addressed. We classify "new" fibers broadly for the purposes of this discussion, including the needs and prospects for improvements in process routes and performance of fibers in current commercial production. The discussion is, however, limited to organic and inorganic *polymer-based* fibers. Fibers that are not obtained via polymer-based routes are not included here.

LINEAR ORGANIC POLYMER FIBERS

<u>PET and Nylon 66/6 Fibers:</u>
The most widely used high performance fibers in composites are fibers of semirigid polyesters and polyamides, especially PET and Nylon 66 and 6. A large class of structural composites - reinforced rubber goods (eg., tires) - has derived its critical mechanical performance from these fibers for several decades. However, it would be incorrect to presume these fibers and the processes in which they are made to be "mature," devoid of further potential for significant improvements. Current research in at least two areas holds significant promise in this regard.

(i) <u>Solid-state polymerization (SSP)</u>. The recent success with ultrahigh molecular weight polyethylene fibers in achieving the highest specific tensile properties of any material [13-16] has led to a renewal of studies pertaining to molecular weight-dependent aspects of morphological order and mechanical performance of other polymers. While it has not been possible with other polymers to achieve the degree of enhancement obtained with PE, a significant potential for improvement vis-a-vis the properties of current commercial PET filaments has been demonstrated. An important example in this regard is the work of Ito *et al*. [17] who has shown that the properties of PET can be increased substantially by processing higher molecular weight polymer (I.V=3.6 dL/g), obtained through a sequence of solid state polymerization followed by dissolution and reprecipitation from solution, beginning with a polymer of 0.61 dL/g intrinsic viscosity.

A major drawback in carrying out SSP prior to fiber extrusion is that it often requires solution-based extrusion because melt spinning of high molecular weight polymer often results in considerable degradation and also requires excessively high pressures. Both environmental and cost considerations make the solution-based routes inappropriate for new, large volume, commercial production. It is therefore necessary to explore options that would integrate SSP with fiber production based on melt spinning. A mechanism for accomplishing this has been examined with relative success in the case of Nylon 66, by incorporating SSP between melt spinning and drawing. Substantial increase in molecular weight has been obtained in as-spun fibers, without causing any embrittlement of the fibers. A barrier to following a similar path with PET has been due to the fact that as-spun PET fibers often become brittle upon annealing above T_g. It has been demonstrated, however, that such embrittlement can be eliminated by melt spinning at high speeds, causing oriented crystal nucleation to occur in the spinning process [18].

(ii) High stress drawing. While it has been well recognized and demonstrated that the development of order in relatively flexible polymers can be influenced significantly by anisotropic stress field, most of the fiber formation processes to date have consisted essentially of deformation-dictated steps. Experiments have shown that it is possible to increase significantly the mechanical properties of fibers through deformation (drawing) at higher stress levels, without necessarily causing higher levels of net deformation [18,19]. The novel approach of Cuculo *et al*. [19] to create a high stress-dictated, temperature-controlled deformation in melt spinning of PET is an excellent example in this regard. The melt spinning threadline in their experiments passes through a temperature-controlled liquid bath, the viscosity and the length (path length of the filaments in the bath) of which dictate the drag and, thus, the force exerted on the filaments. A simple experiment that demonstrates such an anisotropic stress-controlled phenomenon has been conducted in our laboratories. Here a tubular heater, maintained at a temperature between T_g and T_m, is moved so as to enclose an as-spun PET filament bundle, while it is subjected to a constant tensile force. When the "drawn" fibers from such experiments are analyzed for mechanical properties, one finds a monotonically increasing influence of the applied stress, but the properties are still well within the range obtained in current commercial fibers. It should be noted here that the maximum applied stress level in these experiments is still much less than the ultimate stress that the as-spun fiber can withstand. Experimental refinements are necessary to obtain the limits of anisotropic stress field that can be applied in such experiments so that the consequent generation of mechanical properties can be maximized. It is also necessary to invent appropriate, commercially feasible schemes for the implementation of the combination of temperature and stress fields inferred from such experiments.

Cellulosic Fibers:
Of foremost interest in fiber forming polymers is cellulose, the most abundant renewable organic material for this purpose. High tenacity rayon fibers have a proven record of high performance as tire cord. Cellulose and its derivatives can form anisotropic solutions, offering the potential to form high performance fibers. Cellulose-based fibers of extraordinarily high tensile properties have been produced using new solvent systems, such as trifluoroacetic (TFA) acid [20,21]. However, current experimental and commercial methods of direct solution spinning or regeneration of cellulose filaments from solutions of its derivatives, viz., the different rayon processes, suffer from significant pollution control and other potential hazards. A renewed effort, integrating the chemistry of cellulose derivatives and solutions with fiber formation, is necessary in order to ensure an abundant supply of low cost, high performance fibers from a renewable resource.

Fibers from Ultrahigh Molecular Weight Polymers:
The manufacture of high tenacity high modulus fibers through gel spinning and drawing of ultrahigh molecular weight polyethylene (UHMWPE) represents an important recent development in the quest for new reinforcing fibers for high performance composites [13-16]. Tensile strength and modulus in excess of 6 GPa and 220 GPa, respectively, have been achieved. Such properties, combined with their low density ($\approx$ 1g/cm^3), make these fibers desirable in composites that require high specific strength, stiffness and toughness, especially for applications at relatively low temperatures. Parallel efforts with other linear organic polymers have, however, failed to yield the dramatic properties obtained with polyethylene. Nevertheless, the mechanical properties that have been claimed with fibers of polymers such as poly(vinyl alcohol) and polyacrylonitrile, processed via the "UHMW $\rightarrow$ gel spinning and drawing" route are much higher than those obtained from normal routes. [22,23]

Fibers of Rigid Polymers:
The success in achieving high mechanical performance in fibers of linear aromatic (rigid) polyamides has been accompanied by a large research effort on two aspects that are considered to have been crucial to them:
   (i) anisotropic, especially nematic, melts/solutions of linear polymers;
   (ii) rigid, unidirectional primary molecular architecture.

The synthetic aspects of many of these homopolymers and copolymers have been widely discussed in the literature (see, for example, reference 24). Promising new developments in this regard pertain mostly to increasing the intrinsic rigidity of the primary polymer structure in order to increase properties such as modulus, improve retention of mechanical properties to at least intermediate temperatures and minimize solvent-induced degradation of mechanical performance. Fibers of polybenzobisthiazole and polybenzobisoxazole serve as good examples in this regard. Large scale commercial development of these and other new organic polymer fibers for composites has been impeded by the absence of one or both of the following two favorable factors:
   - a demonstrated major commercial need for the improved properties offered by the new fibers;
   - a significant performance/cost advantage.

A major deficiency in all the high performance organic polymer fibers is their much lower compressive strength in comparison with their strength and modulus in tension. A primary reason for this is believed to be the absence of strong intermolecular (lateral) forces in the linear polymers from which the fibers are processed. As suggested by DeTeresa *et al.* [25], a likely mechanism to improve this aspect is the introduction of a high density of regular lateral covalent linkages between the molecules. The requirement for shaping the polymer into the filament form limits the introduction of such linkages to the post-extrusion stage. Examples of polymerization in the solid-state to produce ordered polymer arrays have been limited mostly to the formation of polymer single crystals from monomer single crystals [26,27]. Large-scale production of continuous filaments of oriented high molecular weight polymer with a high density of regular lateral crosslinks is most likely to require the formation of oriented, crystallized oligomeric or polymeric filaments via conventional fiber extrusion with subsequent crosslinking in the solid-state. No clear mechanisms by which such ordered primary and secondary structures can be produced have been identified.

CARBON FIBERS

Carbon fibers occupy a premier position among high performance fibers for composites in that they combine high stiffness and strength with high temperature performance, especially in inert environments, and also offer relatively inexpensive routes, in comparison with other high temperature fibers, for large scale production. Two major precursor materials, polyacrylonitrile and pitch, are used in commercial production, with fibers such as rayon being used only as precursors for some specialty carbon fiber structures.

Formation, structure and properties of carbon fibers have been studied extensively [28]. Such efforts have led to a substantial increase in the tensile strength and modulus of these fibers within the last decade. Major contributions in this regard have resulted from realization of the critical role of orientation in dictating the modulus and that of flaws, morphological as well as impurity-generated, in limiting the strength. In spite of these advances, the market for these fibers has fallen short of most projections. The following are among the plausible reasons for this shortfall.

(i) The relatively high cost of carbon fibers, especially those of lower linear density bundles in combination with mechanical properties at the high end of the spectrum, has prevented their use so far in most commodity structural composites.

(ii) Pitch, originally thought to be a low cost, high performance precursor option, has failed to meet the expectations in cost as well as properties, especially in axial compression.

(iii) The substantial improvements achieved in the *tensile* properties of carbon fibers and composites have not been accompanied by any enhancement of performance in *axial compression*. Only some of it can be attributed to the combined influence of the fiber-matrix interface, matrix properties and deficiencies in processing composites.

Much of the current research on the structure and properties of carbon fibers has been concerned with compressive properties that determine the limiting conditions for their use in critical structural applications. Process-related modifications have so far failed to yield positive results. It is now generally accepted that any significant improvement is most likely to be led by synthesis of new precursor materials that can cause increased in lateral -C-C- bonding sequences while maintaining a high axial orientational order in the carbon fibers.

CERAMIC FIBERS FROM POLYMER PRECURSORS

The most dramatic current projections regarding high performance fibers pertain to ceramic fibers for reinforcement of metal and ceramic matrices, with potential maximum use temperatures well in excess of 1000°C. A comprehensive review of the manufacture, structure, properties and uses of currently produced metal oxide (mono- and mixed-) and boron-based and silicon-based monoxide fibers has been published by Cooke [3]. "Synthesis and shaping into filament form of polymeric precursors, followed by pyrolysis under controlled environments", "extrusion of ceramic slurries into filament form with subsequent high temperature sintering", and CVD processes and direct melt spinning are among the major routes to most of the current and projected *continuous* ceramic fibers. Structural fibers in current commercial or advanced developmental stages that are based on conversion of polymeric precursors include mono and mixed oxides of silicon, aluminum, titanium, zirconium and boron (polysol-gel routes), carbide and nitride of silicon, carbides of silicon and titanium, and nitride of boron.

Much of the research on polymer-based ceramic fibers has been focused on synthesis of precursor polymers that would yield, upon controlled pyrolysis, the ceramic structures of interest [29-34]. The *integrated* requirements for synthesis and processing into continuous ceramic fibers have been discussed by Varaprasad *et al.* [35]. The major steps in the successful production of any polymer-based ceramic fiber consist of:

(i) controlled synthesis of precursor polymer with the potential for yielding fluids of spinnable rheological characteristics and also capable of rapid transformation to a gel or solid-state structure that can be subsequently converted cohesively to the desired ceramic structure. The synthesis here need not be limited to materials that can be melt-processed. Solution-based processes have been shown to be entirely appropriate and offer the flexibility of controlling the rheological characteristics of the spinning fluid through the incorporation of compatible linear organic polymers [35];

(ii) identification and development of a fiber extrusion route, engineered for the specific precursor polymer. It is necessary to consider each of the different routes that are available for continuous filament extrusion, viz., melt spinning, solution-wet spinning, solution-dry spinning, solution-dry jet-wet spinning, and phase separation spinning. The major advantage of melt spinning over the other methods is that it is the simplest. However, infusibility of precursor polymer, possible uncontrolled degradation at the melt extrusion temperatures and lack of (or limited) spinnability are often major barriers to following this route. In such cases, one of the other solution-based routes can offer a viable alternative. As mentioned above, solution-based filament extrusion can be aided by the addition of a rheological aid. In addition to providing enhanced spinnability at low concentrations, it is also important that the rheological aid be fugitive during post-spinning thermal conversion of the precursor to the ceramic structure, without causing loss of integrity of the fiber. Addition of a high molecular weight polymer as a rheological aid is not new. Silazane precursors to silicon nitride have been spun with added poly(ethylene oxide) [36]. Poly(vinyl acetate) has been added to a partially hydrolyzed aluminum alkoxide precursor for Al_2O_3 fiber and also to a silicon alkoxide precursor for Al_2O_3-SiO_2 fiber in order to adjust the viscosity of the solutions for fiber drawing [37,38]. Spinnability of polyborates has been enhanced by the addition of a low concentration of high molecular weight poly(ethylene oxide), poly(1-vinyl-2-pyrrolidinone), poly(vinyl acetate), or poly(methyl methacrylate) which are compatible in polyborate compositions [39]. However, a comprehensive analysis that considers compatibility with precursor synthesis as well as subsequent thermochemical conversion shows poly(methyl methacrylate) to be the most appropriate among those considered. The significance of such modifications of spinning fluids should be considered in the context of expanding the base of potential precursor polymers for ceramic fibers;

(iii) thermochemical treatments of the precursor fibers for controlled <u>conversion and consolidation</u> to obtain the desired high performance ceramic fiber. This stage can often involve steps, such as curing of the precursor polymer and drying or degradation of solvent, reaction byproducts and rheological aids at lower temperatures, prior to final conversion to the ceramic at high temperatures. In addition to causing the desired chemical conversion, this stage of the process also determines the phase compositions and bulk density of the final product. This stage is often the most crucial and also least amenable to the establishment of generic guidelines.

Other requirements have to be considered in special cases. For example, attaining the desired mechanical properties in boron nitride fibers requires the formation of a highly oriented structure. The choices of precursor, fiber formation and thermal conversion should consider together the necessary deformations to produce the required anisotropy in the final ceramic fiber.

In spite of the significant growth in research on the chemistry of precursor polymers for ceramic structures, commercial production of ceramic filaments via this route has been limited to a few silicon- and aluminum-based fibers [40,41]. Even these fibers suffer from compositional and morphological (primarily related to bulk density and phase composition) deficiencies. Di Carlo [41] has provided a critical analysis of the sources that suppress performance and the opportunities for improvement in the production of some of these fibers. Major efforts to integrate materials chemistry and process engineering are necessary in order to produce the currently desired polymer-based ceramic structures with the required level of purity and refinements in chemical composition and morphology.

SUMMARY

We have presented here a brief analysis of the prospects for advances in the formation of high performance fibers that can result in new fibers or refinements in existing fibers. While this analysis has focused on process-related issues, it should be noted that significant improvements in cost of production and/or properties, especially those pertaining to ceramic fibers, are most likely to arise from *synthesis-led* efforts that are cognizant of the need for *integration* of chemical evolution with plausible process options. The following aspects appear to demand the most attention in this regard.

(i) Solid-state polymerization as an integral part of fiber formation, i.e., subsequent to the shaping of relatively lower molecular weight polymer into the "fiber" form.

(ii) High stress deformation and crystallization of linear organic polymer fibers.

(iii) Synthesis and formation of fibers of intrinsically rigid polymers with the potential in their primary structure and crystalline order for "ordered" crosslinking in the fiber form.

(iv) New solution-based routes for infusible polymers, such as polyacrylonitrile-based precursors for carbon fibers and cellulose, that incorporate solvent systems offering one or more of the following attributes:
- minimum environmental hazards
- thermally-induced phase transitions that would eliminate the need for high volume coagulating systems
- a broad, controlled range of specialty morphologies.

(v) Identification and formation of high compressive strength fiber morphologies.

(vi) Solution-based processing of organic, organometallic and inorganic polymers with rheological aides to form new precursor fibers for carbon and ceramic structures.

(vii) Refinements in synthesis and processing of ceramic structures to eliminate impurities that limit high temperature mechanical performance.

One should acknowledge pitfalls in any attempt to make rational projections regarding the formation of new high performance fibers. Current knowledge of the links between primary material structure, processing conditions, morphology and properties is mostly empirical. Accidental discoveries during the course of rational or empirical investigations can alter substantially the course of events in this field.

REFERENCES

1. D. E. Beers and J. E. Ramirez, **Journal of the Textile Institute**, 81(4), 1990, p. 561.
2. A. R. Bunsell, **Journal of Applied Polymer Science, Applied Polymer Symposia,** 47, 1991, p. 87.
3. T. F. Cooke, **Journal of American Ceramic Society,** 74(12), 1991, p. 2959.
4. G.W. Davis and J.R. Talbot, in Encyclopedia of Polymer Science and Engineering, ed. J.I. Kroschwitz, John Wiley & Sons, New York, 12, 1990, p.125.
5. A. Dhingra, *ibid.*, 6, 1990, p. 768.
6. L. Rebenfeld, *ibid.*, 6, 1990, p. 696.
7. W.W. Adams, and R.K. Eby, **Materials Research Society Bulletin**, December, 1987, p. 22-26.
8. "Boron Nitride Fibers and Felts," Product Information from Electroceramics, Toronto, Canada.
9. F. K. Ko, and P. Fang, in Enclyclopedia of Polymer Science and Engineering, Index Vol., ed. J. I. Kroschwitz, John Wiley & Sons, New York, 1990, p. 130.
10. J. M.Prandy and H. T. Hahn, **SAMPE Quarterly,** 22(2), 1991, p. 47.
11. H. G. Sowman, in Sol-Gel Technology for Thin Films, Fibers, Preforms, Electronics, and Specialty Shapes, ed. L. C. Klein, Noyes Publications, Park Ridge, NJ, 1988, p. 162.
12. J. K. Weddel, **Journal of the Textile Institute,** 81(4), 1990, p. 333.
13. A.J. Pennings, J. Smooks, J. de Boer, S. Gogolewski, and P.F. van Hutten, **Pure and Applied Chemistry,** 55(5), 1983, p. 777.

14. D.C. Prevorsek, H.B. Chin, Y.D. Kwon, and J. E. Field, **J. Applied Polymer Science, Applied Polymer Symposia,** 47, 1991, p. 45.
15. S. Kavesh and D.C. Prevorsek, U.S. Patent 4,413,110 (1983).
16. V.A. Marichin, L.P. Myasnikova, D. Zenke, R. Hirte, and P. Weigel, **Polymer Bulletin**, 12, 1984, p. 287.
17. M. Ito, Y. Wakayama and T. Kanamoto, Abstracts of the VIII Annual Meeting of the Polymer Processing Society, 1991, p. 159.
18. U.S. Agarwal, P. Asher, W.W. Carr, F. Pinaud, P. Desai and A.S. Abhiraman, in <u>Polymer and Fiber Science - Recent Advances</u>, R.E. Fornes and R.D. Gilbert, Eds., VCH Publishers, N.Y., 1992, p. 43.
19. J.A. Cuculo, P.A. Tucker and G.-Y. Chen, **Journal of Applied Polymer Science, Applied Polymer Symposia,** 47, 1991, p. 227.
20. J.P. O'Brien, U.S. Patent 4,464,323 (1984).
21. J.P. O'Brien, U.S. Patent 4,501,886 (1985).
22. Y.D. Kwon, S. Kavesh and D.C. Prevorsek, U.S. Patent 4,440,711 (1984).
23. Y.D. Kwon, S. Kavesh and D.C. Prevorsek, U.S. Patent 4,883,628 (1989).
24. H.H. Yang, <u>Aromatic High-Strength Fibers</u>, Wiley Interscience, 1989.
25. S.J. DeTeresa, S.R. Allen and R.J. Farris, in <u>Composite Applications</u>, Eds. T. Vigo and B.J. Kinzig, VCH Publishers Inc., 1992, p. 67.
26. G. Wegner, **Pure and Applied Chemistry**, 49, 1977, p. 443.
27. R.J. Young, in <u>High Technology Fibers - Part B</u>, Eds. M. Lewin and J. Preston, Marcel Dekker, New York, 1989, p. 133.
28. S. Damodaran, P. Desai and A.S. Abhiraman, **J. Textile Institute**, 81(4), 1990, p. 384.
29. S. Sakka, **J. Non-Crystalline Solids,** 121, 1990, p.417.
30. S. Sakka and K. Kamiya, **J. Non-Crystalline Solids**, 42, 1980, p. 42.
31. B.J. Zelinski and D.R. Uhlmann, **J. Physical Chemistry of Solids**, 45, 1984, p. 1069.
32. K. Okamura, **Composites**, 18(2), 1987, p. 107.
33. R.W. Rice, **American Ceramic Society Bulletin**, 62(8), 1983, p. 889.
34. K.J. Wynne and R.W. Rice, **Annual Reviews of Materials Science**, 14, 1984, p. 297.
35. D.V. Varaprasad, B. Wade, N. Venkatasubramanian, P. Desai and A.S. Abhiraman, **Indian Journal of Fibre & Textile Research**, 16, 1991, p. 73.
36. G. Winter, W. Verbeek, and M. Mansmann, U.S. Patent 3,892,583 (1975).
37. S. Horikiri, K. Tsuji, Y. Abe, A. Fukui, and E. Ichiki, Japan Patent 49108325 (1974).
38. S. Horikiri, K. Tsuji, Y. Abe, A. Fukui, and E. Ichiki, Japan Patent 5012335 (1975).
39. N. Venkatasubramanian, B. Wade, P. Desai, A.S. Abhiraman, and L.T. Gelbaum, **J. Non-Crystalline Solids**, 130, 1991, p.144.
40. J. Lipowitz, **J. of Inorganic and Organometallic Polymers**, 1(3), 1991, p.277.
41. J.A. DiCarlo, NASA Technical Memorandum #105174 (1991).

Table 1. Mechanical Properties of High Performance Fibers (From refs. 1-12).

Fiber	Tensile Strength GPa	Tensile Modulus GPa	Tensile Strain at Break (%)	Compressive Strength GPa	Max. Use Temp. (°C) Oxidizing	Max. Use Temp. (°C) Non-oxidizing
Flexible Chain Polymer Fibers:						
UHMWPE(S900)	2.5	60	6			
UHMWPE (S1000)	3	170	3.8	0.07		
Nylon 66	1	6	18	0.1		
Polyester (High Tenacity)	.85–1.2	14.2–14.7	10–20			
Viscose Rayon (HWM, High Tenacity)	0.6–1.35	5.5–9.0	6.5–10			
Rigid-Rod Polymer Fibers:						
Kevlar 49	3.6	130	2.6	0.48		
Kevlar 149	3.5	185	2	0.45		
PBZT	4.2	330	1.3	0.4		
PBO	5.8	365	1.6	0.4		
LCP (Vectran HS)	2.85	65	3.3			
Carbon Fibers:						
PAN-based						
	2–7	250–500		1–3	400	2500
Type I: high strength	2.5–3.5	220–250	1.2–1.4			
High Strain Type I	4.0–4.7	240–270	1.7–1.8			
Type II: high modulus	2.2–2.4	340–380	0.6–0.7			
Type III: ultrahigh modulus	1.8–1.9	520–550	0.3–0.4			
Pitch-based						
	1.4–2.3	200–830		0.5–1.2	400	2500
P-75	2.41	520	0.25			
HM	3	520	0.35			
UHM	3.3	660	0.16			
HS	2.7	245	0.98			
P-100	2.2	720	0.3	0.48		
Ceramic Fibers:						
Oxide						
E glass	3.45	72	4.8			600
S2 glass	4.6	87	5.4			760
Quartz	0.9	69	1.3		1060	1060
Nextel (silica-	1.5–1.7	150–220	0.8–1		1300	1200-1425
Alumina (Fiber FP)	1.4	380	0.4			2045
Alumina-Zirconia (PRD-166)	1.9	380	0.5			
Alumina (Sumica)	1.8	210				1250
Alumina (Saffil RF)	2	295			1540	1600
Alumina (Nextel 312)	1.7–2.1	138–152				1200
Alumina (Nextel 440)	2.1	190				1425
Alumina (Nextel 480)	1.7–2.4	207–242				1425
Non-oxide						
SiC (Avco)	2.4	430	0.56			1150
SiC (Nicalon)	2.62	190	1.1			1200
SiTiCO (Tyranno)	2.8	210	1.4			1300
Silicon Nitride (TNSN)	2.5	295				1200
Boron Nitride	0.86	40–70			1000	2500
Boron	3.8	400	1	>5	560	1200

AN EXPERIMENTAL INVESTIGATION INTO PITTING OF HOLE SURFACES WHEN DRILLING GRAPHITE/EPOXY MATERIALS

K. Colligan
Boeing Defense and Space Group
Boeing Company
Seattle, Washington

M. Ramulu
Department of Mechanical Engineering
University of Washington
Seattle, Washington

ABSTRACT

An experimental study of the formation of macroscopic pitting or fiber pullout caused by drilling graphite/epoxy materials is investigated. The effect of several drilling process variables on the formation of pits on the drilled hole surface is presented. Drilling tests were performed to assess the influence of drill feed rate, cutting speed, the application of coolant, drill point initial geometry, and cutting edge wear on the formation of pits. Fiber orientation relative to the path of the cutting edge, drill point geometry, feed rate and cutting edge wear were found to have a significant effect on the formation of pits on the surfaces of drilled holes in graphite/epoxy.

INTRODUCTION

Many aircraft structural components have traditionally been joined using fasteners such as screws or rivets. This is still generally true of modern aircraft which are constructed of composite materials. The assembly process usually involves trial-fitting the components to be joined using an assembly jig, measuring the gaps between the components, constructing custom-made shims, drilling a large number of holes for fasteners, and installing the fasteners. Literally thousands of holes are drilled in load-bearing members for the installation of fasteners. Specific problems encountered in fiber reinforced plastic materials include delamination of exit surface plies, smearing of the resin on the hole surfaces, burning of the resin on the hole surface, fiber breakout, and cracking on the hole surface. Since the drilled holes and their quality often affects the static failure strength of composite materials [1–4], it is important to understand the different aspects of drilled hole quality and to have a drilling process that will consistently produce superior quality holes.

Drilled hole quality, within the context of laminated composite materials, is generally defined by the geometry of the hole and by the microstructure of the hole surface and surrounding surface area [5]. It is in the microstructure of the hole surface and surrounding area that composite materials differ most from homogeneous materials. In fact, the surface quality of holes drilled in fiber reinforced composite materials has been the most significant problem in drilling these materials. Fiber breakout is extremely common and is either detected by measuring the surface finish of a drilled hole or by visual inspection. Fiber breakout has also been described as macroscopic pitting in the literature [5]. This phenomenon is caused primarily by the dependence of the chip formation mechanism on the fiber orientation relative to the path of the cutting edge. Figure 1 shows a cross section of a typical micrograph of a pitted hole surface. Although it may not always be as well defined as in the micrograph, the pits occurring in drilled holes are often arranged in vertical bands.

Past investigations into machining of fibrous composite materials indicate that different chip formation mechanisms occur depending on the orientation of the fibers of the composite relative to the path of the cutting edge [5-9]. Since laminated composite materials are most often constructed of plies with alternating fiber orientations, the chip formation mechanisms that are active will generally differ from one ply to another which can result in uneven machined edge surfaces and poor surface finishes [10,11]. In the drilling process, the situation is slightly different in that the fiber orientation relative to the path of the cutting edge varies continuously around the circumference of the hole, as well as varying from one ply to another.

Although some of the causes of the pitting of hole surfaces in fiber reinforced composite materials are documented in the literature [12-14], there is very limited information on how to suppress the formation of drilled hole surface pitting in graphite/epoxy. Therefore an experimental investigation was undertaken to study the pit formation in graphite/epoxy material by drilling process. The variables studied included the application of coolant, cutting speed, feed rate, cutting edge initial geometry and cutting edge wear. This paper presents the preliminary results obtained in investigating the drilling process influence on the formation of pitting in a drilled hole.

EXPERIMENTAL PROCEDURE

Test Materials

Test panels constructed from graphite/epoxy unidirectional tape and fabric were used for all of the tests performed. The test panels for each of the tests were approximately 25.4 mm (1") thick. The outer plies on each of the panels were made of bias weave fabric and the internal plies were of unidirectional tape laminated in a +/- 45° alternating pattern, as shown schematically in Figure 2. Fabric was intentionally used on the outer plies to suppress exit and entrance ply delamination. However, the focus of the study was on the internal plies made of unidirectional tape. The resin used was 3501-6 with IM-6 graphite fibers.

Tooling

The drills used in these tests were 12.7 mm (0.5") diameter polycrystalline diamond (PCD) tipped drills fabricated by Manufacturer A and 13.9 mm (0.5494") diameter PCD tipped drills fabricated by Manufacturer B. The drills sizes were not expected to influence the outcome of the study and since the difference in the sizes of these two drills is small they will both hereafter be referred to as 12.7 mm (0.5") diameter drills. The major difference between the drills from the two manufacturers was in the drill point geometry, so the drills will hereafter be referred to as drill point A and drill point B. The drill point geometries are shown in Figure 3. Drill A was constructed with an axial rake angle at the outer diameter of the cutting edge of 27°. Drill B was constructed with a 7° axial rake angle at the corner of the insert. Other than differences in the axial rake angle, the two drills were essentially the same.

Test Machine

A Mazak 3-axis CNC milling machine equipped with a 10,000 RPM spindle was used for all tests performed. The machine is equipped with flood coolant capability. The coolant used was Daracool 706LF, a water soluble synthetic coolant. The test panels were securely fastened to the bed of the machine using a sheet of polyurethane foam that was 25 mm thick as a sacrificial layer between the graphite and the bed of the machine.

Experimental Design

Three series of tests were performed. In the first series of tests, the influence of cutting speed, feed rate, and the application of coolant was studied. These tests consisted of a once replicated full factorial experiment with three variables and two levels for each variable. The parameters for the experiment are outlined in Table 1. Drill point geometry A was used for all of the holes drilled in this test. The second series of tests compared the performance of the two drill point geometries as a function of feed rate as given in Table 2. Holes were drilled at feed rates from 0.025 mm/rev (0.001 in/rev) to 0.25 mm/rev (0.01 in/rev) in increments of 0.025 mm/rev (0.001 in/rev). Two replicates were performed at each condition. Flood coolant was used for all holes drilled and a spindle speed of 4550 RPM was used throughout. After the holes were drilled the workpiece was sectioned through the drilled holes perpendicular to the plane of the laminate so that the hole surfaces could be easily examined under an optical microscope. A water-cooled radial arm saw equipped with an aluminum oxide resin-bond saw blade was used to section the samples. The saw produced a 0.4 micrometer average surface finish (15 micro inches). The sectioned samples were then photographed using an optical microscope and examined to determine the condition of the drilled hole surfaces.

RESULTS AND DISCUSSION

Test 1

The first series of tests performed was designed to determine the influence of cutting speed, feed rate and the application of coolant on the formation of pits on the hole surfaces. Figure 4 shows typical optical micrographs of the sectioned drilled hole surfaces produced by two different cutting conditions. These photomicrographs are representative of the results from this test. The micrograph in Figure 4A shows the hole surface produced by a high feed rate condition (0.25 mm/min), a high speed condition (4550 RPM) and with application of coolant. The pits are generally one ply thickness deep and each individual pitted area in a plane of the laminate and involved 1/8 of the hole circumference. The hole surface had macroscopic pitting over much of the hole surface, as can be seen in the photograph. The hole surface shown in Figure 4B is from a hole drilled at low feed rate condition (0.025 mm/min), high speed condition, and with the application of coolant. As can be seen in the figure, these holes were dramatically improved over the holes drilled at a high feed rate.

The complete results from the first series of experiments along with the test conditions are summarized in Figure 5. Inspection of the holes indicated that the holes produced at a low feed rate (0.025 mm/rev) were far superior to the holes produced at the high feed rate (0.25 mm/rev), regardless of the cutting speed and the application of coolant. Similar results were obtained upon replication of the test matrix. Based on these experimental results it was found that the cutting speed and the application of coolant were not significant in determining the formation of pits on the hole surfaces. However, the feed rate was found to be significant in the formation of pitting in Graphite/Epoxy material.

The chip formation mechanism in composite material varies depending on the orientation of the fibers relative to the path of the cutting edge [5,8]. As a result, the way in which the fibers are cut or fractured will vary as the cutting edge travels around the hole as well as from ply to ply. As an illustration, consider a laminate consisting of alternating +45° and -45° plies. Two adjacent plies are shown schematically in Figure 6. When the cutting edge is cutting the +45° ply it encounters fibers oriented at 0°, then -45°, 90°, 45°, 0° and so on until it returns to the original point on the hole circumference. When the cutting edge is cutting the -45° ply, with fibers that are clocked 90° from the +45° ply, the cutting edge encounters the same fiber orientations except that the contact with the 0° fibers for example will occur 90° from where it occurred in the previous ply. Since the way in which the chips are formed will depend upon the fiber orientation, the chip formation mechanism due to changes in torque [5] will therefore vary around the hole and from ply to ply.

Three main modes of fiber fracture were associated with chip formation when machining fiber reinforced composite materials [6,10] as shown in Figure 7. It has been noted that the orientation of the fibers relative to the cutting edge path is primary in determining which chip formation mode would occur. The relationship between the chip formation mechanism and the occurrence of drilled hole pitting can be understood by noting that fiber fracture in Figure 7A and 7B occurs directly at the cutting edge, whereas fracture in Figure 7C occurs at some distance into the workpiece from the cutting edge. Based on this observation, one would expect that fibers being cut as in Figure 7C would be cut deeper than fibers being cut by the other modes. Also, since it was noted above that the orientation of the fibers relative to the cutting edge varies continuously, it would be reasonable to expect that the occurrence of pitting in drilled hole surfaces would be isolated to where the fibers are oriented -45° to the edge of the hole, using the sign convention as shown in Figure 6.

Test 2

Since the first test series had determined that good holes could be produced at the low feed rate using with Drill A, it was desirable to determine at what feed between the high and low feed rate pitting would begin for each drill geometry. The second test series was intended to compare the performance of the two drill point geometries, A and B, at various feed rates. This test would also determine if the relationship between feed rate and pitting identified in the first test had any interaction effect with the drill point geometry. Figure 8 shows a comparison of hole surfaces drilled with the two drill geometries, both obtained at a speed of 4550 RPM, a feed rate of 0.025 mm/min, and with the application of coolant. The difference between the performance of the two drill geometries was clearly apparent. While drill geometry A produced good holes at or below feed rates of 0.127 mm/rev (0.005 in/rev), drill point geometry B heavily pitted the holes drilled at all feed rates. Figure 9 shows magnifications of the cross section of the surface of the same pair of holes, one from drill geometry A and one from drill geometry B. As can be seen in the photographs, there was a striking difference in the performance of the two drills. Examination of the cross sections of all of the pitted holes revealed that the depth of the pits did not increase with feed rate over the range of feed rates tested. In addition, it was observed that all of the holes produced by drill geometry A were superior to all of the holes produced by drill geometry B.

Indeed, the chip formation mechanism shown in Figure 7C has been directly observed, as shown in Figure 10. A hole drilled in a laminate of graphite/epoxy that contained pits on the surface of the hole was smeared with epoxy and allowed to dry. The epoxy served to stabilize the surface of the hole so that it would not be disturbed during the process of sectioning. The hole was then cut in the plane of the laminate using a water-cooled radial arm circular saw with an aluminum oxide, resin bond blade. The photograph of the cross section shows that fibers that were inclined into the path of the advancing cutting edge (Figure 8A) were fractured by bending. In fact, some fibers were captured in the photograph that were in the process of fracture, showing the bending of the fibers. Figure 8B shows fibers that were inclined away from the advancing cutting edge cut cleanly. In inspecting the cross sections of the holes produced in the second series of tests evidence of a fourth chip formation mode was identified. This mode may be viewed as a variation of the fiber fractures shown in Figure 7C and was produced by drill point geometry B. The chip formation mode noted results from a combination of failure of the matrix in shear and failure of the fibers in bending, shown in Figure 11. The means by which this occurs is shown schematically in Figure 12. As shown in the schematic, as the advancing cutting edge comes into contact with fibers inclined into its path the fibers are bent in the direction of the cutter motion. When the fibers are inclined at a shallow angle, as shown in Figure 12A, the fiber fracture as shown in Figure 7C, leaving the "stair step" microstructure that is visible in Figure 11. However, as the angle between the fibers and the path of the cutting edge approaches 45^0, the matrix between the lateral surfaces of the fibers fails first. The fibers then bend as individual structures rather than as a bundle. The fibers then will individually fail in bending, resulting in a more uniform microstructure. The point of this failure is below the path of the cutting edge, resulting in the characteristic "thumbnail" pitting of the surface of the hole.

The results of the drilling tests suggest that the use of drill geometry A somehow alters the chip formation mechanism which is active in the -45° fiber orientation which was described in the Introduction. Recall that the difference between the two drill point geometries was the axial rake angle at the outer diameter of the cutting edge. The effect of the high axial rake angle (27°) used in drill geometry A was to give that drill a "scooping" cutting action. In contrast, the effect of the low axial rake angle (7°) used in drill geometry B was to give that drill a "scraping" cutting action. In other words, drill point geometry A imparts a significant out-of-plane force to the chips whereas drill point geometry B generates cutting forces that are largely within the plane of the laminate. It is reasonable to speculate that the effect of the out-of-plane cutting forces generated by drill geometry A is to cause the fibers to be bent up against the adjacent ply, thus providing a stabilizing force on the fibers. The out-of-plane bending is apparently sufficient to cause the brittle graphite fibers to fracture near the cutting edge, thus preventing the pitting of the hole surface.

Increasing the feed rate was found to promote the formation of pits on the hole surfaces of holes produced by drill geometry A. It is expected that high feed rates cause pitting of the hole surfaces since the out-of-plane bending caused by drill geometry A would then tend to promote tensile failure of the fibers in bending, which is similar to the chip formation mode that caused the pits in the first place. Therefore, by keeping the feed rate per revolution small, fracture of the fibers is promoted near the cutting edge, producing a smooth drilled hole surface.

SUMMARY AND CONCLUSIONS

The surfaces of holes drilled in graphite/epoxy materials are prone to fiber breakout, which is also commonly referred to as pitting of the hole surfaces. This effect is strongly related to the orientation of the fibers relative to the path of the cutting edge and has been observed to occur at those points where the fibers are inclined into the advance of the cutting edge at approximately 45°. Three drilling process variables have been identified as influencing the formation of pits on the surfaces of drilled holes:

1. Drill Geometry – A polycrystalline diamond (PCD) drill with inserts that have a high axial rake at the outer diameter is preferred to prevent the formation of pits in the hole surface. Drills with a small axial rake at the outer diameter will produce widespread pitting. In the tests performed, an axial rake angle of 27° produced good quality holes while a drill with an axial rake of 7° produced extremely poor holes over a variety of feed rates.

2. Feed rate – The feed rate, expressed in inches/revolution, should be less than one half of the ply thickness. A feed rate no greater than 0.051 mm/rev (0.002"/rev) with a ply thickness of 0.2 mm (0.0078") is recommended in order to produce superior holes through the life of the drill.

REFERENCES

[1] Porter, T.R., "Evaluation of Flawed Composite Structure Under Static and Cyclic Loading," *Fatigue of Filamentary Composite Materials, ASTM STP 636*, K.L. Reifsnider and K.N. Lauraitis, Eds., American Society for Testing and Materials, 1977, pp. 152-170.

[2] Ramkumar, R.L., "Compression Fatigue Behavior of Composites in the Presence of Delaminations," *Damage in Composite Materials, ASTM STP 775*, K.L. Reifsnider, Ed., American Society for Testing and Materials, 1982, pp. 184-210.

[3] Wood, R.E., "Graphite/Epoxy Composite Hole Quality Investigation," *Proceedings of the 10th SAMPE Technical Conference*, Vol. 10, 1978, pp. 636-650.

[4] Tagliaferri, V., Caprino, G. and Diterlizzi, A., "Effect of Drilling Parameters on the Finish and Mechanical Properties of GFRP Composites," *Int. J. Machine Tools Manufacture*, Vol. 30, 1990, pp. 77-86.

[5] Konig, W. and P. Grass, "Quality Definition and Assessment in Drilling of Fibre Reinforced Thermosets," *Annals of the CIRP*, Vol. 38, 1/1989, pp. 119-124.

[6] Santhanakrishan, G., Krishnamurthy, R., and Malhotra, S.K., "A Study on the Cutting Process, Chips, Cutting forces and Surface Morphology in Machining Composites," *Proceedings of the Seventh International Conference on Composite Materials*, Vol. 1, W. Yunshu, G. Zhenlong, and W. Renjie, Eds., Pergamon Press, 1989, pp 171-176.

[7] Koplev, A., Lystrup, A., and Vorm, T., "The Cutting Process, Chips, and Cutting Forces in Machining CFRP," *Composites*, Vol. 14, No. 4, Oct. 1983, pp. 371-376.

[8] Koplev, A., "Cutting of CFRP With Single Edge Tools," *Proceedings of the Third International Conference on Composite Materials*, Paris, Aug. 1980, pp 1597-1605.

[9] Ramulu, M., Faridnia, M., Garbini, J.L., and Jorgensen, J.E., "Machining of Graphite/Epoxy Composite Material with Polycrystalline Diamond Tools," *ASME J. Engineering Materials and Technology*, Vol. 113, No. 4, 1991, pp. 430-436.

[10] Wang, D.H., Ramulu, M., and Wern, C.W., "Orthogonal Cutting Characteristics of Graphite/Epoxy Composite Material," *Transactions of NAMRI/SME*, Vol. XX, 1992, pp. 159-165.

[11] Colligan, K., and Ramulu, M., "A Study of the Delamination in Surface Plies of Graphite/Epoxy caused by the Edge Trimming Process," *Manufacturing Review* (in press).

[12] Radhakrishnan, T. and Wu, S.M., "On-line Hole Quality Evaluation for Drilling Composite Materials Using Dynamic Data," *ASME J. Engineering for Industry*, Vol. 113, No. 1, 1981, pp. 119-125.

[13] Gindy, N.N.Z. "Selection of Drilling Conditions for Glass Fiber Reinforced Plastics," *Int. J. of Production Research*, Vol. 26, No. 8, 1988, pp. 1317-1327.

[14] Malhotra, S.K., "Some Studies on Drilling of Fibrous Composite," *Journal of Materials Processing Technology,* Vol. 24, 1990, pp. 291-300.

Table 1 - Drill Test Parameters - Test 1

Hole No.	Fedrate (mm/rev)	Speed (RPM)	Coolant
A1, B1	0.25	4550	Yes
A2, B2	0.25	4550	No
A3, B3	0.025	4550	Yes
A4, B4	0.025	4550	No
A5, B5	0.25	650	Yes
A6, B6	0.25	650	No
A7, B7	0.025	650	Yes
A8, B8	0.025	650	No

Table 2 - Drill Test Parameters - Test 2

Drills Used: PCD Drills - Point Geometries A and B

Feed rates:

0.025 mm/min	(0.001 in/min)
0.051 mm/min	(0.002 in/min)
0.077 mm/min	(0.003 in/min)
0.102 mm/min	(0.004 in/min)
0.128 mm/min	(0.005 in/min)
0.153 mm/min	(0.006 in/min)
0.179 mm/min	(0.007 in/min)
0.204 mm/min	(0.008 in/min)
0.300 mm/min	(0.009 in/min)
0.255 mm/min	(0.01 in/min)

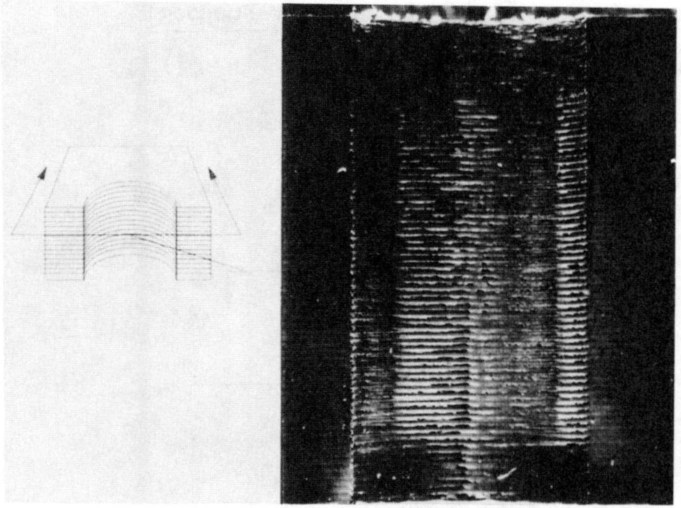

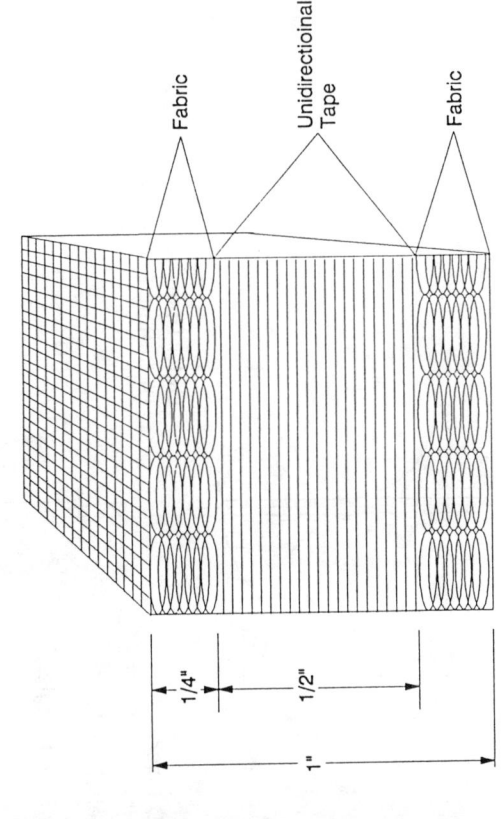

Figure 2
Test Panel Configuration

Figure 1
Cross Section of Typical Pitted Hole

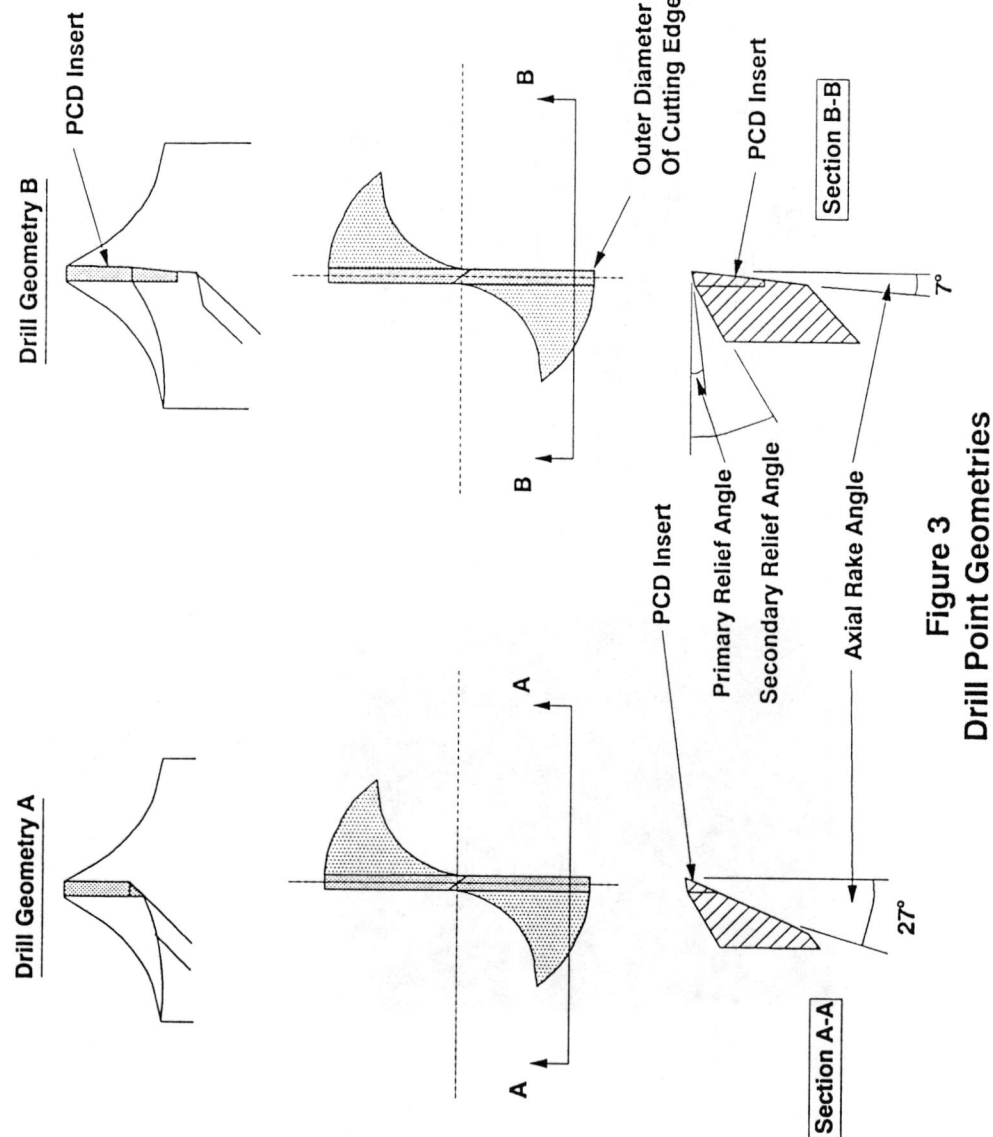

Figure 3
Drill Point Geometries

4A	4X	4B
High Feedrate		Low Feedrate
High Speed		High Speed
With Coolant		With Coolant

Figure 4
Effect of Feedrate on Drilled Hole Surfaces - Drill Geometry A

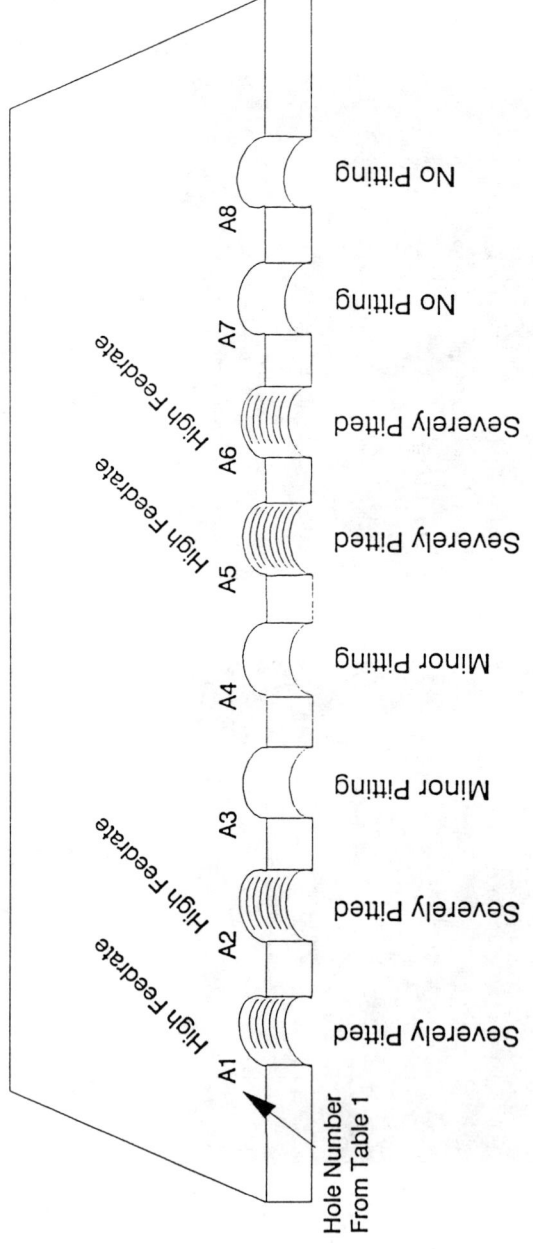

Figure 5
Test Results Summary - Test 1

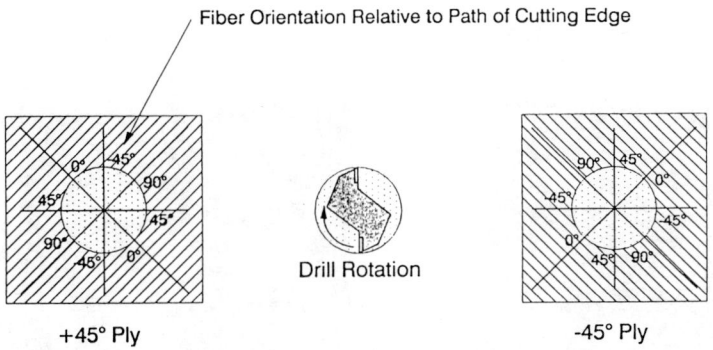

Figure 6
Fiber Orientation Around Circumference of a Drilled Hole

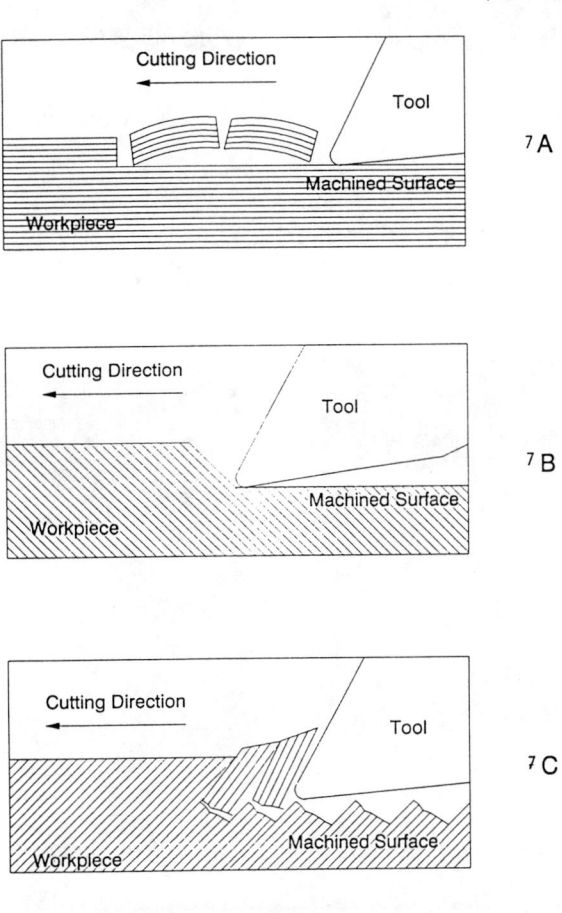

Figure 7
Chip Formation Mechanisms

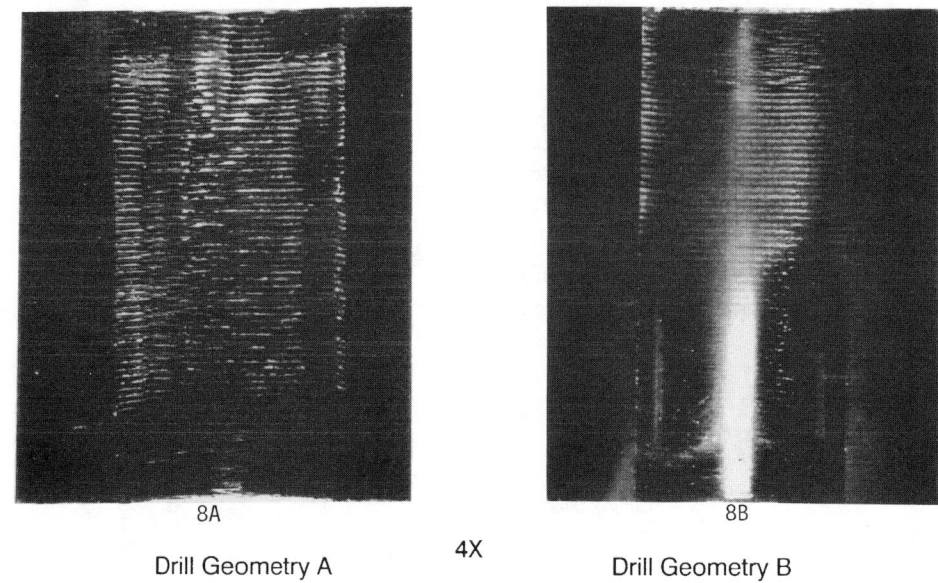

8A   4X   8B
Drill Geometry A   Drill Geometry B

Figure 8
Effect of Drill Geometry on Hole Surface Quality

9A   62.5X   9B
Drill Geometry A   Drill Geometry B

Figure 9
Effect of Drill Geometry on Hole Surface Quality

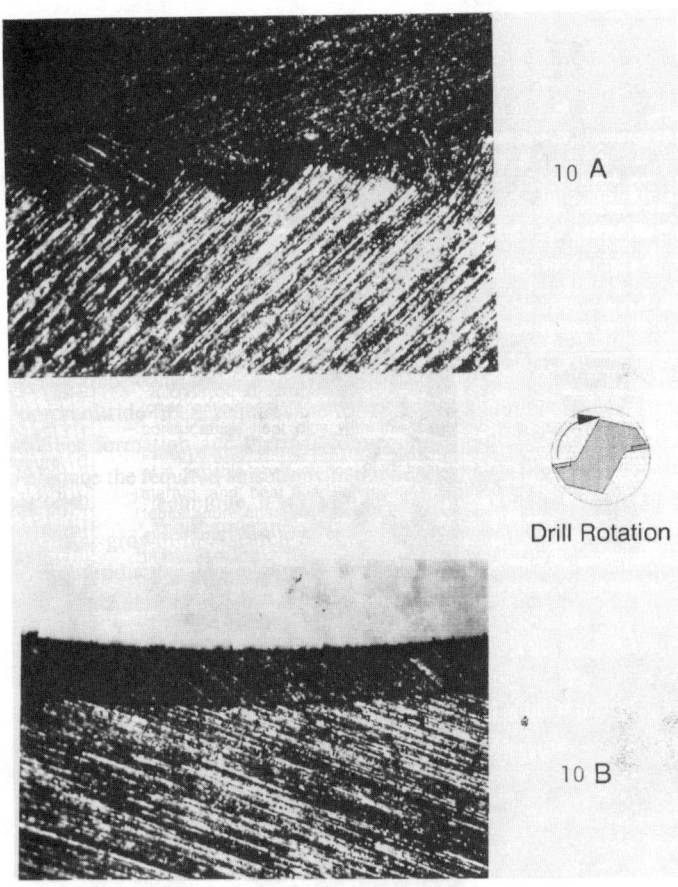

Figure 10
Cross Section of Drilled Hole Surfaces

Drill Geometry A
4,550 RPM
0.128 mm/rev (0.005 in/rev)
80X Magnification

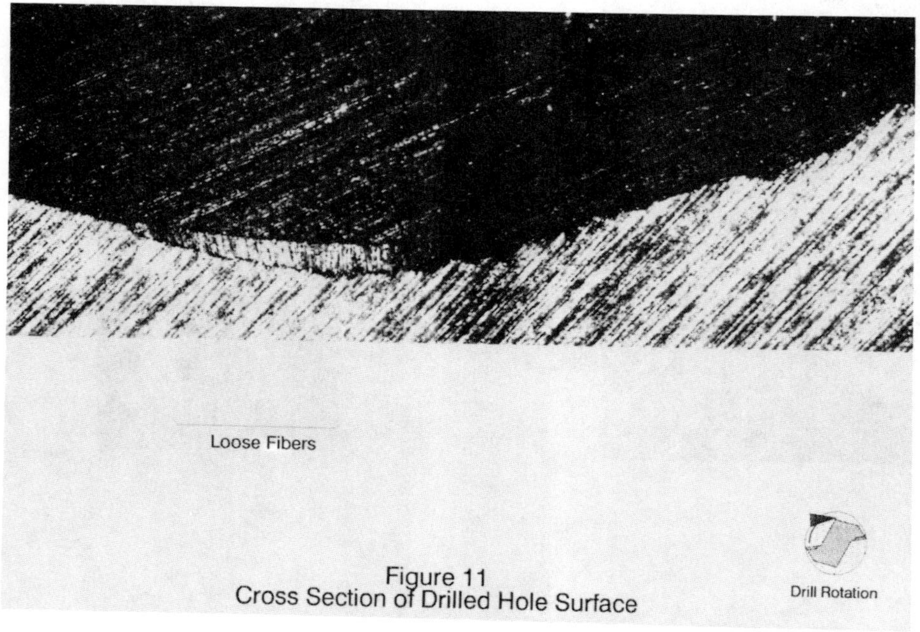

Drill Rotation

Loose Fibers

Figure 11
Cross Section of Drilled Hole Surface

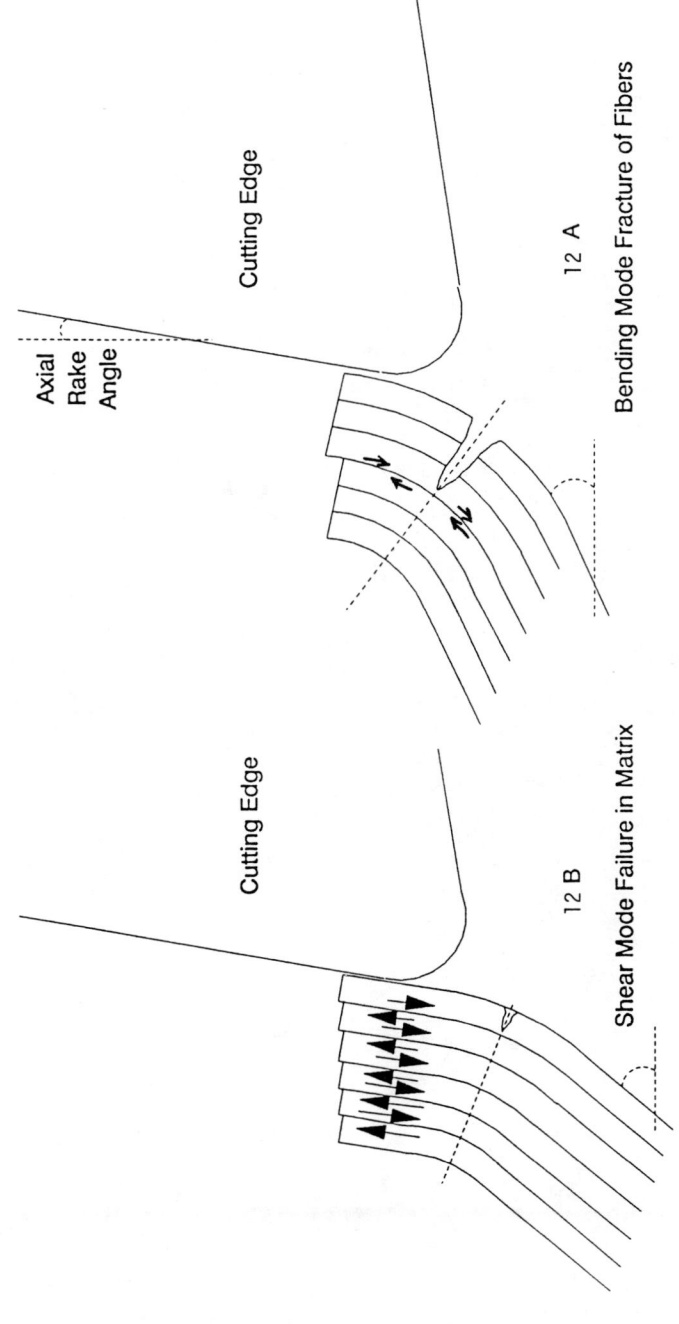

Figure 12
Chip Formation Modes

MD-Vol. 35, Processing, Fabrication, and Manufacturing
of Composite Materials
ASME 1992

AXIAL RESPONSE OF A MULTILAYER STRAND COMPOSED OF VISCOELASTIC FILAMENTS

T. A. Conway
Department of Mechanical Engineering
University of Akron
Akron, Ohio

G. A. Costello
Department of Theoretical and Applied Mechanics
University of Illinois at Urbana-Champaign
Urbana, Illinois

ABSTRACT

A method is presented in which the axial viscoelastic response of a multiple filament strand, constrained by a no end rotation boundary condition, may be predicted. This method is an initial attempt to describe the time-dependent response of the multilayer strand by incorporating the stress relaxation data for a linearly viscoelastic construction material. Specifically, a strand consisting of a core filament, six filaments in the second layer and twelve filaments in the outer layer is analyzed. This analysis could, however, include any number of layers of filaments where each layer has a concentric helix radius. The particular material used in this paper is polymethyl methacrylate (PMMA). The stress relaxation for PMMA is modeled analytically using the Schapery collocation method which determines the constant coefficient values for the elements of a Wiechert response model. Since this is a first approximation model, the approach is limited to linear viscoelasticity. The geometric effects of the strand are then combined with the Wiechert response model to develop a system of convolution integrals which satisfy the equilibrium and imposed boundary conditions for the multiple filament strand construction. The solutions for these integrals are approximated numerically using a modified Newton's iterative method combined with a numerical technique which takes into account the material's stress-strain history.

1. INTRODUCTION

Multiple filament cords composed of twisted polymer filaments are utilized in many of todays manufactured goods. They are the main components in most textile products and the reinforcing components in many composite structures (i.e. tires, biomedical devices, etc.). The advantage of the multiple filament cord over a solid filament with an equivalent radius is the lower bending stiffness while maintaining a high axial stiffness.

Since the simple strand (see Fig. 1) is a fundamental building component of many cords with multiple layered cross-sections (see

Fig. 2), an understanding of the simple strand's mechanical response is necessary. Conway and Costello (1991) analyzed the elastic axial mechanical response of a simple strand. They demonstrated, analytically, that a significant reduction in contact stresses between strand wires could be achieved by a slight modification of the wire's cross-sectional geometry. Conway and Costello (1992b), further, analyzed the viscoelastic response of a simple strand and developed a set of constitutive equations which described the strand's time-dependent response to a static, axial load.

As is evident from the literature, an extensive development of the elastic strand model presently exists (Costello, 1990). Attempts to develop a model for the response of a multiple layer strand which incorporates the time-dependent material behavior for polymer filaments have also been met with some success (Hearle, et al., 1969). This paper is an initial attempt to continue in the efforts of Hearle and his colleagues by developing a theory capable of predicting, within limits, the time dependent mechanical response of a multilayered strand. Since this is an initial attempt to develop a useful theory, the effects of friction and material compliance are ignored. The authors are aware that these first order effects are quite important in the realistic description of a strand's response to loading and will incorporate them in the subsequent development of the theory for polymer cords. This initial development of the theory is accomplished by using a method similar to that described by Conway and Costello (1992a&b). First, a viscoelastic model that incorporates relatively realistic material properties is generated. The model is, then, integrated into a set of constitutive equations for a strand. Finally, the mechanical response of the strand is evaluated numerically.

2. THEORY

Most polymer strand applications require that the strand not retain any residual strains after a period of time subsequent to the removal of the applied load. For this reason, the analytical model chosen to describe the strand material's response to loading must represent a viscoelastic solid. If it is assumed that the material behaves linearly viscoelastic, a number of analytical models are available to describe the material's mechanical response. This assumption is valid for relatively small strains and will be used throughout this work (Bland, 1960). Further, it is assumed that the material is slightly compressible and functions in an isothermal environment, thus any changes in the mechanical properties due to a temperature change are not considered.

The authors of this paper have chosen to use a modified generalized Maxwell model, known as a Wiechert model (see Fig. 3), because of its relative convenience in modeling the stress relaxation for a given material resulting from an imposed strain. Since linear viscoelasticity is assumed for the material property, the response to loading for each element in the Wiechert model is linear and is described by a constant coefficient. The values for these coefficients are calculated from the experimental stress relaxation data by using the Schapery Collocation method.

The Schapery Collocation method models each of the material's major transitions as a single relaxation time process (see Fig. 4).

Thus the values for the stress-strain relationship

$$\sigma(t) = \left\{ E_0 + \sum_{j=1}^{N} E_j \exp\left(\frac{-t}{\tau_j}\right) \right\} \epsilon_0 \qquad (1)$$

can be determined where E_0 is the rubbery modulus, E_j is the modulus for each successive transition, τ_j is the relaxation time for the corresponding transition and ϵ_0 is the imposed constant strain.

Once the strand's material has been mechanically characterized, the geometric constraints for the strand must be determined. The lateral cross-section view of an undeformed three-layered strand is illustrated in Fig. 2. The core filament has an initial radius, R_1, the second layer of 6 filaments each has a radius, R_2, and the outer layer has 12 filaments each with a radius, R_3. The subscripts refer to the layer of filaments, with the first layer being the core filament.

The initial and time dependent helix radius of the second layer of filaments are

$$r_2 = R_1 + R_2 \qquad (2)$$

and

$$r_2(t) = R_1(t) + R_2(t) \qquad (3)$$

The initial helix radius for the third layer of filaments is

$$r_3 = R_1 + 2R_2 + R_3 \qquad (4)$$

and the time dependent helix radius is

$$r_3(t) = R_1(t) + 2R_2(t) + R_3(t) \qquad (5)$$

The initial helix angle of the second layer of filaments, α_2, is determined from the layer's initial pitch, P_2, of its filament, or

$$\alpha_2 = \tan^{-1} \frac{P_2}{2\pi r_2} \qquad (6)$$

The initial helix angle for the third layer, α_3, of filaments is

$$\alpha_3 = \tan^{-1} \frac{P_3}{2\pi r_3} \qquad (7)$$

where P_3 is the initial pitch of the third layer of filaments.

In a purely elastic material, the time necessary for molecular rearrangement is virtually infinite (Tschoegl, 1989). A comparison of an elastic strand's initial configuration and final configuration can, therefore, be used to determine the strand's mechanical response to loading or deformation. If a load is applied to a polymer strand, however, the helix angles, $\alpha_2(t)$ and $\alpha_3(t)$, each filament radius, $R_1(t)$, $R_2(t)$ and $R_3(t)$, and helix radii, $r_2(t)$ and $r_3(t)$, all vary as a function of time. Thus a description of a viscoelastic solid's final configuration is not applicable. Instead, a description of the strand's configuration at some time, $t > 0$, is necessary.

Conway and Costello (1992a&b) have shown that for a given length of strand, the second layer of filaments will have Q_0 turns in the undeformed state and Q turns after deformation at some time $t>0$. The undeformed length of the strand is

$$L_0 = 2\pi Q_0 r_2 \tan\alpha_2 \tag{8}$$

and the deformed length at $t>0$ is

$$L(t) = 2\pi Q r_2(t) \tan\alpha_2(t) \tag{9}$$

The corresponding undeformed length of a second layer filament is

$$l_0 = 2\pi Q_0 r_2 \sec\alpha_2 \tag{10}$$

and the corresponding deformed length is

$$l(t) = 2\pi Q r_2(t) \sec\alpha_2(t) \tag{11}$$

The axial strain in the strand is, thus,

$$\epsilon_1(t) = \frac{L(t)}{L_0} - 1 = \frac{Q r_2(t) \tan\alpha_2(t)}{Q_0 r_2 \tan\alpha_2} - 1 \tag{12}$$

and the axial strain in a filament is

$$\xi_2(t) = \frac{l(t)}{l_0} - 1 = \frac{Q r_2(t) \cos\alpha_2}{Q_0 r_2 \cos\alpha_2(t)} - 1 \tag{13}$$

From Poisson's effect

$$\frac{R_2(t)}{R_2} - 1 = -\nu \xi_2(t) \tag{14}$$

where ν is Poisson's ratio and is considered to be constant for relatively small deformations.

A combination of Eqns. (12)-(14) results in

$$\epsilon_1(t) = \frac{\sin\alpha_2(t)}{\sin\alpha_2}\left[1 + \frac{1}{\nu}\left(1 - \frac{R_2(t)}{R_2}\right)\right] - 1 \tag{15}$$

The angle of twist per unit length for the second layer of the strand is

$$\tau_c(t) = \frac{2\pi(Q - Q_0)}{L_0} = \frac{1}{r_2 \tan\alpha_2}\left(\frac{Q}{Q_0} - 1\right) \tag{16}$$

By combining Eqns. (13), (14) and (16), the twist per unit length becomes

$$\tau_c(t) = \frac{1}{r_2 \tan\alpha_2}\left\{\frac{r_2 \cos\alpha_2(t)}{r_2(t) \cos\alpha_2}\left[1 + \frac{1}{\nu}\left(1 - \frac{R_2(t)}{R_2}\right)\right] - 1\right\} \tag{17}$$

Equation (17) shows that if the strand ends are allowed to rotate, the strain will be coupled with a twist. If, however, the ends of the strand are not allowed to rotate, such that $\tau_c(t)=0$, then an additional constrained condition exists. From Eqn. (17), this constraint is

$$\frac{\cos\alpha_2}{r_2} = \frac{\cos\alpha_2(t)}{r_2(t)}\left[1 + \frac{1}{\nu}\left(1 - \frac{R_2(t)}{R_2}\right)\right] \tag{18}$$

By incorporating Eqns. (2) and (18), a relationship between the helix angle, $\alpha_2(t)$, and the second layer filament radius, $R_2(t)$, can be shown to be

$$\cos\alpha_2(t) = \frac{(R_1(t)+R_2(t))\cos\alpha_2}{(R_1+R_2)\left[1+\frac{1}{\nu}\left(1-\frac{R_2(t)}{R_2}\right)\right]} \quad (19)$$

Similarly, the relationship between $\alpha_3(t)$ and $R_3(t)$ in the third layer of filaments is

$$\cos\alpha_3(t) = \frac{r_3(t)\cos\alpha_3}{r_3\left[1+\frac{1}{\nu}\left(1-\frac{R_3(t)}{R_3}\right)\right]} \quad (20)$$

The general design of a strand minimizes the contact stresses between the filaments by requiring the radius of the second layer filament to be slightly less than the core layer, the radius of the third layer filament to be sightly less than the second layer filament and so on. This reduction in radius does not have to be very large, usually on the order of 3% for a polymer with a Poisson's ratio near $\frac{1}{2}$. This now allows a relationship between $R_1(t)$, $R_2(t)$ and $R_3(t)$ to be known.

Once the geometric constraints for the strand are determined, the equilibrium equations can be developed. Before the viscoelastic material properties are incorporated into this configuration, however, a review of the elastic equilibrium equations for a simple strand are necessary.

Costello (1990) has shown that for an axially loaded strand, the filaments in each layer deform from one helical configuration to another. Thus, the equations of equilibrium that must be satisfied for each wire are:

$$N'_2\tau_2 - T_2\kappa'_2 - X_2 = 0 \quad (21)$$

$$G'_2\tau_2 - H_2\kappa'_2 + N_2 = 0 \quad (22)$$

$$N'_3\tau_3 - T_3\kappa'_3 - X_3 = 0 \quad (23)$$

and

$$G'_3\tau_3 - H_3\kappa'_3 + N_3 = 0 \quad (24)$$

where the subscripts 2 and 3 refers to the second and third layer of wires, N'_2 and N'_3 are the binormal components of the shear load, τ_2 and τ_3 are the twist per unit length of the wire's centerline, T_2 and T_3 are the wire's axial load, κ'_2 and κ'_3 are the binormal components of curvature, X_2 and X_3 are the resultant line loads per unit length of the wire's centerline, G'_2 and G'_3 are the binormal components of the bending moment and H_2 and H_3 are the twisting moments. Further, Costello (1990) showed that for naturally curved thin rods

$$G'_2 = \frac{\pi R_2^4}{4}E\left[\frac{\cos^2\bar{\alpha}_2}{\bar{r}_2} - \frac{\cos^2\alpha_2}{r_2}\right] \quad (25)$$

$$H_2 = \frac{\pi R_2^4}{4(1+\nu)}E\left[\frac{\sin\bar{\alpha}_2\cos\bar{\alpha}_2}{\bar{r}_2} - \frac{\sin\alpha_2\cos\alpha_2}{r_2}\right] \quad (26)$$

$$G'_3 = \frac{\pi R_3^4}{4} E \left[\frac{\cos^3 \bar{\alpha}_3}{\bar{r}_3} - \frac{\cos^3 \alpha_3}{r_3} \right] \tag{27}$$

and

$$H_3 = \frac{\pi R_3^4}{4(1+\nu)} E \left[\frac{\sin \bar{\alpha}_3 \cos \bar{\alpha}_3}{\bar{r}_3} - \frac{\sin \alpha_3 \cos \alpha_3}{r_3} \right] \tag{28}$$

where R_2 and R_3 are the wire radii for each layer after loading, E is the wire material's elastic modulus, $\bar{\alpha}_2$ and $\bar{\alpha}_3$ are the helix angles of the loaded wire and $\bar{r}_2$ and $\bar{r}_3$ are the helix radii for each layer of the loaded strand. Also in the elastic case, the axial load in each wire of layers 2 and 3 may be determined from

$$T_2 = \frac{\pi R_2^2}{\nu} E \left[1 - \frac{R_2}{(R_2)_0} \right] \tag{29}$$

and

$$T_3 = \frac{\pi R_3^2}{\nu} E \left[1 - \frac{R_3}{(R_3)_0} \right] \tag{30}$$

Equations (22), (25) and (26) can be used to determine an expression for N'_2 in terms of $\bar{\alpha}_2$ and R_2 given $\nu, \alpha_2, (R_2)_0$ and E. Equations (24), (27) and (28) can be used to determine an expression for N'_3 in terms of $\bar{\alpha}_3$ and R_3 given $\nu, \alpha_3, (R_3)_0$ and E.

The total forces, F_2 and F_3, and moments, M_2 and M_3, for each layer of wires are

$$F_2 = 6[T_2 \sin \alpha_2 + N'_2 \cos \alpha_2] \tag{31}$$

$$F_3 = 12[T_3 \sin \alpha_3 + N'_3 \cos \alpha_3] \tag{32}$$

$$M_2 = 6[H_2 \sin \alpha_2 + G'_2 \cos \alpha_2 + T_2 r_2 \cos \alpha_2 - N'_2 r_2 \sin \alpha_2] \tag{33}$$

and

$$M_3 = 12[H_3 \sin \alpha_3 + G'_3 \cos \alpha_3 + T_3 r_3 \cos \alpha_3 - N'_3 r_3 \sin \alpha_3] \tag{34}$$

The total force, F_1, in the core wire can be calculated by

$$F_1 = -\frac{\pi R_1^2}{\nu} E \left[1 - \frac{R_1}{(R_1)_0} \right] \tag{35}$$

The total moment, M_1, in the core wire is equal to zero as a result of the no end rotation boundary condition imposed on the simple strand.

Finally, the total axial load, F_T, and the total moment, M_T, in the strand are

$$F_T = F_1 + F_2 + F_3 \tag{36}$$

and

$$M_T = 0 + M_2 + M_3 \tag{37}$$

Returning now to the time dependent behavior of the polymer strand, according to the correspondence principle of viscoelasticity, the tensile relaxation modulus can be treated as an integral operation. The correspondence principle allows Hooke's law to be used in the Laplace transform space and is represented as

$$\bar{\sigma}(p) = p\tilde{E}(p)\bar{\epsilon}(p) \tag{38}$$

where p is the transform variable. Inversion of this transform yields an expression for the Boltzmann superposition integral (Tschoegl, 1989). This general solution to the inverse transform of Eqn. (38) is

$$\sigma(t) = \int_0^t E(t-t_1)\frac{d\epsilon(t_1)}{dt_1}dt_1 \tag{39}$$

Now, instead of varying the modulus and the strain, we now hold the strain constant, as in stress relaxation, and allow another parameter of the design to vary. The convolution integral would have the general form

$$\int_0^t E(t-t_1)\dot{f}(t_1)dt_1 \tag{40}$$

If the differential term is denoted as

$$df = \dot{f} \tag{41}$$

then Eqn. (40) can be represented as

$$(E*df)(t) \tag{42}$$

The time-dependent relationships for the binormal components of the bending moment, $G'_2(t)$ and $G'_3(t)$, and the twisting moment, $H_2(t)$ and $H_3(t)$, for the respective filament can now be determined. These components of the moment become

$$G'_2(t) = \frac{\pi R_2^4(t)}{4}\left(E*d\left[\frac{\cos^2\alpha_2}{r_2}\right]\right)(t) \tag{43}$$

$$G'_3(t) = \frac{\pi R_3^4(t)}{4}\left(E*d\left[\frac{\cos^2\alpha_3}{r_3}\right]\right)(t) \tag{44}$$

$$H_2(t) = \frac{\pi R_2^4(t)}{4(1+\nu)}\left(E*d\left[\frac{\sin\alpha_2\cos\alpha_2}{r_2}\right]\right)(t) \tag{45}$$

and

$$H_3(t) = \frac{\pi R_3^4(t)}{4(1+\nu)}\left(E*d\left[\frac{\sin\alpha_3\cos\alpha_3}{r_3}\right]\right)(t) \tag{46}$$

By incorporating Eqns. (22), (43) and (45), the binormal component of shear, $N'_2(t)$, is

$$N'_2(t) = \frac{\pi R_2^4(t)\cos\alpha_2(t)}{4r_2(t)}\left\{\left[\frac{\cos\alpha_2(t)}{1+\nu}\right]\left(E*d\left[\frac{\sin\alpha_2\cos\alpha_2}{r_2}\right]\right)(t) - \sin\alpha_2(t)\left(E*d\left[\frac{\cos^2\alpha_2}{r_2}\right]\right)(t)\right\} \tag{47}$$

By incorporating Eqns. (24), (44) and (46), the binormal component of shear, $N'_3(t)$, is

$$N'_3(t) = \frac{\pi R_3^4(t)\cos\alpha_3(t)}{4r_3(t)} \left\{ \left[\frac{\cos\alpha_3(t)}{1+\nu}\right]\left(E*d\left[\frac{\sin\alpha_3\cos\alpha_3}{r_3}\right]\right)(t) \right.$$
$$\left. -\sin\alpha_3(t)\left(E*d\left[\frac{\cos^2\alpha_3}{r_3}\right]\right)(t)\right\} \qquad (48)$$

The axial load in a second layer filament can be determined from Eqn. (29) to be

$$T_2(t) = -\frac{\pi R_2^2(t)}{\nu R_2}(E*dR_2)(t) \qquad (49)$$

The axial load in a second layer filament can be determined from Eqn. (30) to be

$$T_3(t) = -\frac{\pi R_3^2(t)}{\nu R_3}(E*dR_3)(t) \qquad (50)$$

The total force, $F_2(t)$, and moment, $M_2(t)$, for the second layer of filaments are

$$F_2(t) = 6[T_2(t)\sin\alpha_2(t) + N'_2(t)\cos\alpha_2(t)] \qquad (51)$$

and

$$M_2(t) = 6[H_2(t)\sin\alpha_2(t) + G'_2(t)\cos\alpha_2(t)$$
$$+ T_2(t)r_2(t)\cos\alpha_2(t) - N_2'(t)r_2(t)\sin\alpha_2(t)] \qquad (52)$$

The total force, $F_3(t)$, and moment, $M_3(t)$, for the third layer of filaments are

$$F_3(t) = 12[T_3(t)\sin\alpha_3(t) + N'_3(t)\cos\alpha_3(t)] \qquad (53)$$

and

$$M_3(t) = 12[H_3(t)\sin\alpha_3(t) + G'_3(t)\cos\alpha_3(t)$$
$$+ T_3(t)r_3(t)\cos\alpha_3(t) - N_3'(t)r_3(t)\sin\alpha_3(t)] \qquad (54)$$

The total force, $F_1(t)$, in the core filament is calculated similarly to the axial load in a second or third layer filament where

$$F_1(t) = -\frac{\pi R_1^2(t)}{\nu R_1}(E*dR_1)(t) \qquad (55)$$

The total moment, $M_1(t)$, in the core filament is, again, equal to zero as a result of the no end rotation boundary condition imposed on the simple strand.

Finally, the total axial load, $F(t)$, and the total moment, $M(t)$, in the three layer strand are

$$F(t) = F_1(t) + F_2(t) + F_3(t) \qquad (56)$$

and

$$M(t) = 0 + M_2(t) + M_3(t) \qquad (57)$$

With Eqns. (42)-(57), the mathematical tools are now available for the development of a solution technique for the response of a strand given a specific loading criterion.

3. GENERAL METHOD OF SOLUTION

Now, the parameters used to calculate the various aspects of the strand response must be non-dimensionalized. In the subsequent analysis, R_2, R_3, $R_2(t)$ and $R_3(t)$ will be evaluated in terms of R_1 and $R_1(t)$. Also, the tensile modulus at time, $t=0^+$, is $E(0^+)$ and is denoted by E_0. Thus, all of the variables can be non-dimensionalized in terms of E_0 and R_1. This results in the following variables where the bar indicates a dimensionless quantity.

$$\bar{E} = \frac{E(t)}{E_0}, \quad \bar{R}_1(t) = \frac{R_1(t)}{R_1}, \quad \bar{R}_2(t) = \frac{R_2(t)}{R_1}, \quad \bar{R}_3(t) = \frac{R_3(t)}{R_1}$$

$$\bar{R}_1 = 1, \quad \bar{R}_2 = \frac{R_2}{R_1}, \quad \bar{R}_3 = \frac{R_3}{R_1}, \quad \bar{r}_2(t) = \frac{r_2(t)}{R_1}, \quad \bar{r}_3(t) = \frac{r_3(t)}{R_1}$$

$$\bar{T}_2(t) = \frac{T_2(t)}{E_0 \pi R_1^2}, \quad \bar{T}_3(t) = \frac{T_3(t)}{E_0 \pi R_1^2}, \quad \bar{N}'_2(t) = \frac{N'_2(t)}{E_0 \pi R_1^2} \quad (58)$$

$$\bar{N}'_3(t) = \frac{N'_3(t)}{E_0 \pi R_1^2}, \quad \bar{G}'_2(t) = \frac{G'_2(t)}{E_0 \pi R_1^3}, \quad \bar{G}'_3(t) = \frac{G'_3(t)}{E_0 \pi R_1^3}$$

$$\bar{H}_2(t) = \frac{H_2(t)}{E_0 \pi R_1^3}, \quad \bar{H}_3(t) = \frac{H_3(t)}{E_0 \pi R_1^3}, \quad \bar{F}_1(t) = \frac{F_1(t)}{E_0 \pi R_1^2}, \quad \bar{F}_2(t) = \frac{F_2(t)}{E_0 \pi R_1^2}$$

$$\bar{F}_3(t) = \frac{F_3(t)}{E_0 \pi R_1^2}, \quad \bar{M}_2(t) = \frac{M_2(t)}{E_0 \pi R_1^3}, \quad \bar{M}_3(t) = \frac{M_3(t)}{E_0 \pi R_1^3}$$

$$\bar{F}(t) = \frac{F(t)}{E_0 \pi R_1^2}, \quad \bar{M}(t) = \frac{M(t)}{E_0 \pi R_1^3}$$

For specific values of ν, $F(t)$ and $E(t)$, the viscoelastic deformation of a three layer strand is governed by the following equations:

$$\bar{F}_1(t) = -\frac{1}{\nu}\bar{R}_1^2(t)(\bar{E}*d\bar{R}_1)(t) \quad (59)$$

$$\bar{T}_2(t) = -\frac{1}{\nu}\bar{R}_2^2(t)(\bar{E}*d\bar{R}_2)(t) \quad (60)$$

$$\bar{N}'_2(t) = \frac{\bar{R}_2^4(t)\cos\alpha_2(t)}{4\bar{r}_2(t)} \left\{ \left[\frac{\cos\alpha_2(t)}{1+\nu}\right]\left(\bar{E}*d\left[\frac{\sin\alpha_2\cos\alpha_2}{\bar{r}_2}\right]\right)(t) \right.$$
$$\left. -\sin\alpha_2(t)\left(\bar{E}*d\left[\frac{\cos^2\alpha_2}{\bar{r}_2}\right]\right)(t) \right\} \quad (61)$$

$$\bar{G}'_2(t) = \frac{\bar{R}_2^4(t)}{4}\left(\bar{E}*d\left[\frac{\cos^2\alpha_2}{\bar{r}_2}\right]\right)(t) \quad (62)$$

$$\bar{H}_2(t) = \frac{\bar{R}_2^4(t)}{4(1+\nu)}\left(\bar{E}*d\left[\frac{\sin\alpha_2\cos\alpha_2}{\bar{r}_2}\right]\right)(t) \quad (63)$$

$$\bar{T}_3(t) = -\frac{1}{\nu}\bar{R}_3^2(t)(\bar{E}*d\bar{R}_3)(t) \quad (64)$$

$$\overline{N}'_3(t) = \frac{\overline{R}_3^4(t)\cos\alpha_3(t)}{4\overline{r}_3(t)} \left\{ \left[\frac{\cos\alpha_3(t)}{1+\nu}\right]\left(\overline{E}*d\left[\frac{\sin\alpha_3\cos\alpha_3}{\overline{r}_3}\right]\right)(t) \right.$$
$$\left. -\sin\alpha_3(t)\left(\overline{E}*d\left[\frac{\cos^2\alpha_3}{\overline{r}_3}\right]\right)(t) \right\} \tag{65}$$

$$\overline{G}'_3(t) = \frac{\overline{R}_3^4(t)}{4}\left(\overline{E}*d\left[\frac{\cos^2\alpha_3}{\overline{r}_3}\right]\right)(t) \tag{66}$$

$$\overline{H}_3(t) = \frac{\overline{R}_3^4(t)}{4(1+\nu)}\left(\overline{E}*d\left[\frac{\sin\alpha_3\cos\alpha_3}{\overline{r}_3}\right]\right)(t) \tag{67}$$

$$\phi = \overline{F}(t) - \{\overline{F}_1(t) + 6[\overline{T}_2(t)\sin\alpha_2(t) + \overline{N}'_2(t)\cos\alpha_2(t)]$$
$$+ 12[\overline{T}_3(t)\sin\alpha_3(t) + \overline{N}'_3(t)\cos\alpha_3(t)]\} \tag{68}$$

$$\overline{M}(t) = 6[\overline{H}_2(t)\sin\alpha_2(t) + \overline{G}'_2(t)\cos\alpha_2(t) + \overline{T}_2(t)\overline{r}_2(t)\cos\alpha_2(t)$$
$$- \overline{N}'_2(t)\overline{r}_2(t)\sin\alpha_2(t)] + 12[\overline{H}_3(t)\sin\alpha_3(t) + \overline{G}'_3(t)\cos\alpha_3(t)$$
$$+ \overline{T}_3(t)\overline{r}_3(t)\cos\alpha_3(t) - \overline{N}'_3(t)\overline{r}_3(t)\sin\alpha_3(t)] \tag{69}$$

When the external force, $F(t)$, and a boundary condition such as no end rotation are applied at $t = 0$, the instantaneous response at $t = 0^+$ is elastic. The governing equations for this elastic response are:

$$\overline{F}_1(0^+) = \frac{1}{\nu}\overline{R}_1^2(0^+)[1 - R_1(0^+)] \tag{70}$$

$$\overline{T}_2(0^+) = \frac{1}{\nu}\overline{R}_2^2(0^+)[\overline{R}_2 - R_2(0^+)] \tag{71}$$

$$\overline{T}_3(0^+) = \frac{1}{\nu}\overline{R}_3^2(0^+)[\overline{R}_3 - R_3(0^+)] \tag{72}$$

$$\overline{N}'_2(0^+) = \frac{\overline{R}_2^4(0^+)\cos\alpha_2(0^+)}{4\overline{r}_2(0^+)} \left\{ \left[\frac{\cos\alpha_2(0^+)}{1+\nu}\right]\left[\frac{\sin\alpha_2(0^+)\cos\alpha_2(0^+)}{\overline{r}_2(0^+)}\right.\right.$$
$$\left.\left. -\frac{\sin\alpha_2\cos\alpha_2}{\overline{r}_2}\right] - \sin\alpha_2(0^+)\left[\frac{\cos^2\alpha_2(0^+)}{\overline{r}_2(0^+)} - \frac{\cos^2\alpha_2}{\overline{r}_2}\right] \right\} \tag{73}$$

$$\overline{N}'_3(0^+) = \frac{\overline{R}_3^4(0^+)\cos\alpha_3(0^+)}{4\overline{r}_3(0^+)} \left\{ \left[\frac{\cos\alpha_3(0^+)}{1+\nu}\right]\left[\frac{\sin\alpha_3(0^+)\cos\alpha_3(0^+)}{\overline{r}_3(0^+)}\right.\right.$$
$$\left.\left. -\frac{\sin\alpha_3\cos\alpha_3}{\overline{r}_3}\right] - \sin\alpha_3(0^+)\left[\frac{\cos^2\alpha_3(0^+)}{\overline{r}_3(0^+)} - \frac{\cos^2\alpha_3}{\overline{r}_3}\right] \right\} \tag{74}$$

$$\phi = \overline{F}(0^+) - \{\overline{F}_1(0^+) + 6[\overline{T}_2(0^+)\sin\alpha_2(0^+) + \overline{N}'_2(0^+)\cos\alpha_2(0^+)]$$
$$+ 12[\overline{T}_3(0^+)\sin\alpha_3(0^+) + \overline{N}'_3(0^+)\cos\alpha_3(0^+)]\} \tag{75}$$

In order to find the correct value for $\overline{R}(0^+)$ in Eqns. (70)-(74), an iterative method must be used to satisfy Eqn. (75). As stated previously, the analysis of the strand is limited to a no end rotation boundary condition. Thus, a modified Newton's iterative method is used. This method uses three values of $\overline{R}(0^+)$. These are $\overline{R}_i(0^+)$, $\overline{R}_i(0^+)-\Delta$ and $\overline{R}_i(0^+)+\Delta$, where Δ is a small number. From these values the corresponding terms ϕ_1, ϕ_2 and ϕ_3 can be calculated. The derivative $d\phi/d\overline{R}$ at $\overline{R}(0^+)=\overline{R}_i(0^+)$ can be approximated by using a central difference equation, such as

$$\phi' = \left.\frac{d\phi}{d\overline{R}}\right|_{\overline{R}=\overline{R}_i} = \frac{1}{2\Delta}(\phi_3 - \phi_2) \tag{76}$$

Now, Newton's iterative formula can be used to determine a new value for $\overline{R}(0^+)$ which is

$$\overline{R}_{i+1} = \overline{R}_i - \frac{\phi_1(0^+)}{\phi'} = \frac{2\Delta\phi_1(0^+)}{\phi_3(0^+)-\phi_2(0^+)} \tag{77}$$

This iteration technique is used until $\phi_1(0^+) \approx 0$. Since this technique is used on a computer, a minimum value for $\phi_1(0^+)$, such as 1×10^{-9}, must be set. Once this value is arrived at, the corresponding value for $\overline{R}(0^+)$ is used to determine the loads and moments at $t=0^+$.

Once the filaments in a strand are deformed into a new helical configuration by this elastic response, they continue to vary with respect to loading and deformation over time, depending on the viscoelastic properties of the strand's construction material. This time-dependent behavior can be closely approximated by using a numerical integration technique developed by Lee and Rogers (1963) which evaluates the hereditary integral introduced in Eqn. (39). This technique uses the equation

$$\int_0^t \overline{E}(t-t_1)\dot{f}(t_1)dt_1 = S_N + \frac{1}{2}[1+\overline{E}(t_N-t_{N-1})]f(t_N) \tag{78}$$

where

$$S_N = \overline{E}(t_N)[f(0^+)-f_0] - \frac{1}{2}[1+\overline{E}(t_N-t_{N-1})]f(t_{N-1})$$

$$\frac{1}{2}\sum_{i=0}^{N-2}\{[\overline{E}(t_N-t_{i+1})+\overline{E}(t_N-t_i)][f(t_{i+1})-f(t_i)]\} \tag{79}$$

In this numerical scheme the time, t, is divided into N intervals where t_0 is the time of the initial elastic response and t_N is the time when the viscoelastic response is to be evaluated. Also, f is the function that is differentiated with respect to time in Eqns. (43)-(50). At each step in the time interval, a value for the convolution integral is determined from this method. This value is used to determine values for $\overline{N}'_2(t)$, $\overline{N}'_3(t)$, $\overline{T}_2(t)$ and $\overline{T}_3(t)$ from Eqns. (47)-(50). $\overline{N}'_2(t)$, $\overline{N}'_3(t)$, $\overline{T}_2(t)$ and $\overline{T}_3(t)$ are then placed in Eqn. (68). If ϕ is not less than some prescribed value, the modified Newton's iterative method is used to evaluate a $\overline{R}_i(t)$ which will satisfy the conditions for ϕ. Once a value for $\overline{R}_i(t)$ is determined, $\alpha_2(t)$ and $\alpha_3(t)$ can be calculated from Eqn. (19) and (20). These can then be used to determine $F_1(t)$, $\overline{T}_2(t)$, $\overline{T}_3(t)$, $\overline{N}'_2(t)$, $\overline{N}'_3(t)$, $\overline{G}'_2(t)$, $\overline{G}'_3(t)$, $\overline{H}_2(t)$, $\overline{H}_3(t)$, $\overline{M}_2(t)$, $\overline{M}_3(t)$ and $\epsilon_1(t)$ from Eqns. (59)-(67)

and (69), for the specific time interval from which $\overline{R}_i(t)$ was evaluated. This iterative technique must then be used at each time interval in order to describe, satisfactorily, the viscoelastic response for the strand.

4. STRAND RESPONSE FOR A PARTICULAR VISCOELASTIC MATERIAL

In order to determine the time-dependent response of a polymer strand, a viscoelastic material must be chosen. This material must have a linear stress-strain relationship in the Laplace transform space since the correspondence principle in viscoelasticity is used in the subsequent analysis. A material which meets this criterion is polymethyl methacrylate (PMMA). The modulus relaxation is shown in Fig. 5. Also in this figure is the corresponding theoretical curve determined from the Schapery collocation method. As discussed previously, the Wiechert model is used in the prediction of this theoretical curve. For good correlation between the experimental data for PMMA and the theoretical results, ten Maxwell elements are required along with a spring in parallel. The modulus relaxation equation is thus

$$E(t) = 2.24 \times 10^7 + 1.60 \times 10^{10} e^{(-t/0.01)} - 7.01 \times 10^9 e^{(-t/0.1)} + 9.65 \times 10^9 e^{(-t)}$$
$$+ 4.36 \times 10^9 e^{(-t/10)} + 4.12 \times 10^9 e^{(-t/100)} + 2.47 \times 10^9 e^{(-t/1000)} + 4.98 \times 10^8 e^{(-t/10000)}$$
$$+ 1.27 \times 10^8 e^{(-t/100000)} + 5.85 \times 10^7 e^{(-t/1000000)} + 1.94 \times 10^7 e^{(-t/100000000)} \tag{80}$$

where at $t > 10^9$ hours, $E(t) \sim 10^{7.35} dynes/cm^2$.

Once the modulus relaxation is closely approximated numerically, the creep response for a solid cylindrical rod and a three layer strand with equal diameters can be computed. Figure 6 shows this comparison where both the rod and the strand have an overall, initial diameter of 9.6463 cm, a Poisson's ratio of 0.45 and a total axial load of $2 \times 10^9 dynes$. The second and third layers of filaments for the strand have helix angles of $75°$ and $100°$, respectively. Also, the second and third layer filament diameters are 1.94 cm and 1.8818 cm, respectively. This difference in diameters ensures that during deformation the resultant contact loads are not influenced by a filament significantly touching another in the same layer. The no end rotation boundary condition is imposed on the strand. The instantaneous jump in strain at $t = 0^+$ is, again, caused by the initial elastic response of the material, as described in Eqns. (70)-(75). Both strains asymptotically approach an equilibrium value corresponding to the delayed time modulus for PMMA of $10^{7.35} dynes/cm^2$, shown in Fig. 5.

Other time-dependent geometric relationships including helix angles and helix radii in the three layer strand also show the asymptotic approach equilibrium values. The helix angle of the second layer of filaments increases while the helix angle of the third layer decreases. The helix radii for each layer of filaments decreases with respect to time. The results are consistent with those expected for the creep response of the strand.

The shear loads for each filament decrease over time, while the axial loads remain relatively constant. This is due to the large helix angles for each layer resulting in relatively small shear load components for the filaments in those layers. As the filaments straighten out during the creep response, the shear load components are transferred to the axial load components. The shear loads are, however, three to four orders of magnitude less than the axial loads and produce no significant addition to the axial load components with respect to time. This time dependent behavior is also observed with respect to the bending and twisting components of the imposed moments for the filaments in each layer of the strand.

5. SUMMARY AND CONCLUSIONS

An analytical method is presented for determining various mechanical properties of a multilayer strand. Specifically, a three layer strand is analyzed, however, the technique developed can be used for any strand with concentric layers of filaments with circular cross-sections. The mechanical properties that are determined include the overall axial, time-dependent strain of the strand as well as the internal time-dependent, geometric and loading variations. The internal construction parameters of the strand are also incorporated in this model.

The mechanical stress-strain relationship for a linear viscoelastic solid is modeled by combining linear springs and dashpots into what is known as the Wiechert model. The parameters for this time-dependent model are determined for a specific linear viscoelastic material, polymethyl methacrylate (PMMA), by using the Schapery collocation method. A numerical integration technique is then introduced to solve a convolution integral which developed, as a result of the correspondence principle of linear viscoelasticity, from the inverse Laplace transform of the elastic solution for a multilayer strand. This integration technique is combined with a modified Newton's iterative technique to solve the time-dependent, geometric and loading relationships in the strand.

ACKNOWLEDGEMENTS

TAC acknowledges, with thanks, the State of Ohio: Board of Regents (Grants FRG #1194 and FRG #1208) for its support.

REFERENCES

Bland, D. R. (1960), The Theory Of Linear Viscoelasticity, Pergamon Press, New York, Chapter 1.

Conway, T. A. and Costello, G. A. (1991), "Response Of A Strand With Elliptical Outer Wires", International Journal Of Solids And Structures, Vol. 28, No. 1, pp. 33-42.

Conway, T. A. and Costello, G. A. (1992a), "Time-Dependent Response Of A Three Filament Cord", Journal Of Engineering Mechanics, ASCE, to be published.

Conway, T. A. and Costello, G. A. (1992b), "Viscoelastic Response Of A Simple Strand", International Journal of Solids and Structures, to be published.

Costello, G. A. (1990), Theory of Wire Rope, Springer-Verlag, New York, Chapters 1-4.

Hearle, J. W. S., Grosberg, P. and Backer, S. (1969), Structural Mechanics of Fibers, Yarns, and Fabrics, Wiley, New York, Chapter 4.

Lee, E. H. and Rogers, T. G. (1963), "Solution Of Viscoelastic Stress Analysis Problems Using Measured Creep Or Relaxation Functions", Journal Of Applied Mechanics, **33**, pp. 127-133.

Tschoegl, N. W. (1989), The Phenomenological Theory Of Linear Viscoelastic Behavior, Springer-Verlag, New York, Sections 2.0, 3.6 and 11.4.

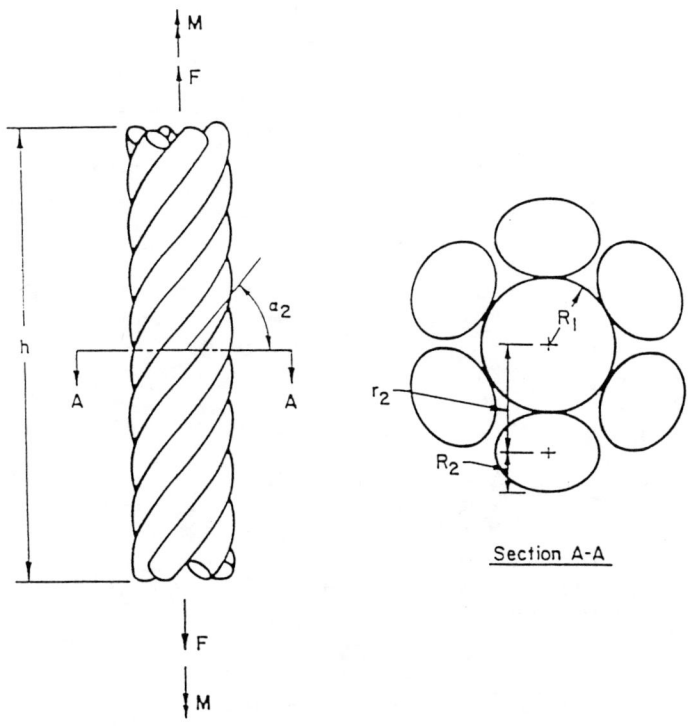

Fig. 1: Side view & cross-section of a simple strand.

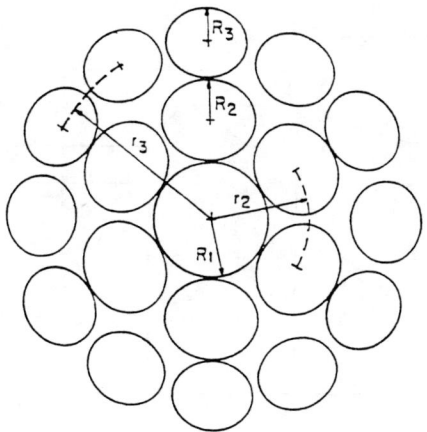

Fig. 2: Cross-section of a three layer strand.

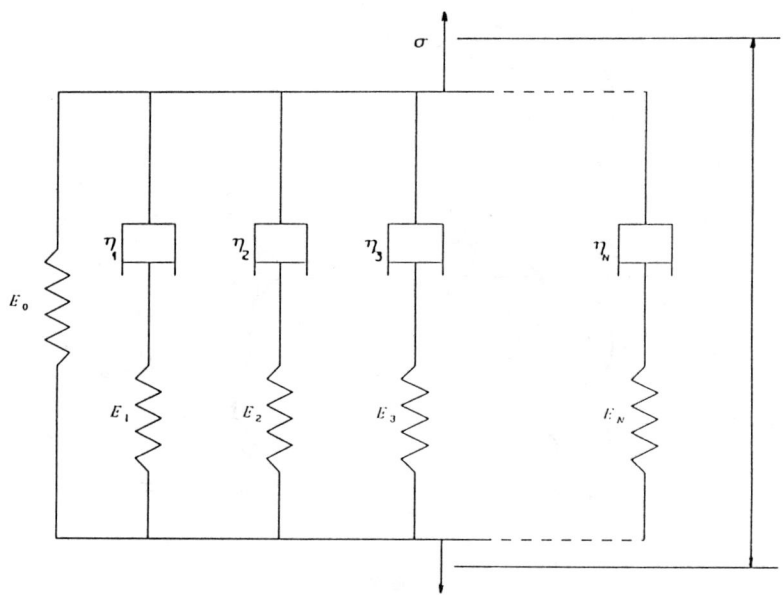

Fig. 3: Wiechert model

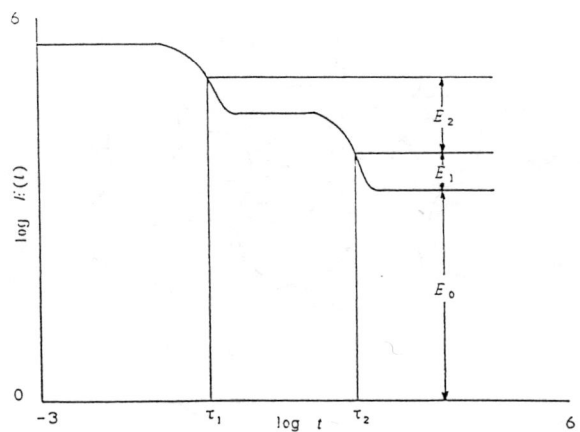

Fig. 4: Theoretical stress relaxation response.

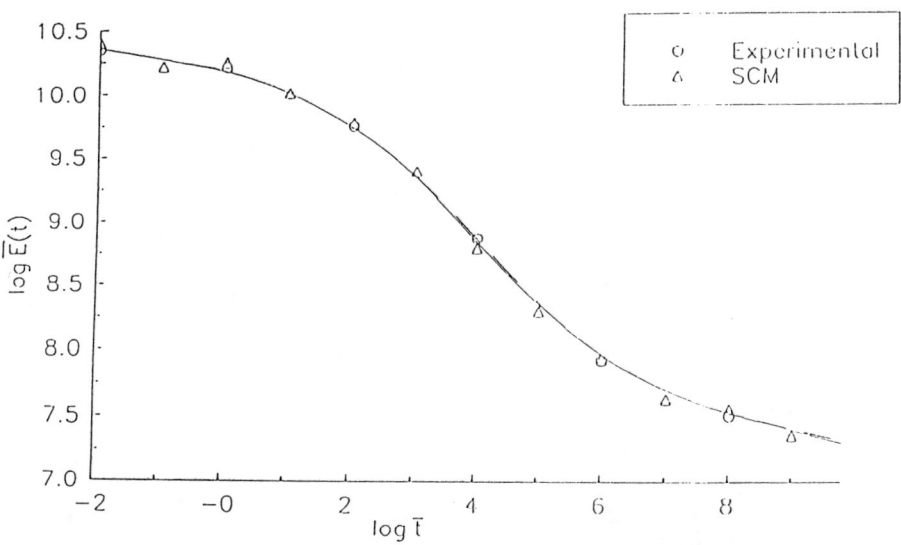

Fig. 5: Stress relaxation for polymethyl methacrylate. (Lee & Rogers, 1963)

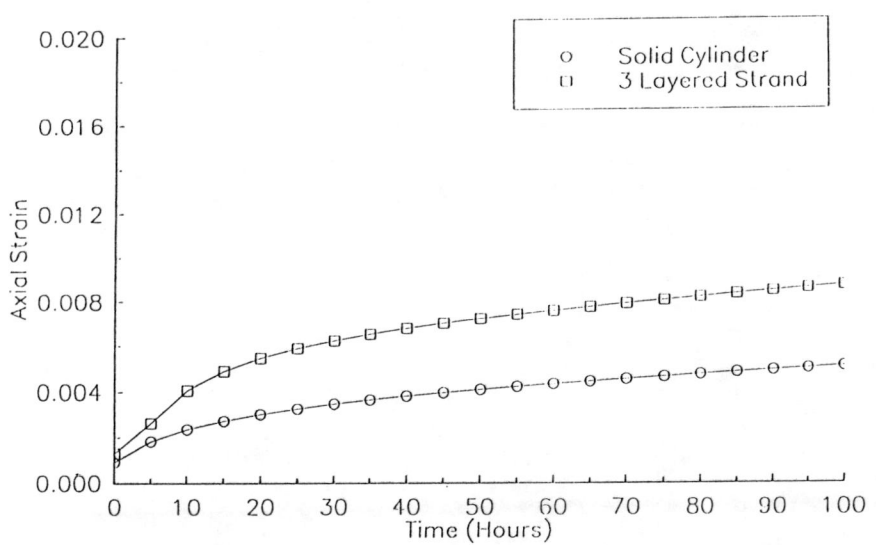

Fig. 6: Creep response for three layer strand of PMMA.

DELAMINATION-FREE DRILLING OF COMPOSITE LAMINATES

Sanjeev Jain* and Daniel C. H. Yang
Department of Mechanical, Aerospace and Nuclear Engineering
University of California, Los Angeles
Los Angeles, California

Abstract

Composite laminates in significant numbers are rendered unacceptable due to delamination that occurs during the drilling operation. Thrust generated during the drilling operation is identified as responsible for delamination. Expressions developed for critical thrusts and critical feed rates, by modeling the delamination zone as an elliptical plate in unidirectional laminates, appear to be fairly accurate. It has been demonstrated that the critical thrusts and feed rates obtained for unidirectional laminates can be conservatively used for multi-directional laminates. With regard to the tool geometry, the chisel edge width appears to be the single most important factor contributing to the thrust force and hence delamination. A hollow grinding drill tool was designed and tested. This tool resulted in a much smaller thrust and much better hole quality as compared with the standard twist drills.

Nomenclature

a, b   Half major axis and minor axis lengths respectively of the elliptical delamination

a/b   Ellipticity ratio

P   Thrust force generated in the drilling operation

P*   Critical thrust force when the crack just begins to propagate (a function of ellipticity ratio)

Pc^* Minimum critical thrust force corresponding to a particular ellipticity ratio

Pc^*, Pc^{**}, Pc^{***} Critical thrusts corresponding to delamination occurring at the bottom ply, second ply and third ply from bottom respectively

G   Crack propagation energy

G_{IC} Critical crack propagation energy in mode I

h   Thickness of the plate below the delamination crack

---

*Currently at National Institute of Standards and Technology, Gaithersburg, Maryland

D_{11} Bending stiffness along the fiber direction
D_{22} Bending stiffness in the direction transverse to the fibers
E_{11}, E_{22} Young's moduli along fiber and transverse directions respectively
f   Feed of the drilling operation (mm/rev)
fc* Feed corresponding to Pc*
fc*, fc**, fc*** Critical feeds corresponding to delamination occurring at the bottom ply, second ply and third ply from bottom respectively
d   Diameter of the drill tool
H_B Brinell hardness number (Kg/mm^2)
G_{66} Shear modulus of the laminate
U   Strain energy stored in a plate when it is bent

1. Introduction

The real impetus for manufacturing advanced composites has come from the aircraft and aerospace industries, where lightweight design and engineering have become increasingly important. The prime objective of this effort is to improve the performance to weight ratio. In addition to light weight, these composites offer high strength, high modulus and, most importantly, excellent fatigue performance. The properties of these materials can be tailored to suit one or more engineering goals.

Over the last decade or so, the use of fiber-reinforced polymers has increased dramatically, continually leading to new applications. Initially the cost of these materials was very high, justified only for specialized, low volume applications such as aerospace, defense etc. As these materials and their manufacturing methods are becoming cheaper, they are finding an increasing use in consumer-oriented applications. As confidence in composites technology builds up, a greater fraction of commercial aircraft will be constructed with composites. As stricter mileage requirements are imposed on automotive industry, composites will inevitably become viable for automobiles. Computer industry will increasingly use composite laminates to tailor the thermal properties of printed wire boards. Medical industry is considering the use of composites for prostheses and implants.

As composites become more and more popular, an increasing emphasis is placed on manufacturing and fabricating them better, cheaper and faster. Composites pose additional difficulty owing to their inhomogeneity and anisotropy. While some studies have focused on general aspects of composites machining [1-4], several recent studies [5-16] have emphasized drilling of composites. These studies on drilling report empirical data and observations. The production of holes in composite materials presents different problems from those encountered in drilling metals. Most of the problems are quality-related. Delamination has been recognized as one major problem. It is generally regarded as a resin or matrix dominated failure behavior, which usually occurs in the interply region. It appears as peeling away of the bottom ply or plies and is attributed to the thrust of the drill which pushes the layers apart rather than cutting through them.

The presence of delamination reduces the stiffness and strength of a laminate and hence its load carrying capacity. Delamination can often be the limiting factor in the use of composite materials for structural applications, particularly when subjected to compressive, shear and fatigue type of loads and when exposed to moisture and other aggressive environments over a long period of time. If delamination is so detrimental, an attempt should be made to avoid delamination in the first place.

By design, each composite lamina possesses large strength and stiffness along the fiber direction and low strength and stiffness in the transverse direction. This is because, the properties in the transverse direction are governed by matrix, and matrix is weaker of the fiber and matrix components. Depending on the strength requirements, several of these laminas (or plies) can be stacked together to possess large in-plane strength, but their out-of-plane strength is inevitably very poor. Depending on the tool geometry, material being drilled, machining parameters such as feed rate, rotational speed etc., a thrust force is generated during the drilling operation. This out-of-plane thrust force is responsible for occurrence of delamination during drilling.

While thrust force has been cited as the cause of delamination by several researchers and some [3] have even reported the value of thrust force at which delamination just begins to occur, no realistic model has been proposed to analytically correlate thrust force (or feed rate) with the onset of delamination. The present paper represents authors' continued effort to better understand and solve the problem of delamination accompanied with the drilling operation [18-19].

2. Delamination in Unidirectional (UD) Laminates

Initial results from experiments on UD laminates were reported earlier [18-19]. A considerable number of experiments have been performed since then and that has helped refine the model to its present form.

Experimental Setup: A commercial strain-gage based sensor was used to measure the thrust force. The signal was then fed into a HP-54501A digitizing oscilloscope. A dedicated printer permitted printing of the waveforms. The data was acquired as an ASCII data file into an IBM-PC-AT, using an HP-IB interface bus. The ASCII data was then plotted using a graphics package. No coolant was used during the drilling experiments. A vacuum pump was, however, used to suction away the powdery chips. The schematic of the setup is shown in Fig. 1.

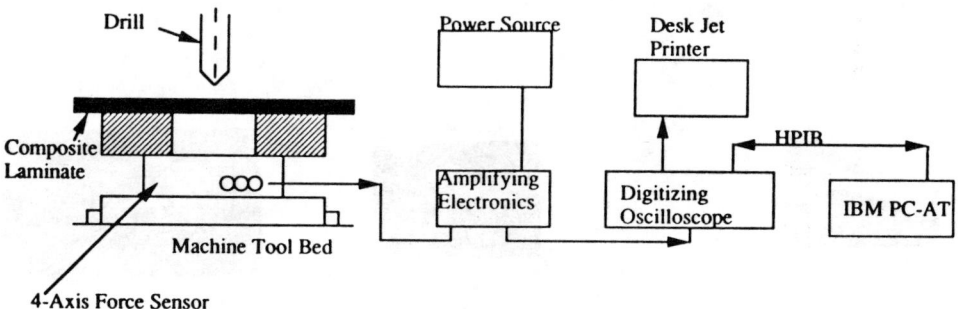

Fig. 1 Schematic of the experimental setup

2.1 Critical Thrusts and Feed Rates in UD Laminates

Fig. 2 shows the schematic of delamination during a drilling operation. For the case of a unidirectional laminate, the realistic shape of delamination is elliptical, with the directions parallel to and transverse to fibers also being principal directions as shown in Fig. 3. This shape has been verified by experiments. Fig. 4 shows the photograph of an actual delamination and a C-scan of a hole drilled in a UD laminate.

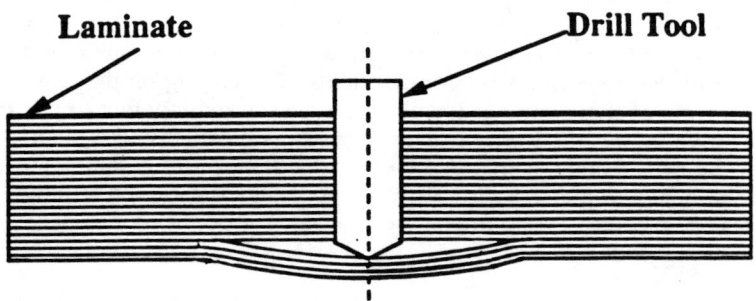

Fig. 2 A schematic depiction of delamination in laminated composites

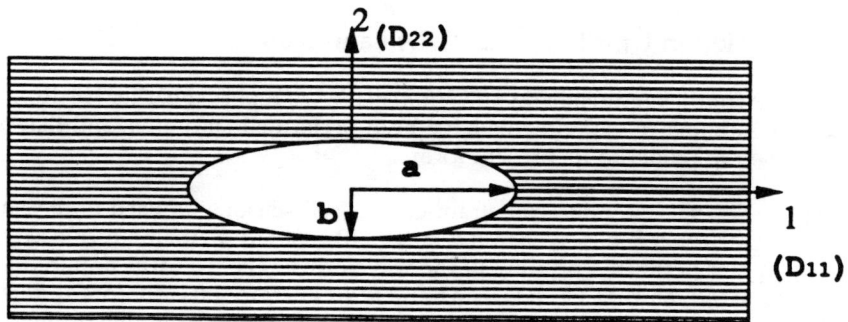

Fig. 3 Schematic shape of the delamination zone in a (unidirectional) laminate

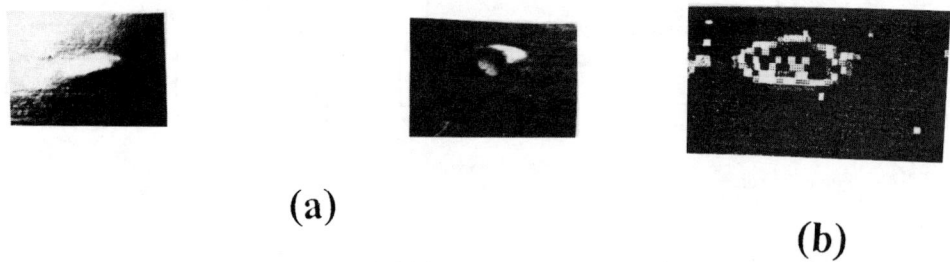

(a)   (b)

Fig. 4 (a) Actual shape of the delamination in a drilled unidirectional laminate, and (b) a C-scan of a delamination around a drilled hole

After incorporating the assumption of (i) clamped boundary condition, (ii) concentrated load at the center of the plate, (iii) delamination crack propagation via mode I, (iv) in a self-similar manner, and (v) applicability of linear elastic fracture mechanics, the following expression for critical thrust was obtained.

$$P^* = 3\pi \left(\frac{b}{a}\right)\sqrt{2G_{IC}D^*} \quad \text{where } D^* = D_{11} + \frac{2(D_{12}+2D_{66})}{3}\left(\frac{a}{b}\right)^2 + D_{22}\left(\frac{a}{b}\right)^4 \quad (1)$$

and in turn $D_{11} = \dfrac{E_{11}h^3}{12(1-\nu_{12}\nu_{21})}$, $D_{22} = \dfrac{E_{22}h^3}{12(1-\nu_{12}\nu_{21})}$, $D_{12} = \dfrac{\nu_{12}E_{22}h^3}{12(1-\nu_{12}\nu_{21})}$ and $D_{66} = \dfrac{G_{66}h^3}{12}$

In the above equations D's are bending stiffnesses. Subscript '11' signifies the fiber direction and '22' the transverse direction. G_{IC} is the critical crack propagation energy and represents the minimum energy that needs to be supplied to propagate crack of unit area quasi-statically.

The critical thrust is a function of ellipticity ratio 'a/b'. Minimizing P^* with respect to (a/b) results in a value of $a/b = (D_{11}/D_{22})^{.25}$. For higher values of thrust, a higher a/b ratio results. Since our goal is to avoid delamination altogether, the absolutely safe critical thrust is given by substituting $a/b = (D_{11}/D_{22})^{.25}$ in the above expression for thrust, i.e.,

$$P_c^* = 3\pi \sqrt[4]{\frac{D_{22}}{D_{11}}}\sqrt{2G_{IC}D_c^*} \quad \text{where } D_c^* = 2D_{11} + \frac{2(D_{12}+2D_{66})}{3}\sqrt{\frac{D_{11}}{D_{22}}} \quad (2)$$

Having established critical thrust, the next step will be to establish critical feed rate. It is important to correlate feed rate with thrust forces because once this relation is established, one can directly program feed rates into machining commands so that delamination is avoided. This will eliminate the need to monitor thrust.

Depending on feed rate and other cutting conditions, a certain thrust force will be generated. Thrust has been found to be a function of feed per revolution 'f' rather than the linear feed rate in inches per minute.

A set of experiments was performed to correlate thrust with feed rate for composites. Several holes were drilled using standard High Speed Steel (HSS) drills of 6.25mm and 12.5mm diameter. Thrusts were measured for different feed rates while drilling the unidirectional T300/5208 laminate. Figs. 5 and 6 show the plots. Curve fitting through our data points yielded the following power relation:

$$P = 56.3 \, (f)^{.40}$$

where P is thrust force in Newtons and f is the feed rate in (10^{-3}mm/rev). As shown in Fig. 5, a set of experiments with 12.5mm standard HSS drills revealed a similar conformance to the above power law dependence of thrust on feed ($P=120.7(f)^{0.39}$).

In another set of experiments, holes were drilled with 1.63mm, 2.1mm, 3.15mm, 6.25mm and 12.5mm HSS tools at the same feed rate. The following relationship was obtained upon curve-fitting (Fig. 7):

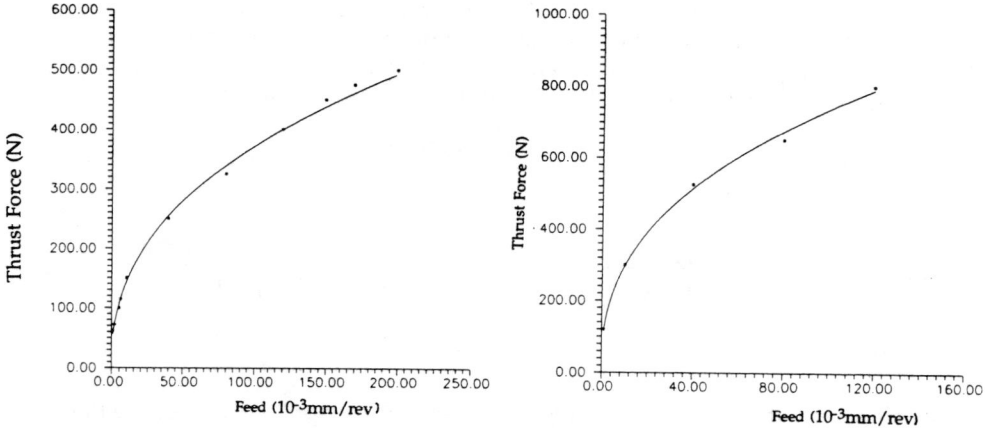

Fig. 5 Feed versus thrust for a 6.25mm standard, HSS drill   Fig. 6 Feed versus thrust for a 12.5mm standard, HSS drill

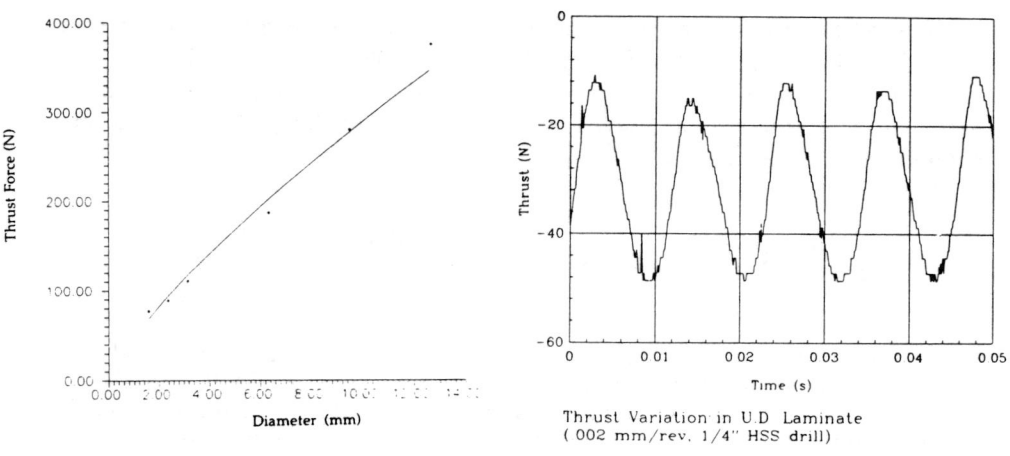

Fig. 7 Thrust versus drill diameter (with identical feed rate)

Thrust Variation in U.D Laminate
(.002 mm/rev, 1/4" HSS drill)

Fig. 8 Thrust variation with time in Unidirectional laminate

$P = 48.0 \, (d)^{0.78}$

where drill diameter d is in mm's. Combining the diameter and feed dependence equations and incorporating the assumed proportionality of thrust to H_B (as is typically done for metals, [20]), we get:

$$P = 0.136 \, H_B (d)^{0.78} (f)^{0.4} \qquad (3)$$

Substituting for P in the relation for critical thrusts, we can obtain corresponding critical feeds, given by the following equation

$$f_c^* = \left\{ \frac{3\pi}{0.136 \, H_B \, (d)^{0.78}} \left[\sqrt[4]{\frac{D_{22}}{D_{11}}} \sqrt{2 G_{IC} D_c^*} \right] \right\}^{2.5} \qquad (4)$$

Based on the above model, a variable feed rate strategy was devised, which helped drill in a time optimal fashion while avoiding delamination [18]. According to this strategy, one would drill as fast as practically permissible in the beginning and then progressively decrease the feed rate as the tool approached exit. More specifically, if Pc* corresponds to delamination at the first ply from the bottom, Pc** to the second ply, Pc*** to the third ply, then the corresponding feed rates can be calculated by using equation (4) and called fc*, fc** and fc*** respectively. When the tool nears the exit, it can be slowed to fc*** just before reaching the third ply, to fc** before reaching the second ply and to fc* before the bottom-most ply. This strategy when experimentally implemented resulted in a delamination-free hole.

The critical thrust was also found to be consistent with that reported in the literature, and was 26.5 N for 5 mil ply laminate and 64 N for 9 mil ply laminate corresponding to delamination at the bottom ply [19].

3. Delamination in Multidirectional (MD) Laminates

Delamination is also observed in MD laminates. For MD laminates analysis and modeling become very difficult. It has been demonstrated in the following section that the model established for UD laminates can be conservatively used for MD laminates.

3.1 Critical Thrust and Feed Rates in MD Laminates

To an advancing tool the first ply from the bottom always appears to be a UD ply irrespective of whether it is a UD laminate or MD laminate. If the the delamination were to take place at the bottom ply, the shape would always be elliptical with major axis along the fiber direction and minor axis in the transverse direction.

Let us next see what happens for the second, third, etc. plies from the bottom. In a UD laminate, according to classical laminate theory, an out of plane load, results in pure bending of the laminate. For the case of a general, multi-directional laminate, the load will cause bending, twisting, midplane extension and shear of the plate. Moreover, the shape of delamination at different ply locations is likely to be different. This makes the analysis much more complicated.

For instance, let us consider the delamination to take place at the second ply (from bottom) during drilling of a cross-ply laminate. If we make a simplifying assumption that bending is the only mode of plate deformation, and ignore twisting and mid-plane extension, and that the shape of delamination is circular (which is likely since $D_{11} = D_{22}$), an approximate expression for critical thrust can be obtained. From fracture mechanics, the energy balance equation is given as:

$$Pdw_0 = GdA + dU \qquad (5)$$

where G is the crack propagation energy, dU is the infinitesimal strain energy and dA is the increase in the area of the crack.

Using Timoshenko's plate theory [17], the deflection 'w' and strain energy 'U' for a circular plate of radius 'a', clamped and subjected to a central load 'P', are given by:

$$w = w_0 (1 - \frac{x^2}{a^2} - \frac{y^2}{a^2})^2 \qquad (6)$$

where $w_0 = \dfrac{Pa^2}{\pi[6D_{11} + 4(D_{12} + 2D_{66}) + 6D_{22}]}$

$$U = \frac{1}{2} \int \int [D_{11}(\frac{\partial^2 w}{\partial x^2})^2 + 2D_{12}\frac{\partial^2 w}{\partial x^2}\frac{\partial^2 w}{\partial y^2} + D_{22}(\frac{\partial^2 w}{\partial y^2})^2 + 4D_{66}(\frac{\partial^2 w}{\partial x \partial y})^2] dx dy$$

Differentiating, we get

$$dw_0 = \dfrac{Pada}{\pi[3D_{11} + 2(D_{12} + 2D_{66}) + 3D_{22}]}$$

$dA = 2\pi a da$

$$dU = \dfrac{2Pada}{3\pi[3D_{11} + 2(D_{12} + 2D_{66}) + 3D_{22}]}$$

Substituting these in the energy balance equation $Pdw_0 = dU + GdA$, solving for P and replacing G by G_{IC} for critical thrust, we get

$$P_c^* = P^* = \pi \sqrt{6G_{IC}[3D_{11} + 2(D_{12} + 2D_{66}) + 3D_{22}]} \qquad (7)$$

Upon substituting into the above equation various stiffness values corresponding to 2 plies--one 0° and the other 90°--each 0.125mm thick, the value of turns out to be 108.5 N. The value of critical thrust for a 2-ply thick UD laminate comes out to be 75 N. Thus, the critical thrust value arrived for a UD laminate can be conservatively used for a cross-ply laminate.

3.2 Experiments and Results for MD Laminates

Experiments were performed on cross-ply and quasi-isotropic laminates, which fall under the category of multi-directional laminates. The cross-ply laminate was 20-ply thick and the quasi-isotropic was 24-ply thick, with a ply thickness of about 0.125mm. Holes were drilled at varying feed rates and cutting forces were measured and observations regarding occurrence and extent of delamination were made.

Figs. 8, 9 and 10 show thrust variation plots for UD, cross-ply and quasi-isotropic laminates, respectively. The maximum magnitude of the thrust is of interest to us because delamination will propagate at this thrust. From the comparison of Figs. 8 and 9, it is clear that the magnitude of cyclic thrust variation has decreased, but the frequency has increased. However, a further decrease in magnitude and a further increase in frequency is not noticed in Fig. 10 corresponding to quasi-isotropic laminate.

The general decrease in magnitude and increase in frequency of the thrust undualtions can be reasoned out as follows. The two drill lips are 180° apart. In the case of a unidirectional laminate, the two lips simultaneously cut the fibers twice during each rotation. Since fiber is the stronger among the fiber/matrix components, each time the cutting lips cut perpendicularly through the fibers, the force is maximum. The force is minimum when the cutting lips move parallel to the fibers. For a U.D. laminate all the fibers are aligned in the same direction and therefore maximum thrust occurs when the cutting edges cut through them

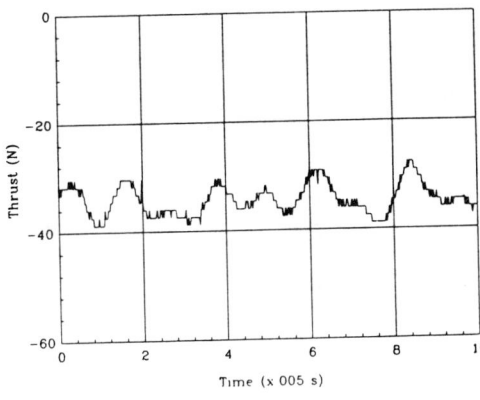
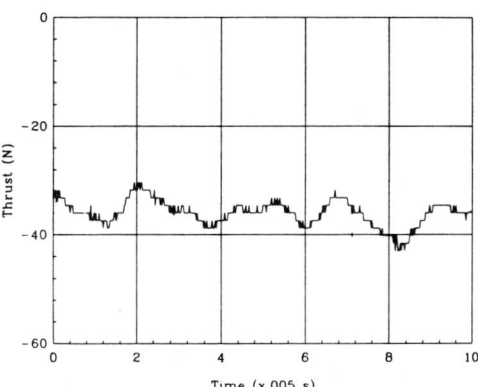

Thrust Variation in Cross-Ply laminate (.002 mm/rev, 1/4" HSS drill)

Thrust Variation in Quasi-isotropic Laminate) (.002 mm/rev, 1/4" HSS drill)

Fig. 9 Thrust variation with time in Cross-Ply laminate

Fig. 10 Thrust variation with time in Quasi-Isotropic laminate

perpendicularly. For a multidirectional laminate, fibers are distributed and therefore, the maximum thrust is not as high as in the case of a U.D. laminate and the minimum is not as low the minimum in U.D. laminates. Thus, some sort of averaging effect takes place in the case of multi-directional laminates.

Fig. 11 shows the plots of thrust variation with feed rate. As stated earlier, the thrusts for cross-ply and quasi-isotropic laminates are consistently lower than the unidirectional laminates. Curve fitting through the data points yielded the following relationships between thrusts and feeds:

$P = 41.8 \, (f)^{.317}$ for cross-ply laminates, and

$P = 43.1 \, (f)^{.224}$ for quasi-isotropic laminates,   (8)

where P is thrust in Newtons and f is feed in 10^{-3} mm/rev.

A diametral relationship, similar to that for the U.D. laminate, could not be established for the cross-ply and quasi-isotropic laminates due to relative thinness of these available laminates. For example, the conical part of a 12.5 mm drill will not fully immerse into these laminates and therefore the recorded thrust would be a function of thickness.

Based on above observations, it can be argued that since (i) for any feed rate the thrust generated in MD laminates is lower, and additionally (ii) the critical thrust for MD laminates is higher, the feed rates obtained for UD laminates can be obtained safely for MD laminates. These feed rates will be somewhat sub-optimal for MD laminates, but they still allow appreciable feed rates while avoiding delamination. The above was verified experimentally. Every time there was no delamination in UD laminates, there was also no delamination seen in MD laminates under identical drilling conditions. However, there were times when some delamination was seen in UD laminates, but no delamination was noticed in MD laminates under the same drilling conditions. Fig. 12 shows the holes drilled in UD, cross-ply and quasi-isotropic laminates at the same feed rate. Whereas the UD laminate exhibits delamination, the other two are delamination-free.

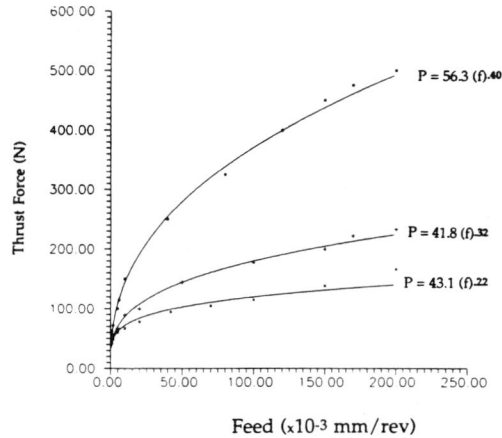

Fig. 11 Thrust variation with feed rate for different laminate configurations.

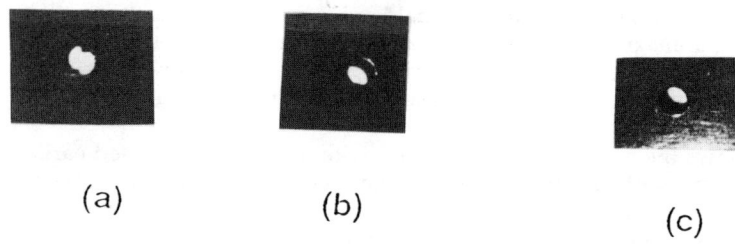

Fig. 12 Appearance of holes drilled in unidirectional, cross-ply and quasi-isotropic laminate (at the same feed rate, 0.1 mm/rev).

4. A Hollow-Grinding Drill

From our analysis and experiments, it became apparent that the conventional HSS drill can be used for drilling holes of diameters less than 6.25mm by exploiting variable feed rate strategy. For larger diameters, variable feed rate strategy is not good enough because even at zero feed rate, these drills will give rise to thrust larger than the critical thrust. This highlights the need to modify tool geometry so that it produces less thrust in the first place.

4.1 Effect of Tool Geometry on Thrust in Drilling

Oxford [21] has pointed out that the thrust is made up of three components. The three components are called--(i) primary cutting at the two lips, (ii) secondary cutting at the outer part of the chisel edge and (iii) indentation (or extrusion) at the center of the chisel. Although the contribution of chisel edge to the torque may not be significant, it has major contribution to thrust. At normal feed rates, the total chisel edge contribution to the thrust may be as high as 40 to 60 percent. For this reason, an in-depth study of the effect of chisel edge width on thrust force was carried out experimentally.

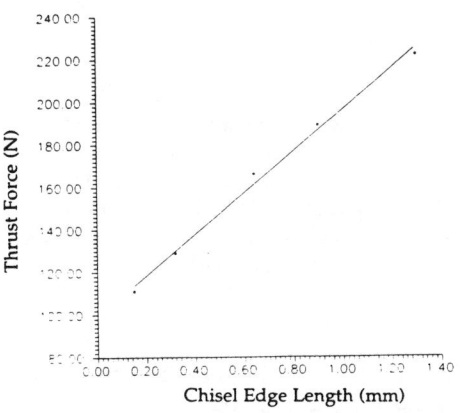

Fig. 13 Plot of thrust force variation with varying chisel edge length

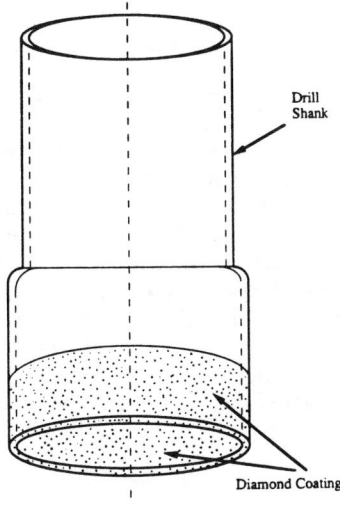

Fig. 14 Schematic of a hollow grinding drill.

In a study of the effect of chisel edge, 6.25mm holes were drilled with drill bits ground to possess different chisel widths, while keeping other parameters constant. Fig. 13 shows profound effect of the chisel edge on the thrust force. Based on these experiments, it became clear that the chisel edge is a major contributor to the thrust force and hence the occurrence of delamination in laminated composites. Experiments to study the effect of point angle revealed it to be of minor significance.

4.2 Rationale behind a Hollow Grinding Drill

From our study of milling operation [22], it became clear that one way to decrease cutting forces is to increase the number of teeth (for a given feed rate and rpm). Although the average force remains about the same, when cutting with a larger number of teeth the magnitude of undulations decreases while their frequency increases. Thus, increasing the number of teeth decreases the magnitude of maximum cutting force generated. Using this idea a cutting tool was conceived which is tubular in appearance with one end of the tube coated with diamond particles of a certain grit size. The choice of grit size will depend on the feed/rev and on the quality of hole desired. The mechanism of material removal is grinding. These small angular diamond particles offer numerous cutting edges and therefore, the thrust force variation is very smooth. This type of drill can be called a hollow-grinding (HG) drill and appears as shown in Fig. 14.

Since the drill is hollow, there is no contribution of chisel edge as in the case of a conventional drill. Moreover, in conventional drilling, the entire volume under the tool diameter is cut down into small chips and for the case of HG drill, depending on the wall thickness of the drill, only a fraction of the entire volume is ground away. This also results in thrust reduction. Finally, a conventional drill due to its point angle needs to travel an extra distance in order to completely clear the laminate, whereas due to its flat end, the HG drill can stop right after it reaches the bottom of the laminate. This results in time savings.

4.3 Results from the HG Drilling Experiment

A 1/2" outer diameter HG drill was designed and custom-ordered from an Ohio manufacturer of diamond coated tools. A 12.5mm hole was drilled on the cross-ply laminate at 12.5mm/min at 600 rpm (about .02 mm/rev). This resulted in a fairly steady thrust variation with a maximum at about 45 N, which resulted in a near perfect hole, without any delamination or any other apparent hole defect. Figs. 15 and 16 show the thrust variations obtained under identical conditions with an HG drill and a standard HSS drill. The thrust for the case of the HSS drill is nearly three times as much and the hole quality is very poor as compared with that from the HG drill. Fig. 17 shows the photographs of the two holes.

Experiments with HG drill on the same laminate at 6.5 mm/min and 12.7 mm/min revealed thrusts to be 28 N and 70 N and again yielded holes of good quality.

5. Recommendations for Delamination Prevention while Drilling Composite Laminates

Use variable feed rate for diameters upto 6.25mm, when only conventional, standard twist drills are available.

Modify tool geometry especially chisel edge in combination with newer, harder materials such as HSS drills brazed with poly crystalline diamond bits.

Use the HG drill, designed and tested to exert much smaller thrust in the first place. This becomes especially viable for holes of diameter larger than 12.5mm and when the laminate thickness is 12.5mm or less.

6. Conclusions

The results obtained from our experiments agree with the critical thrust and critical feed rate models proposed. Gradually lowering the feed rate in the last few plies will result in a delamination-free hole. But for 12.5mm diameter drill, the limiting thrust, obtained by extrapolating to zero feed rate, is over 100 N, far above the critical thrust. In this case, delamination is inevitable if conventional drill point tool is used. This highlights the need to modify tool geometry to obtain sub-critical thrust at a reasonable feed rate. From the limited experiments on chisel edge, it has become clear that it is possible to reduce the thrust force significantly by reducing the chisel edge width. In practice, a combination of feed rate strategy and modified tool geometry can be used to avoid delamination.

In conclusion, a model has been proposed to predict critical thrusts and feed rates at different ply levels. The model, though developed for unidirectional laminates, can be conservatively used for other multi-directional laminates. While small chisel edge is desirable, the ability of the tool material to provide sufficient strength and hardness at those small dimensions may be a limiting factor. The need to combine small chisel edge with newer materials such as carbides and poly-crystalline diamond is emphasized. A cross-section of conventional and modified geometry tools are tested. Lastly, a tool called hollow grinding drill has been developed which can cut clean holes.

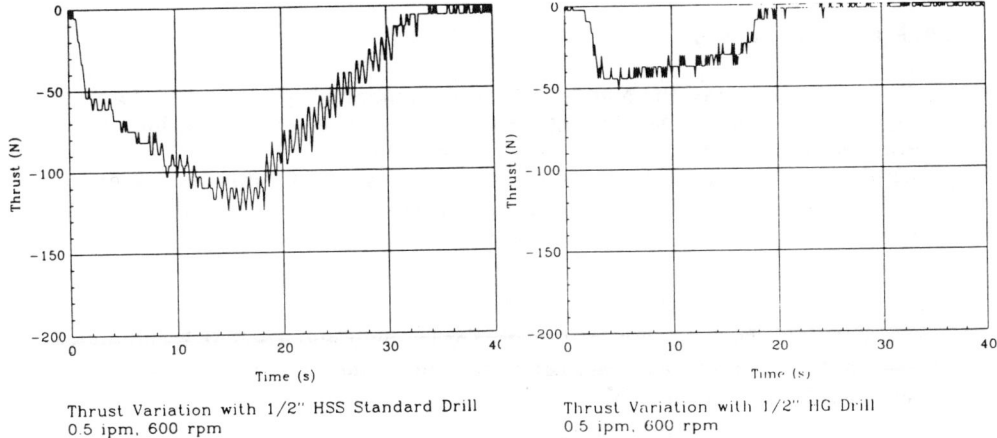

Fig. 15 Thrust force variation in drilling with standard HSS, 1/2" dia tool.

Fig. 16 Thrust force variation in drilling with the 1/2" dia, hollow-grinding drill tool.

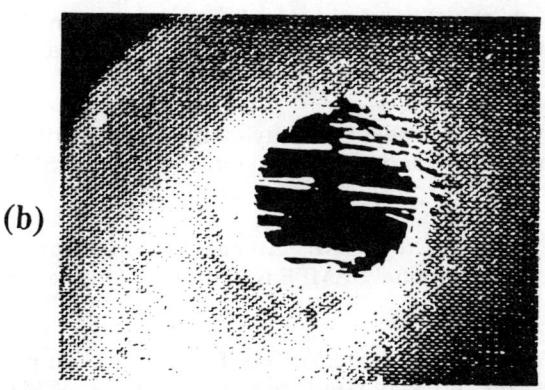

Fig. 17 Photographs of holes drilled with 12.5 mm (a) HG and (b) HSS drill tools

7. Acknowledgements

The financial supports of the National Science Foundation (IUCRC for Intelligent Manufacturing Systems CDR-8717322) and the Institute for Manufacturing and Automation Research (IMAR) are greatly appreciated. Both the center (IUCRC) and the institute (IMAR) are jointly hosted by the University of California, Los Angeles (UCLA) and the University of Southern California (USC).

8. References

[1] Doerr R. et al,"Development of Effective Machining and Tooling Techniques for Kevlar Composite Laminates", Fabrication of Composites Source Book, ed. Mel M. Schwartz, pp. 339-384, ASM, 1985.

[2] Koplev A., Lystrup Aa. and Vorm T.,"The cutting process, chips, and cutting forces in machining CFRP", Composites, v.14, no. 4, October 1983, pp.371-376.

[3] Friend C.A., Clyne R.W. and Valentine G.G,"Machining graphite composite materials", Composite materials in engineering design, ed. Noton, B.R., pp.217-225.

[4] Koenig W., Wulf Ch., Grass P. and Willerscheid H.," Machining of fiber-reinforced plastics", Annals of the CIRP, vol.34, no.2, 1985, pp.538-548.

[5] Doran J.H. and Maikish C.R.,"Machining boron composites", Composite materials in engineering design, ed. Noton, B.R., pp.242-250.

[6] Cutting Advanced Composite Materials, Manufacturing Engineering, v. 79, November 1977, pp.63-65.

[7] Petrof R.C.,"On the dynamics of drilling glass reinforced plastics with different drill point geometries", Autocom '86 conference, June 9-12, Dearborn, Michigan.

[8] Koenig W.and Grass P.,"Quality definition and assessment in drilling of fiber reinforced thermosets", Annals of the CIRP, vol.38, no.1, 1989, pp.119-124.

[9] Koenig W., Grass P., Heintze A., Okcu, F. and Schmitz-Justin C.,"Developments in drilling, contouring composites containing Kevlar", Production Engineer, Sept. 1984, pp. 56-61.

[10] Sakuma K., Yokoo, Y. and Seto M., "Study on drilling of reinforced plastics - relation between tool material and wear behavior", Bulletin of JSME, pp. 1237-1244, v.27, no.228, June 1984.

[11] Wong T.L., Wu S.M. and Croy, G.M., "An analysis of delamination in drilling composite materials", 14th national SAMPE technical conference, October 12-14, 1982, pp. 471-483.

[12] Tagliaferri V., Caprino G. and Diterlizzi A.,"Effect of drilling parameters on the finish and mechanical properties of GFRP composites", Int. J. of Machine Tools and manufacture, v.30, no.1, 1990, pp.77-84.

[13] Ho-Cheng H. and Dharan C.K.H., "Delamination during drilling in composite laminates", J. of Engineering for Industry, August 1990.

[14] Bishop G.R. and Gindy N.N.Z.,"An investigation into the drilling of ballistic kevlar composites", Composites Manufacturing, Sept. 1990, pp. 155-159.

[15] Hickey J.,"Drilling graphite composites", Modern machine shop, March 1987, pp.84-90.

[16] Beall, R.T.,"Drilling composites with Gun Drills", Fabrication of Composites Source Book, ed. Mel M. Schwartz, pp. 320-326, ASM, 1985.

[17] Timoshenko S. and Woinowsky-Keiger S., Theory of plates and shells, 2nd ed., McGraw Hill, 1959, pp.51-59, 376-377.

[18] S. Jain & D. Yang, "Variable Feedrate Drilling for Delamination Prevention in Composites", Presented at the ASME/ETCE conference held in Houston, Texas, Jan.21-23, 1991.

[19] S. Jain & D. Yang, "Effects of Chisel Edge and Feed Rate for Delamination Prevention in Drilling of Composites", 1991, ASME Winter Annual Meeting, PED. vol 49, pp. 37-52.

[20] Shaw M.C., Metal cutting principles, Clarendon press, Oxford, 1984, pp.475-480.

[21] Oxford C.J., "On the drilling of metal - Basic mechanics of the process", Transactions of ASME, Feb. 1955, pp.103-114.

[22] S. Jain & D. Yang, "A Systematic Force Analysis of the Milling Operation", 1989, ASME Winter Annual Meeting, PED. vol 40, pp.55-63.

ELASTIC PROPERTIES OF TWO-DIMENSIONAL COMPOSITES CONTAINING POLYGONAL HOLES

I. Jasiuk
Department of Materials Science and Mechanics
Michigan State University
East Lansing, Michigan

J. Chen and M. F. Thorpe
Department of Physics and Astronomy
Center for Fundamental Materials Research
Michigan State University
East Lansing, Michigan

ABSTRACT

Evaluation of the effective elastic moduli of two dimensional materials containing polygonal holes reveals that the area Young's modulus is independent of the Poisson's ratio of the host material. In addition, the elastic response of materials containing elliptical holes and planar cracks is presented.

1. INTRODUCTION

We discuss some new exact results in two dimensional (plane) elasticity that have applications in studies of the elastic response of two dimensional composite materials. These results are related to a recently proven theorem (Cherkaev, Lurie and Milton, 1992) which reduces a parameter space of the local stress fields and the effective elastic properties. This theorem is most powerful for materials containing holes. Recently, we have shown that the effective elastic modulus of a sheet containing circular holes is independent of the Poisson's ratio of the matrix (Day et al., 1992; Thorpe and Jasiuk, 1992). Now we consider a material with polygonal holes and also show that the effective elastic modulus depends on the shape of holes but not on the Poisson's ratio of the matrix. In addition, we discuss the results available in literature for the elastic response of materials containing circular and elliptical holes, and planar cracks, which relate directly to the subject of this paper.

2. TWO-DIMENSIONAL ELASTICITY

The stress-strain equations for a linear elastic and isotropic material in three dimensions (3d) are given by (Timoshenko and Goodier, 1951)

$$\varepsilon_{ij} = \frac{1}{E'}[(1+\nu')\sigma_{ij} - \nu'\sigma_{kk}\delta_{ij}] \qquad i,j,k = 1,2,3 \qquad (1)$$

where ε_{ij} and σ_{ij} are the strain and stress tensors respectively, and E' is the 3d Young's modulus and ν' is the 3d Poisson's ratio. We use the primes for the elastic constants in three-dimensions (3d) so that we may use unprimed quantities in two-dimensions (2d).

In 2d elasticity the constitutive equations have a form similar to (1), but only involve two coordinates

$$\varepsilon_{ij} = \frac{1}{E}[(1+\nu)\sigma_{ij} - \nu\sigma_{kk}\delta_{ij}] \qquad i,j,k = 1,2 \qquad (2)$$

The unprimed quantities E and ν are the area Young's modulus and the area Poisson's ratio, respectively. Note that the area Poisson's ratio is bounded by $-1 < \nu < 1$, in contrast to the bounds $-1 < \nu < 1/2$ for the 3d Poisson's ratio. The area bulk modulus K and the shear modulus G are expressed in terms of E and ν as (Sen and Thorpe, 1985),

$$K = \frac{E}{2(1-\nu)}$$
$$G = \frac{E}{2(1+\nu)}$$

so that

$$\frac{4}{E} = \frac{1}{K} + \frac{1}{G}. \qquad (3)$$

In this paper we will use equations (2) and the various 2d elastic moduli (3). The 2d constitutive equations (2) are usually derived from (1) by assuming either <u>plane strain</u> or <u>plane stress</u>. We summarize the mappings from 3d to 2d for both plane strain and plane stress in Table 1. The 2d elastic constants discussed in this paper may represent the effective in-plane elastic constants of a transversely isotropic material or a sheet containing inclusions, for example.

Table 1. The connection between the 2d elastic constants (unprimed) and the 3d elastic constants (primed) for both plane strain and plane stress.

Plane Strain	Plane Stress
$K = K' + \dfrac{G'}{3}$	$K = \dfrac{9K'G'}{3K' + 4G'}$
$G = G'$	$G = G'$
$E = \dfrac{E'}{1-\nu'^2}$	$E = E'$
$\nu = \dfrac{\nu'}{1-\nu'}$	$\nu = \nu'$

3. CLM TRANSFORMATION AND THEOREM

A new result in plane elasticity which we refer to as the CLM theorem has recently been proved for 2d composite materials (Cherkaev, Lurie and Milton, 1992). It is based on an earlier work of Lurie and Cherkaev (1986). The CLM theorem applies to linear elastic materials with general anisotropy and holds for an arbitrary geometry. In this paper we are only concerned with the case where the components are isotropic, when the CLM theorem can be stated as follows.

Suppose that a 2d composite material has spatially varying bulk and shear moduli given by $K(\mathbf{r})$ and $G(\mathbf{r})$ respectively, and that the effective moduli of the material are K^* and G^*, then if

$$\frac{1}{K^t(\mathbf{r})} = \frac{1}{K(\mathbf{r})} - C \quad \text{and} \quad \frac{1}{G^t(\mathbf{r})} = \frac{1}{G(\mathbf{r})} + C \qquad (4)$$

then,

$$\frac{1}{K^{t*}} = \frac{1}{K^*} - C \quad \text{and} \quad \frac{1}{G^{t*}} = \frac{1}{G^*} + C \qquad (5)$$

where the superscript t denotes the transformed system and C is a constant. Note that the constant C is restricted in order to ensure that the elastic moduli are positive everywhere in the transformed system. The vector **r** lies in the plane of the 2d material. We will refer to (4) as the CLM transformation, leading to the CLM theorem (5). Under the CLM transformation the stress field is the same in both the original and the transformed material, for given external tractions, even though the elastic constants differ. Cherkaev, Lurie and Milton (1992) refer to such materials as <u>equivalent</u>. The CLM theorem contains most other previously known exact results in 2d for composite systems, as special cases (Cherkaev, Lurie and Milton, 1992; Thorpe and Jasiuk, 1992). In this paper we are interested in the case where the elastic moduli are piecewise constant and the composite contains two phases only (which is a special case of (5) and (6)). The CLM thereom applies to arbitrary geometries, but in this paper we are only concerned with polygonal and elliptical holes. The constraint on the allowed values of C means that the CLM theorem is most powerful in the limit when the inclusions are holes, and becomes useless in the limit where the sample contains rigid inclusions which require that C=0. In this paper we focus our attention on the case when the inclusions are holes.

It is useful to rewrite the CLM transformation (4) in terms of the Young's modulus using (3) to give

$$E^t(\mathbf{r}) = E(\mathbf{r}) \tag{6}$$

which indicates that the Young's moduli of the material are invariant under the CLM transformation and to rewrite (5) as

$$E^{t*} = E^* \tag{7}$$

which states that the area Young's modulus of the composite is invariant under the CLM theorem.

4. DUNDURS CONSTANTS

The CLM transformation is related to an earlier result which is due to Dundurs (1967, 1970). Dundurs showed that if a composite material consisting of two linearly elastic and isotropic phases is subjected to specified tractions and undergoes plane deformation, then the stress has a reduced dependence on the elastic constants. The stress depends on only <u>two</u> dimensionless parameters, as opposed to <u>three</u> dimensionless combinations of elastic constants in 3d. This can by written as

$$\sigma_{ij} = \sigma_{ij}(\mathbf{r}, \alpha_{12}, \beta_{12}) \tag{8}$$

where the vector $\mathbf{r}$ lies in the plane of the 2d material. The Dundurs constants α_{12} and β_{12} are defined as

$$\alpha_{12} = \frac{\frac{1}{E_1} - \frac{1}{E_2}}{\frac{1}{E_1} + \frac{1}{E_2}}$$

$$4\beta_{12} = \frac{\frac{1}{K_1} - \frac{1}{K_2}}{\frac{1}{E_1} + \frac{1}{E_2}} \tag{9}$$

where the subscripts 1 and 2 denote the matrix material and inclusions, respectively. The Dundurs result follows from the CLM transformation (4), which reduces the number of parameters by one. The parameters α_{12} and β_{12} are appropriate as they are clearly invariant under the CLM transformation. The Dundurs constants are not unique and other representations can be used.

A particularly important special case occurs when one of the components, say 2, becomes holes (of any size or shape). Then $K_2=E_2=0$, leading to $\alpha_{12}=-1$ and $4\beta_{12} = -E_2/K_2 = -2(1-\nu_2)$. For holes, the dependence on Poisson's ratio remains in the Dundurs constants, but drops out in the expressions for stresses as expected from the result of Michell (1899). Michell showed that the stresses in a 2d multiply-connected body, induced by specified tractions, are independent of the elastic constants if the resultant force over each boundary vanishes and there are no body forces. This result led to the development of the photoelasticity method, an experimental technique of using optical birefringence in transparent materials to measure the stress fields. These stress fields depend only on the geometry of the holes and <u>not</u> on the elastic constants of the material.

5. A MATERIAL CONTAINING HOLES

If a 2d composite material is made by removing material to form holes (of any size, shape, area fraction etc.), then <u>the relative area Young's modulus</u>

E^*/E of a 2d material containing holes is the same for all materials, independent of Poisson's ratio, for a prescribed geometry (Day et al., 1992). This result is easily proved by the CLM result (Cherkaev, Lurie and Milton, 1992; Thorpe and Jasiuk, 1992). The CLM transformation leaves holes as holes, and therefore any matrix material can be reached by using the CLM transformation. Also, the effective area Young's modulus remains unaffected by the change in the Poisson's ratio as seen from (7).

To illustrate this special case of the CLM theorem, we recently considered a sheet with circular holes in various regular and random arrangements and used a computer simulation (Day et al., 1992). For simplicity these holes were all of the same size. We showed that the results all lie on a single curve, independent of the Poisson's ratio of the matrix material, so that

$$E^{t*}/E^t = E^*/E \qquad (10)$$

where E is the area Young's modulus of the matrix. The area Poisson's ratio of the composite can be written as

$$\nu^{t*} - \nu^* = (\nu^t - \nu) \, E^*/E \qquad (11)$$

where we have used the values of the area Poisson's ratio ν and Young's modulus E of the matrix at the area fraction of holes $f=0$ to eliminate the (unknown) constant C. This equation provided the <u>first rigorous proof</u> of the conjectured flow (Garboczi and Thorpe, 1985; Schwartz el al., 1985; Garboczi and Thorpe, 1986) of the Poisson's ratio to a fixed point at the percolation threshold (Day et al., 1992; Thorpe and Jasiuk, 1992). The result (11) shows explicitly that $\nu^{t*} = \nu^*$ as $E^*/E \to 0$. We have found numerically that for randomly centered circular holes with $\nu=1/3$, the Poisson's ratio remains as 1/3 to within numerical accuracy for all values of f (Day et al., 1992). It should be emphasized that the universal value of the area Poisson's ratio is only with respect to the elastic constants of the material; different geometries (arrangement and shape of inclusions) will lead to different universal values of the Poisson's ratio as percolation is approached.

6. A MATERIAL WITH POLYGONAL HOLES

Next, we consider a 2d material with regular polygonal holes (as shown in Fig. 1 for triangles) and investigate the relation between the shape of polygon and the effective elastic moduli in the dilute limit.

Fig. 1   A sketch of a composite containing randomly positioned triangular inclusions.

When the area fraction of polygons f is small, the effective Young's modulus E^* can be written as

$$\frac{E^*}{E} = 1 - \alpha f \tag{12}$$

where E is the Young's modulus of the matrix and α is some positive constant which is a function of the shape of polygon only. From the CLM theorem, we know that for a given shape of polygon holes in a matrix, E^*/E is a universal curve which is independent of the Poisson's ratio of the matrix. The relation (11) holds, so that to leading order in the area fraction f we have

$$\nu^* = \nu + \beta f \tag{13}$$

where β is some constant depending upon the shape of the polygon only. By using (11), (12), and (13), we have

$$\nu^{t*} = \nu^t + f[\beta - \alpha(\nu^t - \nu)] \tag{14}$$

Therefore, if we choose

$$\nu^t = \nu_0 = \nu + \frac{\beta}{\alpha} \tag{15}$$

then $\nu^* = \nu$ to the leading order in f so that the term linear in f vanishes. We shall refer to this value of the Poisson's ratio ν_0 as the <u>invariant Poisson's ratio</u>, and the general result for the Poisson's ratio for small f can be written

$$\nu^* = \nu - \alpha(\nu-\nu_0)f \tag{16}$$

For the rest of the paper we will use the notation of (12) and (16). This means that if the matrix has an invariant Poisson's ratio given by (15), then the effective Poisson's ratio does not change (to the leading order in f) when a small concentration of polygon holes is cut in the matrix. If $\nu > \nu_0$, then the Poisson's ratio decreases with f, and vice versa, so that the Poisson's ratio tends to "flow towards" ν_0 as f increases.

The purpose of this work is to study the dependence of constants α and ν_0 on the shape of polygon. One of the extreme cases is when the polygon is a circular hole. In this case, $\alpha = 3$ and $\nu_0 = 1/3$, which can be calculated exactly using the result for a single circular inclusion. In the other cases, no analytical formulas are available yet. But from the Hashin's (Hashin, 1965) second order bounds on E^*, we can show that $3 \leq \alpha < \infty$. We note that these bounds are attained for the circular and very thin elliptical inclusions respectively, and are therefore optimal.

7. SIMULATION RESULTS

We do the numerical simulation by using a digital-image-based method (Day et al., 1992), in which we use a triangular net with central forces to represent the matrix. We place a single polygonal hole in a sample which is periodic. In order to calculate α and ν_0 accurately, we have to make the polygonal hole as small as possible. But on the other hand, the smaller the polygonal hole, the larger the finite size error. Therefore, there exists some tradeoff here. Furthermore, the exact area of the polygon is a rather ambiguous quantity due to the finite size effect. If the shape of the polygon is neither triangle nor hexagon, the situation is even worse, because in this case, it is harder to mimic the polygon exactly by cutting bonds in the triangular matrix. Of course, one can always increase the sample size to reduce the finite size error, but the computational time increases dramatically with the sample size. Other numerical methods, such as the finite element method, should give better results than we were able to obtain with the finite difference method used here. Nevertheless our results are accurate enough to be interesting.

In order to get precise results, we do our simulation in two ways. One way is to calculate the area by counting the number of pixels inside the polygonal hole and each pixel corresponds to some fixed area. The other way is that we define the area by calculating the exact area enclosed by the polygon, but the strength of the spring constants for the bonds crossing the hole matrix interface are reduced by a factor of 2. This is due to the fact that each bond at this interface is shared by both the matrix and the hole. The two sets of results of α and ν_0 given by these two methods are coincident when f is large, and the deviations become larger and larger when f becomes smaller and smaller. Fortunately, these deviations move away in opposite directions as the hole becomes smaller and f approaches 0, which unables us to take the arithmetic average of the two sets of data to reduce the finite size effect. After that, we use least-square-fit to interpolate the values of α and ν_0. Three different sample sizes, 102 x 102, 156 x 156 and 210 x 210, are used to estimate the finite size error and ν_0 in (16) is chosen to be 1/3. Because of the finite size error, we can only calculate α and ν_0 up to n = 6 (hexagon), which is converging rapidly on the large n limit of the circle. Note that all simulations were done with the spring constants of the triangular grid α, β and γ all set equal, so that the Poisson's ratio of the host was 1/3 (Day et al., 1992). Because of the CLM theorem, this single value of ν is sufficient to obtain the quantities of interest α and ν_0, which are given in the Table 2.

Table 2. The values of α for use in (12) and the invariant Poisson's ratio ν_0, with estimated errors from the simulations, for regular polygonal holes.

n	α	ν_0
3 (triangle)	4.23 ± 0.02	0.227 ± 0.003
4 (square)	3.39 ± 0.04	0.302 ± 0.006
5 (pentagon)	3.25 ± 0.02	0.325 ± 0.003
6 (hexagon)	3.14 ± 0.02	0.328 ± 0.003
∞ (circle)	3	0.333

One special case is when the regular polygon is a square. In this case, the effective elastic moduli are not isotropic quantities. All other inclusions, with regular polygons, lead to isotropic elasticity equations (2) for the composite, and no angular averaging is required. The square is different as the equivalent directions are at right angles which is not sufficient symmetry to give isotropy, and three elastic constants, rather than two are required. We compute results for the stress parallel to the side of the square, and find that α=3.78±0.02 and the invariant Poisson's ratio ν_0=0.378±0.003. Rotating the stress loading by $\pi/4$ so that it is along a diagonal of the square, we find that α=3.00±0.02 and the invariant Poisson's ratio ν_0=0.205±0.003. Averaging α and β and using (15), we obtain the results shown in Table 2. This is the correct procedure to first order in the area fraction f.

We emphasize that the results in Table 2 are for polygonal holes in continuum materials, and the grid was just a numerical device used in the computation. The results in Table 2 show a smooth evolution from the triangle to the circle, which is the limit of a regular n-sided polygon. The shape effect is surprisingly large, especially on the invariant Poisson ratio. Note that we could have added an n=2 result for a slit, which can be obtained from the ellipse as the aspect ratio goes to zero. This would give α=∞, and zero for the invariant Poisson's ratio as can be seen by taking the limits on the equations for ellipses in the next section. In Fig. 2, we show plots of the results fitted to the following curves,

$$\alpha = 3 + \frac{33.2}{n^3} \quad \text{and} \quad \nu_0 = \frac{1}{3} - \frac{25.8}{n^5} \tag{17}$$

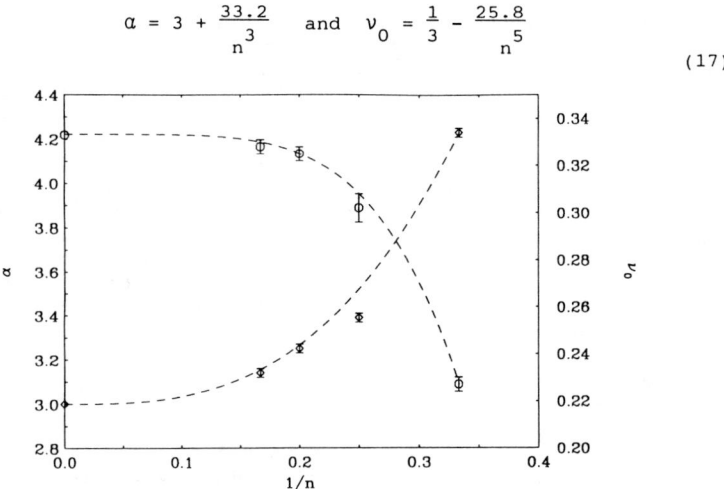

Fig. 2 Showing the results for α and ν_0 for various n-gons, using the results from Table 2, and the power law fits (17).

These fits were made by plotting the results against various power laws and choosing the one that came closest to giving a straight line. The result (17) may be useful in extending our work beyond n=6, to estimate the difference between say an octagon and a circle. Notice that the result for the Young's modulus for the square does not fit onto the curve, which we find surprising as we expected a monotonic progression. The square is different because of the averaging required and this may account for its different behavior.

As a concluding remark here, it is important to comment that for the geometry of polygonal holes with n finite the stress is singular in the matrix at the sharp corners. However, since it is a geometric type of singularity (Muskhelishvili, 1953; Timoshenko and Goodier, 1951) an elastic strain energy is finite (or integrable). The elastic moduli are obtained directly from the elastic strain energy.

8. EFFECTIVE MEDIUM THEORIES

The CLM theorem and the CLM transformation provide an important check on effective medium theories of composite materials in 2d. It is clearly desirable that these approximate theories should be invariant under the CLM transformation. Because the dilute result is exact, it is of course invariant under the CLM transformation. The dilute result is used as the starting point in many effective medium theories. We have tested four commonly used methods, the self-consistent method (Budiansky, 1965; Hill, 1965; Wu, 1966), the differential scheme (McLaughlin, 1977), the generalized self-consistent method (Christensen and Lo, 1979), and the Mori-Tanaka method (Mori and Tanaka, 1973; Benveniste, 1987), and we have found that in <u>any</u> of the above methods the CLM invariance is maintained.

Here, we will use our dilute results (12) and (16) along with Table 2 to predict the effective elastic moduli at higher volume fractions of holes by using two effective medium theories: the differential scheme and the self-consistent method.

The differential scheme is an iteration on the dilute limit such that incremental amounts of larger holes are embedded in a medium containing the previous level as an effective medium. Independent differential equations are obtained for the Young's modulus and the Poisson's ratio via,

$$\frac{dE}{E} = -\frac{\alpha df}{(1-f)} \qquad (18)$$

and

$$d\nu = (\nu_0 - \nu)\frac{\alpha df}{(1-f)} \qquad (19)$$

Equations (18) and (19) can be integrated, with the initial conditions that for small f, the dilute limits given in (12) and (16) are recovered. This is a standard technique that is explained in more detail in Jasiuk, Chen and Thorpe (1992). The results are

$$\frac{E^*}{E} = \frac{\nu^* - \nu_0}{\nu - \nu_0} = (1-f)^\alpha \qquad (20)$$

The results (20) for the differential scheme show that the Young's modulus goes to zero, and the Poisson's ratio $\nu^* \to \nu_0$ as $f \to 1$. The Young's modulus result is shown in Fig. 3 for triangles, squares and hexagons, and it gives an idea of the shape dependence of the result for a given area fraction of holes. This is qualitatively the same for all polygonal inclusions but differs in detail with the parameters α and ν_0 given in Table 2. The differential scheme is set up to describe a composite containing regular polygons with a wide size variation. Equations (20) apply to the situation where there is a single type of inclusion (square etc.). However the result can easily be generalized to an arbitrary weighting of different types of polygons (triangles, squares etc.), by taking a weighted average of α and ν_0.

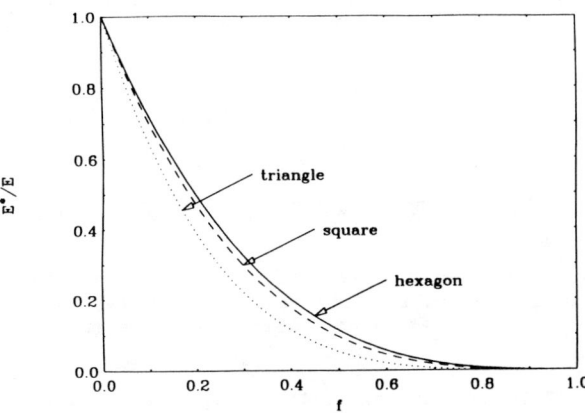

Fig. 3 The Young's modulus E^* using the differential scheme (20) is shown for triangles, squares and hexagons.

Similarly, we can use the self-consistent method to predict the effective elastic moduli of composites containing polygonal holes. Using the dilute result for E^* and ν^* given in (12) and (16) we can obtain the expressions for the effective bulk and shear moduli K^* and G^* which are related to E^* and ν^* via (3)

$$\frac{K^*}{K} = 1 - \alpha \frac{1-\nu_0}{1-\nu} f \qquad (21)$$

$$\frac{G^*}{G} = 1 - \alpha \frac{1+\nu_0}{1+\nu} f \qquad (22)$$

We can achieve self-consistency by replacing ν by ν^* on the right hand sides of equations (21) and (22).

$$\frac{K^*}{K} = 1 - \alpha \frac{1 - \nu_0}{1 - \nu^*} f \tag{23}$$

$$\frac{G^*}{G} = 1 - \alpha \frac{1 + \nu_0}{1 + \nu^*} f \tag{24}$$

We combine these equations via (3) to form equations for E^* and ν^*.

$$\frac{E^*}{E} = \frac{\nu^* - \nu_0}{\nu - \nu_0} = 1 - \alpha f \tag{25}$$

which are the same as the dilute results (12) and (16), but they hold for non-dilute concentration. The result (25) for the Poisson's ratio is plotted in Fig. 4 for the cases of triangles, squares and hexagons. It can be seen that <u>both</u> the invariant value of the Poisson's ratio <u>and</u> the critical value of the area fraction f are sensitive to the type of polygon.

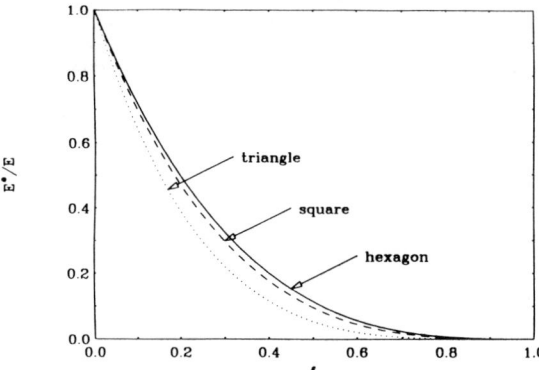

Fig. 4 The Poisson's ratio ν^* using the self consistent scheme (25) is shown for triangles, squares and hexagons. Here, the star * marks the fixed point of the Poisson's ratio which is at ν_0.

Substituting (25) into (23) and (24) we have the following expressions for K^* and G^*

$$\frac{K^*}{K} = 1 - \alpha \frac{1 - \nu_0}{1 - \nu + f\alpha(\nu - \nu_0)} f \tag{26}$$

$$\frac{G^*}{G} = 1 - \alpha \frac{1 + \nu_0}{1 + \nu - f\alpha(\nu - \nu_0)} f \tag{27}$$

The results (23) and (24) for the self-consistent method show that the moduli go to zero and the Poisson's ratio $\nu^* \to \nu_0$ at the critical area fraction $f = 1/\alpha$.

The present result for the self-consistent method complements an earlier study which was done for elliptical inclusions (Sen and Thorpe, 1985). For this case, the area Young's modulus and the area Poisson's ratio are and

$$\frac{E^*}{E} = 1 - f/f_c$$

$$\nu^* = \nu - (\nu - \nu_c)f/f_c \tag{28}$$

where

$$\nu_c = 1/(\frac{a}{b} + \frac{b}{a} + 1)$$

and

$$f_c = 1/(\frac{a}{b} + \frac{b}{a} + 1) \tag{29}$$

Here b/a is the aspect ratio of the randomly oriented and randomly centered ellipses, and the subscript c denotes the quantity at percolation. Here, the subscript c has the same meaning as the subscript 0 used earlier in this section.

We comment that these effective medium equations at least have a structure compatible with the CLM theorem. Although this compatibility is desirable, it is of course not sufficient to ensure a good effective medium theory. A good figure of merit is given by the percolation concentration. For circular holes, equation (29) gives $f_c = 1/3$ which is far from the exact result of $f_c = 0.66$ (Xia and Thorpe, 1988). Note that in any dimension, the percolation concentration is a geometrically determined quantity that is independent of the Poisson's ratio; any effective medium theory that is invariant under the CLM transformation ensures that this property is achieved for holes.

Finally, we comment on the results for a material with planar cracks which exist in the literature. The dilute result is

$$\frac{E^*}{E} = 1 - \eta\pi \tag{30}$$

$$\nu^* = \nu(1 - \eta\pi) \tag{31}$$

where η is a crack density parameter defined by $\eta = Na^2/A$, where N is a number of cracks, the length of the crack is 2a, and A is the area of the sample. Please note that the results (12) and (16) reduce to (30) and (31) if we set $\nu_0 = 0$ and $\alpha f = \eta\pi$.

The self-consistent result is (Gottesman, 1980; Benveniste, 1986)

$$\frac{E^*}{E} = \frac{\nu^*}{\nu} = 1 - \eta\pi \tag{32}$$

Note that these expressions are identical to (30) and (31). However, since the crack interaction has been taken into account here, the equations (32) are also valid in a non-dilute situation.

The differential scheme solution of Sagalnik (1973) gives

$$\frac{E^*}{E} = \frac{\nu^*}{\nu} = e^{-\eta\pi} \tag{33}$$

The Mori-Tanaka result obtained by Benveniste (1986) yields

$$\frac{E^*}{E} = \frac{\nu^*}{\nu} = \frac{1}{1 + \eta\pi} \tag{34}$$

As discussed earlier, the CLM result is particularly powerful for holes, where it shows that the Young's modulus should be independent of the Poisson's ratio of the material, and the Poisson's ratio should flow to a universal value at the fixed point (Day et al., 1992). Please note that in all the results discussed in this section the Young's modulus is indeed independent of the Poisson's ratio.

9. CONCLUSIONS

In this paper, we have drawn together and discussed a number of exact results for 2d composite materials. These results begin historically with the result of Michell (1899) concerning the stresses in plates containing holes, and are all brought together by a recent theorem proved by Cherkaev, Lurie and Milton (1992). We have discussed the special case of a material containing holes and showed that the effective area Young's modulus is independent of the Poisson's ratio of the matrix. We presented the results for the elastic constants of a material containing polygonal holes and also discussed other geometries such as circular, elliptical and slitlike holes (cracks). We have also pointed out that effective medium theories in 2d should reflect the transformation properties in the CLM theorem.

ACKNOWLEDGEMENTS

We acknowledge a partial support from the State of Michigan Research for Excellence Fund.

REFERENCES

Benveniste, Y., 1986, "On the Mori-Tanaka's method in cracked bodies," Mech. Res. Comm., **13**, 193-201.

Benveniste, Y., 1987, "A new approach to the application of Mori-Tanaka's theory in composite materials," Mech. Materials, **6**, 305-317.

Budiansky, B., 1965, "On the elastic moduli of some heterogeneous materials" J. Mech. Phys. Solids, **13**, 223-227.

Cherkaev, A., Lurie, K. and Milton, G. W., 1992, "Invariant properties of the stress in plane elasticity and equivalence classes of composites," Proc. Roy. Soc. A., in press.

Christensen, R.M. and Lo, K. H., 1979, "Solutions for effective shear properties in three phase sphere and cylinder models," J. Mech. Phys. Solids, **27**, 223-227.

Day, A.R., Snyder, K.A., Garboczi, E.J., and Thorpe, M. F., 1992, "The elastic moduli of a sheet containing circular holes," J. Mech. Phys. Solids, **40**, 1031-1051.

Dundurs, J., 1967, "Effect of elastic constants on stress in a composite under plane deformation," J. Comp. Mats., **1**, 310-322.

Dundurs, J., 1970, "Some properties of elastic stresses in a composite," in <u>Recent Advances in Engineering Science</u>, Vol. 5 (ed. A.C. Eringen), Gordon and Breach, 203-216.

Garboczi, E. J. and Thorpe, M. F., 1985, "Effective-medium theory of percolation on central force networks. II Further results" Phys. Rev. **B31**, 7276-7281.

Garboczi, E. J. and Thorpe, M. F. 1986 "Effective-medium theory of percolation on central force networks. III The superelastic problem," Phys. Rev. **B33**, 3289-3294.

Gottesman, T. 1980, Ph.D. thesis, Tel-Aviv University.

Hashin, Z., 1965, "On elastic behavior of fibre reinforced materials of arbitrary transverse phase geometry," J. Mech. Phys. Solids, **13**, 119-134.

Hill, R., 1965, "Self-consistent mechanics of composite materials," J. Mech. Phys. Solids, **13**, 213-222.

Lurie, K. A. and Cherkaev, A. V., 1986, "The effective properties of composite materials and problems of optimal designs of constructions," Uspeckhi Mekaniki (Adv. in Mech.), **9**, 3-81 (in Russian).

McLaughlin, R., 1977, "A study of the differential scheme for composite materials," Int. J. Eng. Science, **15**, 237-244.

Michell, J.H., 1899, "On the direct determination of stress in an elastic solid, with application to the theory of plates," Proc. London Math. Soc., **31**, 100-125.

Mori, T. and Tanaka, K., 1973, "Average stress in matrix and average elastic energy of materials with misfitting inclusions," Acta Metallurgica, **21**, 571-574.

Muskhelishvili, N.I., 1953, <u>Some Basic Problems of the Mathematical Theory of Elasticity</u>, P. Noordhoff Ltd., Gronongen, Holland.

Salganik, R.L., 1973, "Mechanics of bodies with many cracks," Mechanics of Solids, **8**, 149-158.

Schwartz, L.M., Feng, S., Thorpe M. F. and Sen, P. N., 1985, "Behavior of depleted elastic networks: comparison of effective medium theory and numerical calculations," Phys. Rev. **B32**, 4607-17.

Sen, P. N. and Thorpe, M. F., 1985, "Elastic moduli of two-dimensional composite continua with elliptical inclusions," J. Accout. Soc. Am. **77**, 1674-1680.

Thorpe, M.F. and Jasiuk, I., 1992, "New results in the theory of elasticity for two-dimensional composites," Proc. Roy. Soc. A, in press.

Timoshenko, S. and Goodier, J.N., 1951, <u>Theory of Elasticity</u>, 2nd ed. McGraw-Hill, New York.

Wu, T.T., 1966, "The effect of inclusion shape on the elastic moduli of a two-phase material," Int. J. Solids Struct. **2**, 1-8.

Xia, W. and Thorpe, M. F., 1988, "Percolation of random ellipses," Phys. Rev. **A38**, 2650-2656.

FIBRE DIAMETER/MECHANICAL BEHAVIOR CORRELATION IN CARBON FIBRE PROCESSING

S. Ozbek and D. H. Isaac
Department of Materials Engineering
University of Wales Swansea
Swansea, United Kingdom

ABSTRACT

As part of an ongoing programme on the study of high temperature stretching of carbon fibres, a series of experiments has been performed on fibres with nominally similar mechanical performance but different filament diameters. Three types of carbon fibres tows, containing 12000, 6000 and 3000 filaments of nominal diameter 5μm, 7μm and 10μm respectively were stretched at 2600°C with various loads up to 6kg for a dwell time of 5min. It was found that the general response to hot stretching was similar for the three fibre types in that substantial increases in Young's moduli were induced. Tensile strengths generally decreased after fibres were subjected to high temperatures and low loads, but when high strains were induced by using higher stresses, the tensile strengths were restored to their original values. It was found that smaller diameter fibres showed greater improvements in properties, although fibres with a relatively large diameter of 8.3μm were produced with a modulus of ~470GPa and strength of ~3.7GPa.

1.   *INTRODUCTION*

Although a wide range of materials has been used to produce carbon fibres since their introduction in the early sixties, current commercial production is concentrated on two basic precursors, polyacrylonitrile (PAN) fibres and mesophase pitch (MP) fibres. The pitch based fibres have the highest Young's moduli, typically ~500GPa to ~700GPa, but their low strength (~2-3GPa) and consequent poor strain to failure (<0.5%) preclude widespread applications. Hence these fibres are limited to situations in which the ultimate in stiffness is required, and strength and strain to failure are of secondary importance. PAN-based carbon fibres are of greatest commercial significance and a large variety is available. Various different classifications of these fibres have been suggested and one that is in common use considers three main categories. High strength (HS,HT or Type II) fibres are typically treated at ~1300°C during processing, giving rise to a Young's modulus of ~230GPa and tensile strength of ~4GPa. High modulus (HM or Type I) fibres are made by treating to ~2500°C, producing a modulus of ~400GPa but a significantly reduced strength of ~3GPa. The effect of the high temperature treatment is to increase the ordering at the microstructural level and more importantly to increase the degree of preferred orientation of the individual "crystallites" along the fibre axis. Various studies have shown that it is this improved preferred orientation that gives rise to the higher modulus values (Bacon and Schalamon, 1969). However the high temperatures also reduce the strength (Johnson, 1969; Watt and Johnson, 1970), it is suggested, by altering the intrinsic defect population which is thought to control the ultimate tensile strength of the fibres (Moreton *et al.*, 1967). The third category of carbon fibres that has emerged more recently is based on a new generation of intermediate modulus (IM) fibres which possess strengths at least equal to the HT type, but have improved moduli (typically ~300GPa). The processing route for generating these IM fibres has not been published in any detail.

It has been known for some time that carbon fibres derived from rayon could be hot stretched at temperatures above 2500°C (Bacon, 1960). After stretching, the axial modulus was found to rise by an order of magnitude and a high degree of preferred orientation was induced in the "sub-crystalline" domains. Specifically, an axial strain of 400% produced fibres with a modulus of ~620GPa and strength ~3.45GPa (Bacon and Schalamon, 1969). Later, several attempts were made to stretch PAN-based carbon fibres at high temperatures and these resulted in improved mechanical properties (Johnson *et al.*, 1969; Johnson, 1970; Hishiyamo 1973). However, the technology was not adapted to encompass continuous fibre production and the advantages were not realised on a commercial scale, possibly due to limitations in the PAN-based precursors then available. Recent developments in Japan by Toray (Sumida *et al.*, 1989) have shown that PAN-based carbon fibre can now be produced commercially (Torayca M60J) with tensile modulus of ~590GPa and strength ~3.8GPa, although production details for these fibres have not been published.

Previous reports from this laboratory have described hot stretching experiments on PAN-based carbon fibre tows which have been used to determine the combined effects of load, temperature and dwell time on the final mechanical properties of the carbon fibres (Ozbek *et al.*, 1991; Ozbek and Isaac, 1991). The starting materials (tows containing 3000 filaments of 7μm diameter) were stretched with varying loads up to 0.5g/filament, over a range of temperatures between 2600°C and 3000°C and for dwell times up to 30 minutes. All three variables were found to be important in determining the properties of the stretched fibres and significant increases in modulus were recorded. Although a loss in strength, as previously documented, occurred for high heat treatment temperatures and low strains, at higher strains the strength was improved, indicating that simultaneous application of thermal and mechanical work can be used to restore the fibre strength. For example, one particular severe combination of load, temperature and dwell time namely 0.5g/filament at 2800°C with a dwell time of 30 minutes, induced an increase in Young's modulus from ~180GPa to ~710GPa whilst concurrently the tensile strength showed a small increase from ~3.9GPa to ~4.2GPa. In addition to mechanical characterisation of individual stretched fibres, the effects of hot stretching on microstructure were investigated (Francis *et al.*, 1991). X-ray diffraction patterns of numerous individual fibres were taken using a high intensity synchrotron radiation source. Large increases (from ~38° to ~10°) in the preferred orientation parameter Z (FWHM of an azimuthal scan through the (00.2) reflection) were observed and a clear correlation with the enhancement in modulus was noted. The modulus was found to increase approximately linearly with decreasing Z (i.e. increasing preferred orientation).

Recent developments in high performance carbon fibres have resulted in smaller diameter fibres, leading to an increase in fibre strength. This is probably one of the reasons why intermediate modulus (IM), high strength fibres have average diameters of only 5-6μm, compared with the former quality of high tensile (HT) fibres with fibre diameters between 8 and 10μm. Even higher strength would seem to be possible by further reductions in the fibre diameter. However, the toxicity and especially the carcenogenic danger associated with fine fibres will limit the minimum diameter to about 5μm (Fitzer and Heine, 1988).

It is against this background that the experiments in the present report were planned, with the idea of carrying out a systematic study of the effects of fibre diameter on mechanical properties. This report highlights results from three different experimental carbon fibres tows with various thermal-stress histories, specially prepared by Courtaulds to have different fibre diameters but similar mechanical properties. Each of these three tows was subsequently hot stretched in this laboratory under various load/temperature regimes for dwell times of 5 minutes. Individual fibres were then subjected to mechanical testing to determine their Young's modulus and tensile strength.

2. EXPERIMENTAL

A graphite element resistance furnace from the Astro 1000A series, with a hot-zone length of approximately 140mm at 2600°C, was adapted to allow the insertion of a carbon fibre tow subjected to a constant load. Further details of the equipment have been reported previously (Ozbek and Isaac, 1991). Three sets of experiments were carried out using Courtaulds experimental fibre tows with different thermal-stress histories. These starting fibres had nominal diameters of 5μm (12000 filaments), 7μm (6000 filaments) and 10μm (3000 filaments). The three fibre types were designed to have similar mechanical properties and weights per unit length of tow and hence similar cross-sectional areas per tow. They were hot stretched using constant loads of 0.06kg, 1.2kg, 2.4kg, 3.6kg and 6kg at a constant hot-zone temperature of 2600°C for a dwell time of 5 minutes, the process parameters being chosen to be commercially attainable.

For each run, the tow of carbon fibres was cut to a length of ~820mm, the ends of which were embedded in resin. One end was gripped at the top of the furnace, whilst the end at the bottom of the furnace was attached to a dead weight. The hot zone of the furnace corresponded to ~140mm at the centre of the cut tow. In each case the furnace was heated to 2600°C at a rate of 27°C per minute with minimal stretching. During this heating up period (~95mins), a small fixed extension of 1.5mm, including thermal expansion was allowed in order to keep the fibres taut and to ensure that all the fibre tows had the same thermal-stress history prior to the stretching procedure. When the final temperature was reached the load was applied by moving the holding platform away from the loading system. The tow was then allowed to stretch at the fixed temperature of 2600°C with various constant loads for 5 minutes and the extension was monitored as a function of time using a linear voltage displacement transducer (LVDT) attached to the lower end of the cut tow. Thus the measured extension corresponded to the sum of the extension within the hot zone and a smaller extension in the region close to the hot zone where the temperature was still high enough to allow some plastic deformation. Hence an accurate value for the strain is not easily obtained, but an approximate figure can be determined by dividing the recorded extension by 140mm. A more accurate value would require a complex numerical analysis to account for the furnace temperature profile. Therefore, in this paper, only the overall extensions are quoted, in order to provide an idea of the strain without introducing the inevitable inaccuracies associated with the unstretched length.

The Young's modulus and tensile strength of single filaments were measured on specially made single filament testing equipment which was detailed in a previous paper (Ozbek and Isaac, 1991). The equipment was calibrated and checked with well characterised fibres. Mechanical testing was carried out on a fibre gauge length of 27mm. The diameter of each filament was measured using an optical microscope fitted with an image shearing eyepiece. A minimum of five individual fibres was tested for each stretching condition.

3. RESULTS

The complete set of results for the three series of experiments are shown in Table I, and these data are plotted in various forms in Figures 1-8. Figure 1 shows the total extensions for the three fibre types after 5 minutes dwell time at 2600°C as a function of the initial stretching stress applied to the fibre tows. These graphs illustrate the expected trend of an approximately linear increase in extension with stretching stress, the only anomaly being that for the 7μm fibre tow with a tensile stress of ~300MPa (6kg load) a surprisingly large extension was recorded. As mentioned above, these three fibre tows, with different filament diameters, were prepared by Courtaulds to have similar weights per unit length of tow and hence similar total cross-sectional areas for the complete tows. This should have ensured that the application of equal loads to each tow would have produced similar stress levels. However, when the fibre diameters were measured it was found that the nominally 7μm fibre tow actually had an average fibre diameter of ~6.5μm (Table I). The measured fibre diameters of the as-received fibres as quoted in Table I are consistent with the

TABLE I MECHANICAL PROPERTIES OF 5μm (12K), 7μm (6K) and 10μm (3K) DIAMETER PAN-BASED CARBON FIBRES FOLLOWING HOT STRETCHING WITH VARIOUS LOADS AT 2600°C FOR 5 MINUTES

Material	Stretching Load (kg)	Stretching Stress (MPa)	Total Extension* (mm)	Filament Diameter (μm)	Tensile Strength (GPa)	Young's Modulus (GPa)
5μm (12K)	AS	RECEIVED	-	5.1 ± 0.1	4.19 ± 0.35	259 ± 14
"	0.06	2.4	0.156	4.9 ± 0.1	1.76 ± 0.35	306 ± 31
"	1.2	48	1.804	4.8 ± 0.2	3.00 ± 0.10	361 ± 34
"	2.4	96	2.796	4.8 ± 0.2	4.09 ± 1.01	431 ± 31
"	3.6	144	3.977	4.4 ± 0.1	3.47 ± 0.20	450 ± 22
"	6.0	240	6.671	4.3 ± 0.1	5.02 ± 0.50	527 ± 35
7μm (6K)	AS	RECEIVED	-	6.5 ± 0.2	4.11 ± 0.78	238 ± 10
"	0.06	3	0.631	6.1 ± 0.3	2.40 ± 0.42	326 ± 24
"	1.2	59	2.271	6.0 ± 0.4	3.29 ± 1.01	368 ± 60
"	2.4	118	3.649	5.9 ± 0.3	4.13 ± 1.04	422 ± 58
"	3.6	177	5.424	5.7 ± 0.1	3.94 ± 0.74	470 ± 37
"	6.0	296	15.662	5.3 ± 0.3	5.41 ± 1.60	548 ± 91
10μm (3K)	AS	RECEIVED	-	10.0 ± 0.2	4.27 ± 0.84	225 ± 12
"	0.06	2.5	0.471	9.8 ± 0.1	2.39 ± 0.22	286 ± 15
"	1.2	50	2.054	9.4 ± 0.1	3.17 ± 0.50	358 ± 12
"	2.4	100	3.169	9.1 ± 0.2	3.58 ± 0.42	396 ± 18
"	3.6	150	4.457	8.9 ± 0.2	3.51 ± 0.31	406 ± 23
"	6.0	250	7.954	8.3 ± 0.2	3.73 ± 0.51	467 ± 21

* Total extension at 2600°C after 5 minutes. Fibres stretched 1.5 mm including thermal expansion up to 2600°C. Heating rate 27°C/min from room temperature to 2600°C.

TABLE II WEIGHT OF 1 m LENGTH OF 5μm (12K), 7μm (6K) and 10μm (3K) DIAMETER CARBON FIBRE TOWS

5 μm diameter	(12000 filament)	Weight	= 0.455 g/m
7 μm diameter	(6000 filament)	Weight	= 0.380 g/m
10 μm diameter	(3000 filament)	Weight	= 0.430 g/m

weight per unit length measurements of Table II, indicating that the densities of these fibres were similar. Thus the stretching stress (load applied per unit cross-sectional area) was in fact significantly higher for this tow than for the other two sizes of fibres. Even so, this large rise in extension with the 6 kg load is curious.

Figures 2 and 3 show the Young's modulus and ultimate tensile strength of these hot stretched fibres as a function of the stretching stress. All of these fibres showed substantial increases in modulus following stretching at 2600°C compared with their as-received values. Tensile modulus (Figure 2) increased approximately linearly with increasing stretching stress in all three cases. Tensile strength (Figure 3) was reduced sharply from the as-received values (~4GPa) to well below 3GPa when little stretch was induced at 2600°C using the nominal 0.06kg load (~2.5MPa stress). However with higher loads, for which greater extensions were induced, the strength increased, so that with a 6 kg load (~250MPa stress) values greater than the as-received condition were found for the 5μm and 7μm fibres, and two thirds of the untreated value were noted for the 10μm fibres. Although best fit straight lines are included in Figure 3, these are merely to indicate the trends rather than to suggest a linear relationship, since the scatter inherent in these strength results precludes accurate fitting of mathematical equations.

Figures 4 and 5 show the Young's modulus and tensile strength as a function of the extension induced during hot stretching, although for clarity the greatly extended 7μm tow with a stretching stress of ~ 300 MPa has been omitted. A general increase of both modulus and strength with extension is clearly evident.

Figures 6 and 7 show how the modulus and strength are related to the fibre diameters of the three fibre types after the hot stretching process. Best fit straight lines are included on these graphs to show the trend of improving mechanical properties with decreasing fibre diameter.

Figure 8 is a plot of tensile strength against Young's modulus showing how the mechanical properties have been changed by the hot stretching. This diagram graphically illustrates how the tensile strength falls from the as received value of ~4GPa to ~2GPa for all three fibre types when there is just a small increase in modulus to ~300GPa. However it also shows very clearly how, as the modulus rises to ~550GPa following greater stretching, the strength also increases to ~5GPa.

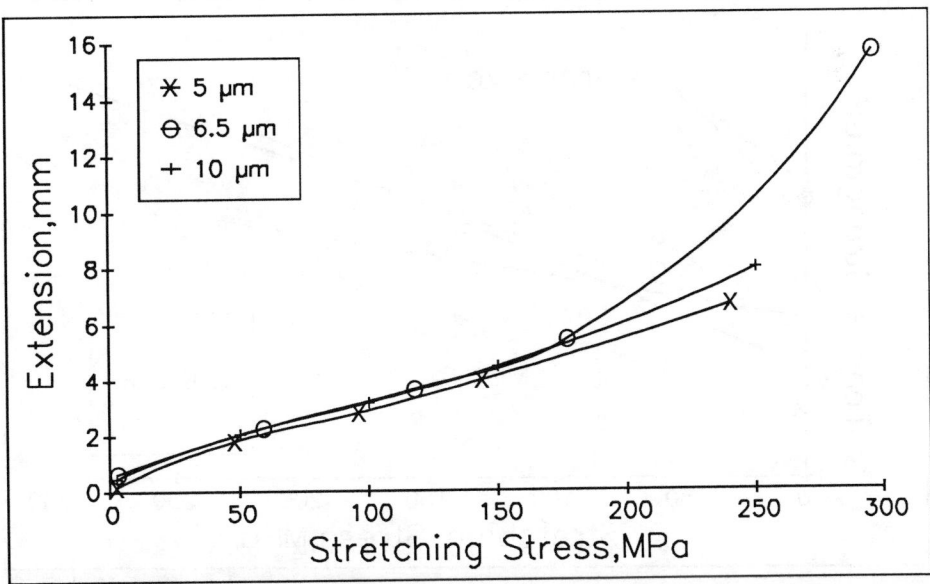

Figure 1   The total extension after 5 minutes stretching at 2600°C as a function of stretching stress for the three different starting fibre types.

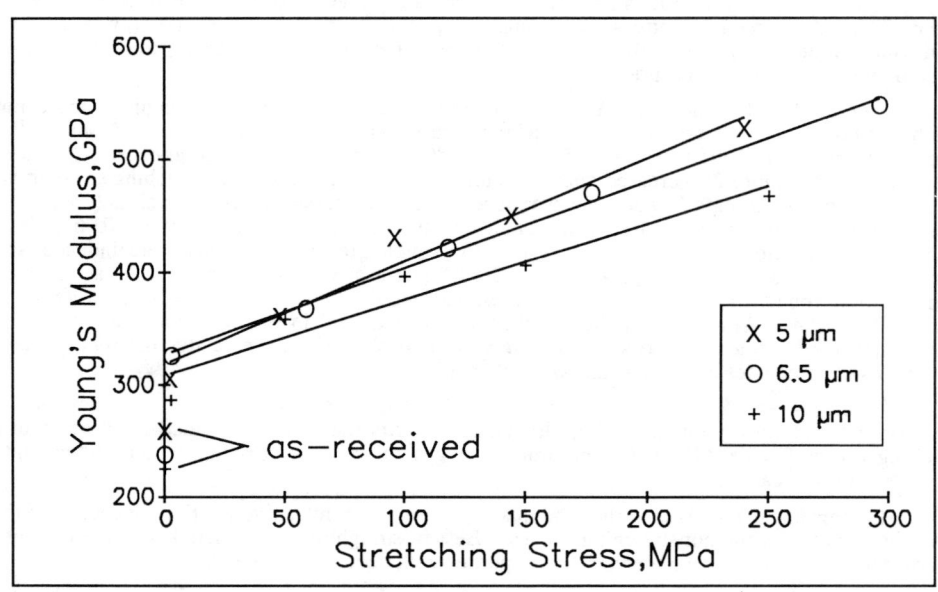

Figure 2   Young's modulus as a function of the stretching stress for the three fibre types, with best fit straight lines for each type. The moduli for the as-received fibres are included for comparison purposes.

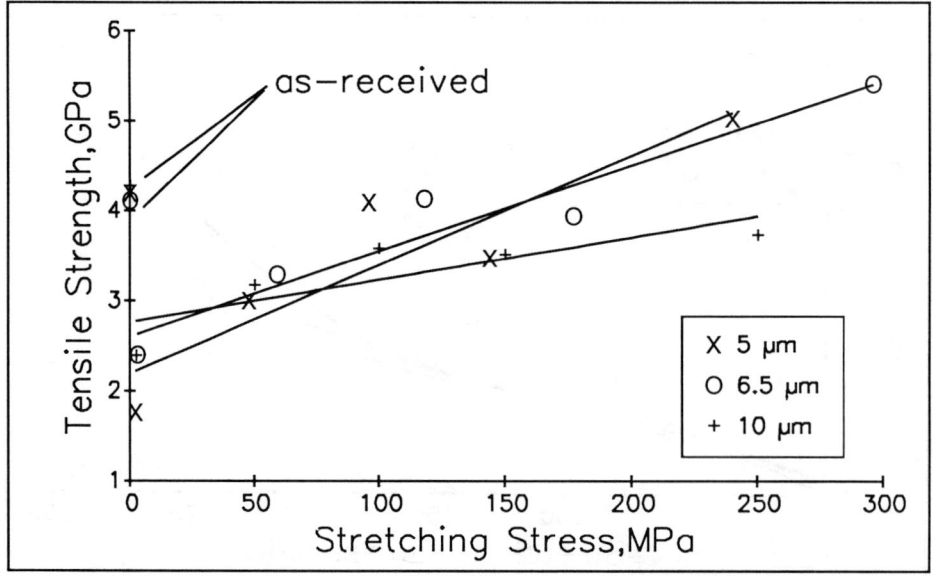

Figure 3   Tensile strength of the three fibre types following stretching at 2600°C as a function of stretching stress. Best fit straight lines are included to show the trends, and the strengths of the as-received fibres are given for comparitive purposes.

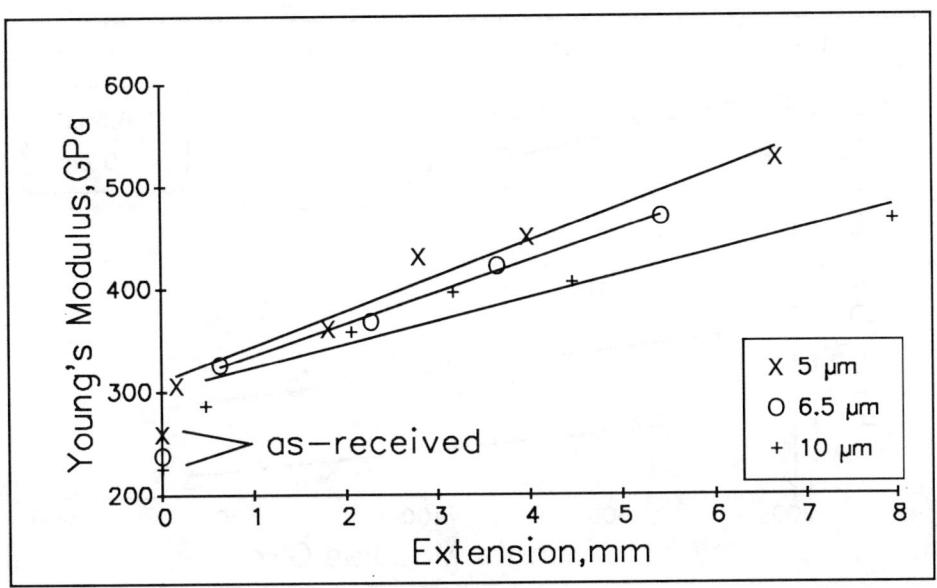

Figure 4   Young's modulus for the experimental fibres as a function of the extension induced by hot stretching.

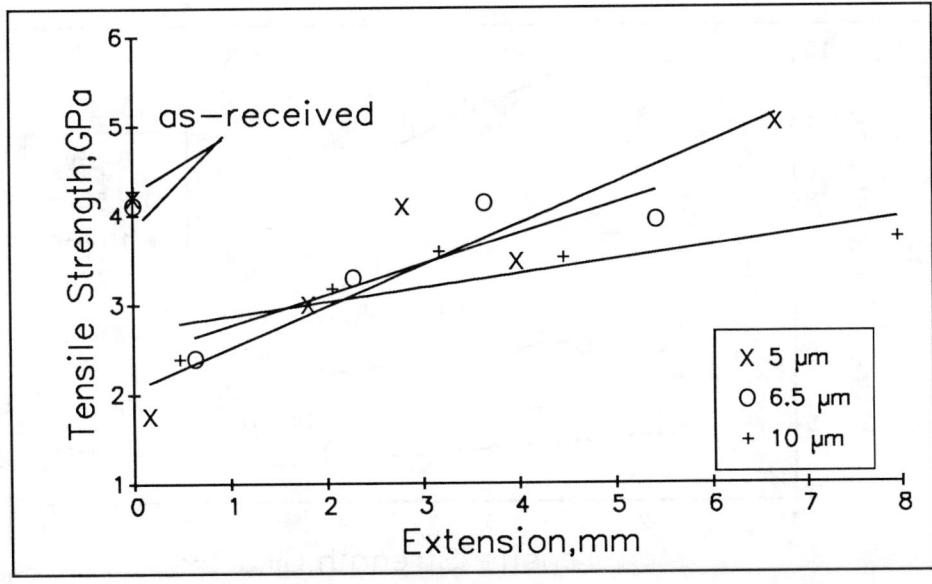

Figure 5   Tensile strength for the experimental fibres as a function of the extension induced by hot stretching. Best fit straight lines are included to illustrate the trend of increasing strength with induced extension.

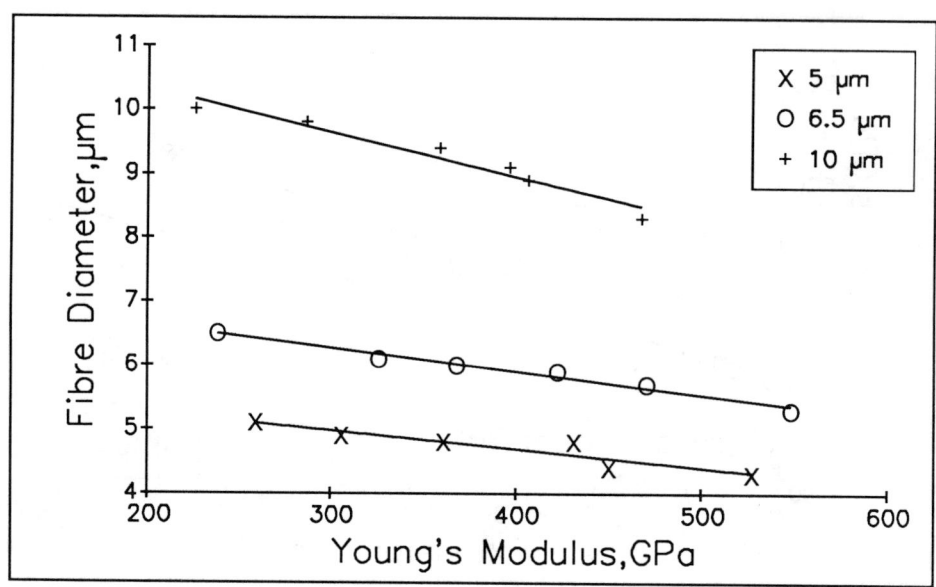

Figure 6   Young's modulus as a function of the filament diameter following hot stretching of the fibre tows. The as-received fibres which are included lie on the best fit straight lines.

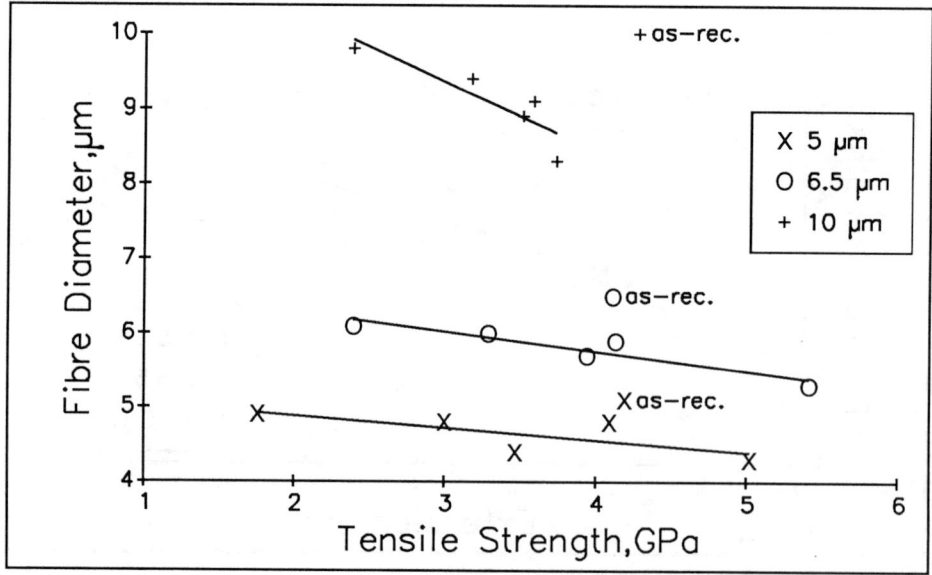

Figure 7   Tensile strength as a function of the filament diameter following hot stretching of the fibre tows. The positions of the as-received fibres clearly show how heat treatment without significant change in fibre diameter change reduces the strength.

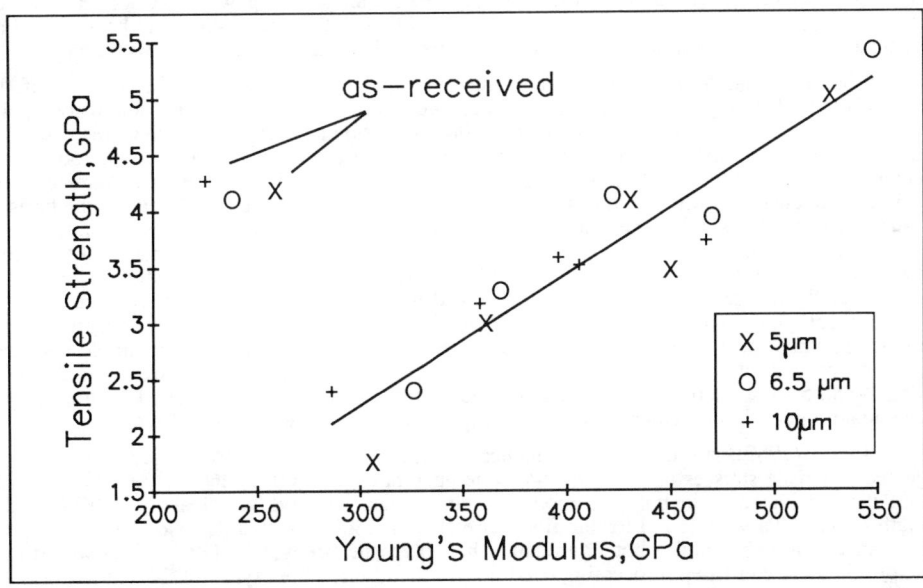

Figure 8 Tensile strength as a function of Young's modulus for the range of fibres produced in the three series of experiments. A single straight line gives a reasonable fit to all the data points.

4. DISCUSSION

Although the results in Figure 1 indicate a similar trend of approximately linear increase in extension with stretching stress for all three fibre types, the extension recorded for the 7μm fibre tow with a stretching stress of ~300MPa was found to be unexpectedly high. A possible explanation for this behaviour could be that at 2600°C with this stretching stress a critical level has been reached and some of the fibres have broken causing the stress on the remaining fibres to increase even further. Since it is impossible to arrange for all the fibres in a tow to be subjected to precisely equal tensions, and the random occurence of defects gives rise to variations in strength between individual fibres, some of the filaments will fracture before others. In these circumstances the overall strain would be expected to increase sharply, but such behaviour would not necessarily lead to rapid catastrophic failure of the tow since, as the fibres extend, their tensile strength increases. Thus although the number of fibres supporting the fixed load is decreasing, their strength is increasing. As would be expected for the most severe stretching conditions reported and the greatest recorded extension, the mechanical tests results showed that the highest Young's moduli and tensile strengths were achieved for this particular fibre tow. However, interestingly, the largest deviations from the mean values were also found for this condition. This would be consistent with some of the fibres breaking prior to the full extension since, not having been subjected to the full stretching treatment, their properties would be somewhat inferior and more variable.

It has been reported in previous studies (Ozbek et al., 1991; Ozbek and Isaac, 1991) that the mechanical properties of hot stretched PAN-based carbon fibres are highly dependent on the process parameters, namely, load, temperature and dwell time. Large increases in modulus have been associated with the overall plastic deformation induced during hot stretching. These variations have also been correlated with the microstructural changes that occur during hot stretching of these fibres (Francis et al., 1991). Large increases (from ~38° to ~10°) in the preferred orientation parameter Z (FWHM of (00.2) reflection arcing) were achieved and direct relationships were established between Z and the stretching conditions. In the limited survey of temperature and dwell time regimes investigated in the present study, large increases in modulus have been achieved for all three different diameter starting fibre tows (Figure 2). Although reductions in tensile strength occurred following lower levels of stretching stress, the strengths recovered to the original values after higher stretching stresses were used, except for the 10μm fibres. More than two thirds of the original strength of the 10μm fibre tow was recovered after straining with 6kg load (~250MPa stress) as seen in Figure 3. There is no reason to doubt that further increases could be achieved by applying a more severe combination of process parameters load, temperature and dwell time.

Figures 4 and 5 show the Young's modulus and tensile strength as a function of the extension induced during the hot stretching procedure. In both these figures the anomalous point (i.e. for the 7μm tow stretched at 6kg load or ~300MPa stress) has been omitted since the excessive extension would have distorted the scales of the graphs, and crowded the other points too closely together. Interestingly though, both modulus (~550GPa) and strength (~5.4GPa) measurements for the 15.7mm extension of this tow fall well below the values that would be predicted on the basis of the best fit straight lines in Figures 4 and 5. Such features are consistent with the explanation given above for the excessive extension since if some of the individual tested fibres had broken before being fully stretched, then their properties would be expected to be below this prediction, leading to a lower overall average. However it should be noted that best fit straight lines have been added to the experimental data points in Figures 4 and 5 to show the trends rather than to confirm a linear relationship, since there is not yet any theoretical reason for anticipating such a simple curve, or indeed any other relationship. One interesting feature of these best fit lines, however, is that both Figures 4 and 5 show increasing gradients for decreasing starting fibre diameter. This implies that mechanical property improvements are more easily induced in smaller diameter fibres.

Fitzer (1989), discussed the influence of heat treatment on PAN-based carbon fibres subjected to tensile stresses of up to 75MPa at temperatures of 2400°C, 2600°C and 2800°C. His results showed increases in both the modulus and strength. A modulus of about 370GPa and a strength of ~1.7GPa were found for the fibres stretched at 2600°C with a stress of 50 MPa. In this present study, similar modulus but significantly higher tensile strength (>3GPa) values were found for approximately equivalent processing conditions with all three fibre types (Table I and Figures 2 and 3). The reason for this higher strength value could be the difference in the process histories of these fibres prior to hot stretching.

Reductions in strength have been attributed to defects formed during heat treatment above 1300°C (Jones and Duncan, 1971; Sharp and Burnay, 1971; Johnson and Tyson, 1969; Tokarsky and Diefendorf, 1975). One of the steps to eliminate these defects is to reduce the filament diameter. According to Weibull statistics, reduction in the material volume can diminish the crack causing notches of critical length. In fact, those commercially available carbon fibres with high strength and modulus have filament diameters in the region of 5-6μm.

Figures 6 and 7 show how the mechanical properties are related to the fibre diameters in the present study. Clearly, for all three starting fibre types an increase in modulus is observed as the fibre diameter is decreased by hot stretching, and the linear relationship illustrated shows a reasonable fit (Figure 6). Although the modulus values achieved by hot stretching the 10μm fibres are not as great as those for the smaller starting diameter fibres Figure 6 does illustrate that there is no direct general relationship between filament diameter and modulus. Furthermore, extrapolation of the curve for the 10μm fibres would suggest that very high moduli could be achieved without recourse to very thin fibres. The situation is not quite so clear cut for the tensile strength results (Figure 7). In this case, for the hot stretched 10μm fibres, the tensile strength values do not increase to the same extent as for the smaller diameter fibres. Combined with a steeper gradient for this line, the indications are that a small fibre diameter is important for achieving adequate tensile strength values. This is consistent with the argument of Weibull statistics that the number of critical size cracks are reduced by decreasing the cross-sectional area.

In Figure 8, an overall summary of the mechanical properties of the fibres is given. From the as-received fibres with a modulus of ~230GPa and strength of ~4GPa the initial drop in strength to ~2GPa with a small increase in modulus to ~300GPa is evident. This is due to the high temperature treatment with little stretching. The effect of stretching is seen as the general curve extending to a modulus ~550GPa accompanied by a strength of ~5GPa. The linear relationship illustrated is a reasonable fit for all three fibre types, suggesting that with even more extreme stretching conditions any of these starting materials could be transformed into very high modulus fibres simultaneously with inducing a much enhanced strength.

5. CONCLUSIONS

It has been demonstrated that carbon fibres can be hot stretched at the commercially viable temperature of 2600°C in the reasonably short time of 5 minutes. Large plastic deformations can be induced with strains of up to approximately 10% resulting from stresses of up to ~300MPa. Such hot stretching induces substantial improvements in mechanical properties, with the most severe conditions transforming the Young's modulus and tensile strength from ~230GPa and ~4GPa to ~550GPa and ~5GPa respectively. A significant feature of this result is that with appropriate control and choice of the hot stretching conditions, strength can be increased above the original value.

The three different fibre types tested (5μm, 7μm and 10μm nominal diameters) showed remarkably similar, approximately linear, extension-stretching stress curves up to about 200MPa. The Young's modulus values following hot stretching also increased approximately linearly with both stretching stress and extension, although the results suggested that the smaller the fibre diameter the greater the response to both stretching stress and induced extension. Scatter in the tensile strength results did not prevent a clear trend of increasing strength with extension being observed. Thus although high temperatures combined with low extensions lead to strengths ~2GPa, when large strains are induced strength can be increased to ~5GPa.

The results suggest that smaller diameter fibres show greater improvements in mechanical properties for similar extensions. This is particularly evident in the tensile strength measurements since, although the tensile modulus of the 10μm fibres could be improved to ~470GPa by hot stretching the corresponding strength was only ~3.7GPa compared with equivalent figures of ~5GPa for the smaller diameter fibres. On the other hand, since such fibres still have a diameter of greater than 8μm, further stretching should lead to fibres of exceptional modulus and strength whilst maintaining a diameter above the 5μm limit at which health and safety problems arise.

6. REFERENCES

Bacon, R., 1960, "Growth, Structure and Properties of Graphite Whiskers", *Journal of Applied Physics*, **31**, 283-290.

Bacon, R. and Schalamon W.A., 1969, "Physical Properties of High Modulus Graphite Fibres Made from a Rayon Precursor", *Applied Polymer Symposia*, ed. J. Preston, American Chemical Society, **9**, 285-292.

Fitzer, E. and Heine M., 1988, "Carbon Fibre Manufacture and Surface Treatment", in *Fibre Reinforcement for Composite Materials*, ed. A.R. Bunsell, Elsevier, Composite Materials Series, **2**, 73-148.

Fitzer, E., 1989, "PAN-Based Carbon Fibres - Present State and Trend of the Technology from the Viewpoint of Possibilities and Limits to Influence and to Control the Fibre Properties by the Process Parameters", *Carbon*, **27**, 621-645.

Francis, J.G., Ozbek S. and Isaac D.H., 1991, "Microstructural Changes in Carbon Fibres During High Temperature Processing", in *Processing and Manufacturing of Composite Materials*, ASME, **PED-Vol. 49 / MD-Vol. 27**, 173-183.

Hishiyama, Y., 1973, "Stress Graphitization in Carbon Fibre", *11th Biennial Conference on Carbon, Oak Ridge National Laboratory, Gatlinburg, Tennessee*, American Carbon Society, 269-270.

Johnson, D.J. and Tyson C.N., 1969, "The Fine Structure of Graphitized Fibres", *Journal of Physics D*, **2**, 787-795.

Johnson J.W., 1969, "Factors Affecting the Tensile Strength of Carbon-Graphite Fibres", *Applied Polymer Symposia*, ed. J. Preston, American Chemical Society, **9**, 229-243.

Johnson J.W., Marjoram J.R. and Rose P.G., 1969, "Stress Graphitization of Polyacrylonitrile-Based Carbon Fibre", *Nature*, **221**, 357-358.

Johnson, W., 1970, "Hot Stretching of Carbon Fibres Made From Polyacrylonitrile Fibres", *Proceedings of the 3rd Conference on Industrial Carbon and Graphite, London*, Society of Chemical Industries, London, 447-453.

Jones B.F. and Duncan R.G., 1971, "The Effect of Fibre Diameter on the Mechanical Properties of Graphite Fibres Manufactured from Polyacryonitrile and Rayon", *Journal of Materials Science*, **6**, 289-293.

Moreton, R., Watt, W. and Johnson W., 1967, "Carbon Fibres of High Strength and High Breaking Strain", *Nature*, **213**, 690-691.

Ozbek, S. and Isaac D.H., 1991, "Carbon Fibres : Effect of Processing Parameters on Mechanical Properties", in *Processing and Manufacturing of Composite Materials*, ASME, **PED-Vol. 49 / MD-Vol. 27**, 307-320.

Ozbek, S., Jenkins G.M. and Isaac D.H., 1991, "Mechanical Properties of Hot Stretched Carbon Fibres", *20th Biennial Conference on Carbon, Santa Barbara*, American Carbon Society, 308-309.

Sharp, J.V. and Burnay S.G., 1971, "High-Voltage Electron Microscopy of Internal Defects in Carbon Fibres", *1st International Conference on Carbon Fibres*, Plastics Institute, London, 68-71.

Sumida, A., Ono K. and Kawazu Y., 1989, "PAN-Based High Modulus Graphitized Carbon Fibres Torayca M60J", *34th SAMPE Symposium*, 2579-2589.

Tokarsky, E.W. and Diefendorf R.J., 1975, "The Modulus and Strength of Carbon Fibres", *12th Biennial Conference on Carbon, University of Pittsburgh*, American Carbon Society, 301-302.

Watt, W. and Johnson W., 1970, "Carbon Fibres from 3 Denier Polyacrylonitrile Textile Fibres", *Proceedings of the 3rd Conference on Industrial Carbon and Graphite, London*, Society of Chemical Industries, London, 417-426.

MICROSTRUCTURE AND TEXTURE DEVELOPMENT DURING CARBON FIBRE PROCESSING

S. Thitipoomdeja, S. Ozbek, and D. H. Isaac
Department of Materials Engineering
University of Wales Swansea
Swansea, United Kingdom

ABSTRACT

The development of texture and microstructure during the processing of PAN-based carbon fibres has been monitored by stretching commercial tows at high temperatures. In particular the interplay between stretching temperature, dwell time and applied load has been studied by producing a series of experimental carbon fibre tows following stretching at 2600°C and 2800°C with loads up to 0.5g/filament and dwell times up to 30 min. The degree of preferred orientation (Z) and apparent crystallite size (L_c) have been determined by wide angle x-ray diffraction of single fibres using synchrotron radiation. For the first time, the development of texture and crystallite size with time has been monitored. Mechanical properties of the same individual fibres have been measured so that it has been possible to correlate variations in texture and properties between single fibres from the same tow. A close correspondence between the degree of preferred orientation and mechanical properties has been confirmed. Electron microscopy and electron diffraction of ground and ultrasonicated fibres qualitatively confirmed the x-ray observations.

1. INTRODUCTION

At the microstructural level, carbon fibres consist of aromatic layer planes that form sets of turbostratic stacks in a highly textured arrangement (Oberlin and Guigon, 1988). Crystallographically this preferred orientation may be described by defining the [00.1] direction as perpendicular to the fibre axis. This ensures that the very strong intralayer covalent bonds lie along the fibre direction, thus endowing carbon fibres with their well known and exceptional mechanical properties. The detailed arrangement of these stacked layers in space clearly controls the mechanical behaviour and two particular factors that have been widely discussed are the degree of preferred orientation which is closely related to the modulus, and the effects of defects, which are thought to control the strength.

The texturing is impressed into the carbon fibres during manufacture and it is therefore dependent on the type of precursor used and on the processing conditions applied to the fibre tow. Studies of the texture have usually been carried out using either x-ray diffraction or electron microscopy and diffraction or sometimes a combination of both (Brydges et al., 1969; Johnson and Watt, 1967; Shindo, 1961). Slightly different results arise from the two techniques due to the inability of the electron beam to penetrate the carbon fibre beyond about 100nm (Butler and Diefendorf, 1969), whereas the x-ray beam can pass through the complete thickness (~5-10μm) of the fibre. Hence electron microscopy is more suited to surface studies or in conjunction with TEM of thin sections. Various reports have related the quantitative data collected from these techniques, such as mean crystal sizes, interlayer spacings and preferred orientations to the mechanical properties of the carbon fibres. From the earliest work of Shindo (1961) and Watt et al.(1966) on PAN-based carbon fibres and Bacon and Schalamon (1969) on rayon-based fibres a correlation between the preferred orientation and Young's modulus has been established.

A previous report from this laboratory discussed the hot stretching of PAN based carbon fibres at a series of temperatures between 2700°C and 3000°C and showed that temperature and load are both important in determining the subsequent properties, and significant increases, particularly in modulus, were found (Ozbek et al., 1991). In the limited survey of load, temperature and time regimes investigated, modulus increases from ~180GPa to ~690GPa were recorded, but there were clear indications that even further enhancement could be achieved by application of more severe processing treatments. The results also suggested that the adverse effects of high temperature, namely the decrease in tensile strength, could be overcome by using the right processing conditions. Although the greatest improvements in mechanical properties were achieved by heating to 3000°C, it was found that substantial deformation and property enhancement could be induced at 2700°C.

In addition to mechanical characterisation of the above fibres, microstructural changes that occurred during the hot stretching process were investigated using individual fibres (Francis et al., 1991). An exploratory study was carried out and demonstrated the feasibility of assessing two of the major microstructural parameters on a reasonable timescale using the synchrotron radiation source (SRS) at SERC Daresbury Laboratory, U.K. It was found that exposure times for x-ray diffraction patterns of single fibres could be reduced by at least 3 orders of magnitude from ~10 days using a laboratory generator to less than ~20 minutes on the SRS. Thus in a 24 hour shift at Daresbury substantial data were generated on a series of individual fibres which had previously been subjected to various thermal/loading processing regimes. The preferred orientation parameter Z (FWHM of an azimuthal scan through the (00.2) arc) and the "apparent" crystallite size perpendicular to the basal planes, L_c (derived from the FWHM of a radial scan through the (00.2) arc) were related to the processing conditions, temperature and load and to the modulus of the stretched fibres. It was found that large increases in preferred orientation (Z changed from ~38° to between ~19° and ~10°) were achieved by hot stretching, and a direct relationship was established between this degree of preferred orientation and the stretching conditions. The orientation parameter, Z, decreased with both heat treatment temperature and load, and was closely related to the extension induced by hot stretching. An approximately linear relationship between Z and Young's modulus was also noted. Furthermore it was found that the "apparent" crystallite size normal to the basal planes, L_c, increased with increasing heat treatment temperature, although L_c appeared not to vary significantly with either hot stretching load or extension.

In further studies (Ozbek and Isaac, 1991), PAN-based carbon fibre tows were hot stretched particularly to determine the effects of dwell time in addition to load and temperature on the final mechanical properties. In these experiments, the starting material was stretched at 2600°C or 2800°C with fixed loads of up to 0.5g/filament and a range of dwell times between 2 and 30 minutes. All three variables were found to be important in determining the mechanical properties of the hot stretched fibres and again significant increases particularly in modulus were reported. At the most severe combination of load, temperature and dwell time, namely 0.5g/filament at 2800°C with a dwell time of 30 minutes, increases in the modulus from ~180GPa to ~710GPa were recorded whilst the tensile strength remained at ~4GPa. In this present report, the results of detailed texture and microstructural studies on these series of fibres are given. In particular, the synchrotron radiation source has been used to obtain x-ray diffraction patterns of individual fibres, from which the microstructural parameters Z and L_c have been determined, and for the first time the same individual fibres have been mechanically tested. Thus it has been possible to correlate variations in microstructure and mechanical behaviour between fibres which have been identically treated. Also variations in these properties have been monitored as a function of the process parameters, with a particular interest in the effect of stretching time at high temperatures.

2. EXPERIMENTAL

A graphite element resistance furnace from the Astro 1000A series, with a hot-zone length of 150 mm at 3000°C was adapted to allow the insertion of a carbon fibre tow subjected to a constant load. The hot stretch processing was performed on carbon fibre tows of length 820mm cut from PAN-based 3000 filament tows (average filament diameter of 7μm) supplied by Courtaulds that had been heat treated to ~1300°C only during manufacture. The quoted tensile strength and modulus were checked and confirmed to be ~3.9GPa and ~180GPa respectively, as measured by single fibre testing. Initially a set of twelve experimental fibre tows was produced by stretching 820mm long cut tows using loads of 0.10, 0.17, 0.20, 0.24, 0.31 and 0.5g/filament at constant hot-zone temperatures of 2600°C and 2800°C for a dwell time of 30 minutes. A further set of ten fibre tows was produced in the second series of experiments using dwell times of 2, 5, 10, 20 and 30 minutes at constant hot-zone temperatures of 2600°C and 2800°C with a fixed load of 0.31g/filament. Mechanical properties were measured on single fibres with a gauge length of 28mm. Further details of the experimental arrangements for hot stretching and mechanical testing have been reported elsewhere (Ozbek and Isaac, 1991).

Specimens for transmission electron microscopy (TEM) were prepared by grinding carbon fibres cut from the centre of the hot-zone, with an agate pestle and mortar, followed by ultrasonifiction in a mixture of alcohol and water (2:1). Ground fibres in solution were then placed on copper grids coated with a very thin carbon supporting film, and the liquid allowed to evaporate. The TEM was operated at 100kV and selected area electron diffraction patterns (SAD) were taken with a camera length of 76cm. Micrographs of bright field images and diffraction patterns of the fibre samples were taken from the same area in each case.

Several x-ray diffraction patterns of single fibres for each condition were taken with the synchrotron radiation source at SERC Daresbury Laboratory, although not all of the photographs proved suitable for subsequent analysis. A pinhole collimated flat plate film camera was used and the camera was flushed with helium to reduce background scatter. Radiation of wavelength ~0.1488nm and a sample to film distance ~60mm were used and fibres were dusted with calcite for calibration. Data from the x-ray diffraction patterns were digitised with a SCANDIG microdensitometer. Subsequent analysis was performed on a VAX 8700 using the GENS suite of programs originally written at Daresbury Laboratory. All the individual fibre samples that had been x-rayed were subsequently subjected to mechanical testing in a purpose built apparatus as described elsewhere (Ozbek and Isaac, 1991).

The preferred orientation parameter Z (in degree) was taken to be the full width at half maximum height (FWHM) of an azimuthal scan through the (00.2) reflection. The measure of "apparent" crystallite size (L_c) perpendicular to the basal planes was obtained using the Scherrer formula $L_c = K.\lambda/B.\cos\theta$ where λ is the x-ray wavelength, K is the Scherrer geometric factor taken to be 1, θ is the Bragg angle and B is the FWHM of a radial scan through the (00.2) reflection (Johnson, 1987a).

3. RESULTS AND DISCUSSION

Individual carbon fibres from the as-received tow exhibited no diffraction pattern after being exposed for 60 minutes to synchrotron radiation. Hence a bundle of as-received fibres was used for comparative purposes. An x-ray diffraction pattern was taken using an evacuated camera on a laboratory source and a Z value of ~38° was found from the x-ray pattern shown in Figure 1. The (00.2) reflection was so diffuse that it was not possible to measure a realistic value of L_c. It is not unreasonable to assume that the Z value for such fibres is close to that measured using bundles since fibres in a bundle would be out of alignment by only a few degrees. Whilst having a large effect on very highly oriented samples this would have little effect on the relatively large measured angle of 38° for as-received fibres (e.g. Johnson, 1987b). Thus for all experimental conditions except as-received fibres, individual hot stretched fibres were used for the x-ray diffraction

Figure 1. X-ray fibre difraction pattern of a bundle of as-received carbon fibres taken using a laboratory generator with a pinhole collimated, flat plate vacuum camera.

Figure 2. X-ray diffraction pattern of a single carbon fibre following hot stretching for 30 minutes at 2600°C with a load of 0.5 g/filament. This photograph was taken on the SRS using a pinhole collimated, helium filled, flat plate camera.

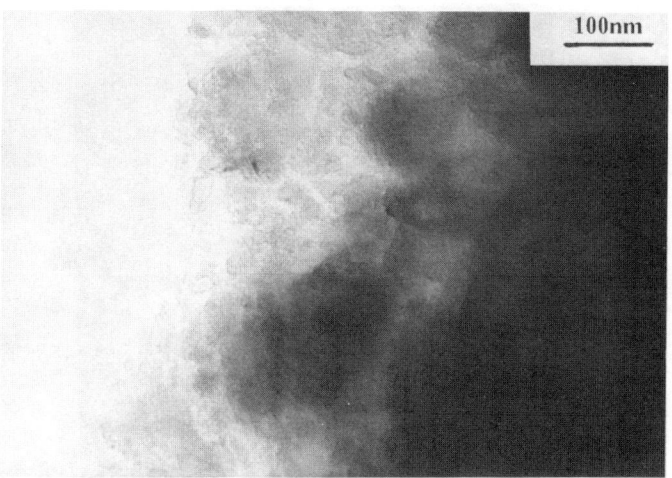

Figure 3. Electron micrograph of a ground and ultrasonicated carbon fibre as-received from Courtaulds. It had been subjected to an HTT of ~ 1300°C only during production.

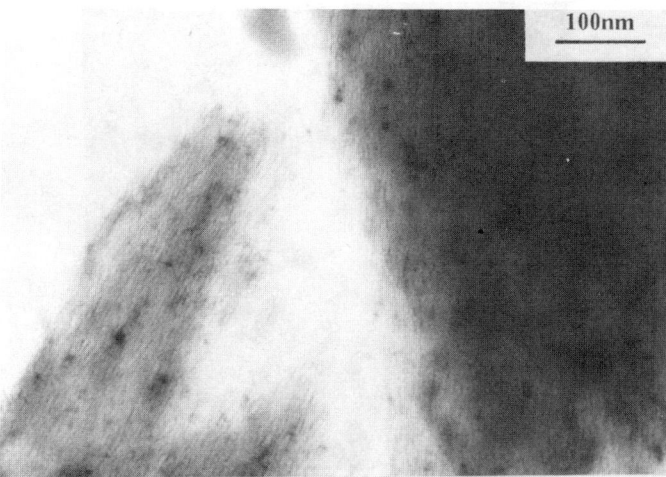

Figure 4. Electron micrograph of a ground and ultrasonicated carbon fibre that had been hot stretched at the same conditions as in Figure 2, (i.e. 30 minutes at 2600°C with a load of 0.5 g/filament).

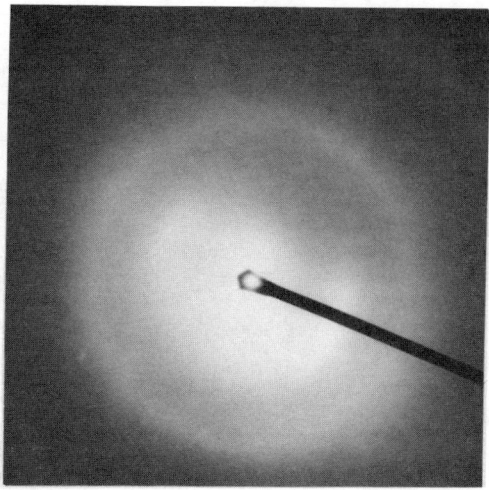

Figure 5.   Selected area electron diffraction pattern (SAD) of part of the sample in Figure 3, (i.e. an as-received fibre).

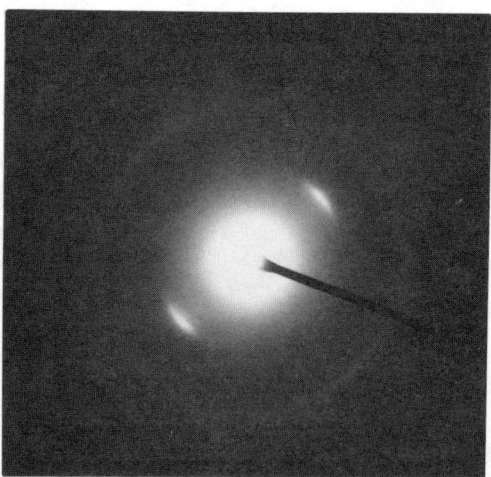

Figure 6.   SAD of the hot stretched fibre of Figure 4, (i.e. stretched for 30 minutes at 2600°C with a load of 0.5 g/filament)

experiments. Figure 2 shows a typical x-ray diffraction pattern of a single hot stretched carbon fibre taken at the SRS. Clearly, for the stretched fibre, not only is the (00.2) reflection much sharper in a radial direction, indicating a substantial increase in "apparent" crystallite size (L_c), but also the arc length is far smaller, confirming a significant increase in the degree of preferred orientation (i.e. a decrease in Z).

Figures 3 and 4 illustrate the bright field TEM images of ground, ultrasonicated as-received fibres and a typical example of hot stretched fibres respectively. Clear differences in morphology are evident between the untreated fibres and those subjected to hot stretching. Very little variation was observed in the micrographs of fibres that had been subjected to different hot stretching conditions so that Figure 4 is representative of the morphologies of the treated carbon fibres for the whole range of processing parameters used in these experiments. No doubt with a wider range of stretching conditions, variations would be found and indeed investigations are being extended to lower temperatures so that the development of morphological form and quantitative microstructural parameters can be controlled and monitored.

The starting fibres (Figure 3) have an irregular and poorly defined "crystallite" arrangement and give rise to electron diffraction patterns with broad highly arced (00.2) reflections as shown in Figure 5. However, during the hot stretching treatment, fibres develop the more regular fibrous morphology of Figure 4 which gives rise to electron diffraction patterns such as that shown in Figure 6. In this case a very sharp (00.2) reflection is seen and it shows very little arcing. Furthermore, this diffraction pattern provides evidence of further ordering within carbon layers since both (10.0) and (11.0) reflections are just discernible. Such reflections however have not been observed on any of the x-ray diffraction patterns. Quantitative data cannot be obtained from these electron diffraction patterns since the grinding up of the fibres for electron microscopy causes disruption of the microstructure, particularly affecting the orientations and positions of individual "crystallites" within the original fibre. However, qualitatively the electron diffraction patterns of Figures 5 and 6 are consistent with the x-ray diffraction pictures of Figures 1 and 2 respectively.

Fortunately, the x-ray diffraction patterns retain the information on preferred orientation and "apparent" crystallite size and so these measurements form the basis of the present paper. However, since the x-rays penetrate the complete carbon fibre, each pattern is an average of the "crystallites" through the complete cross section of the fibre and no detail is available concerning the presence of features such as sheath-core arrangements, or indeed any microstructural variations that might occur across the radius.

The complete set of data from the mechanical testing and x-ray diffraction experiments of the hot stretched individual fibres are shown in Tables I and II. In Figures 7-14, these data are plotted in a variety of different formats to highlight specific aspects. Figures 7 and 8 indicate how the mechanical properties of these individual fibres are related to the "best fits" determined for larger numbers of fibres taken from a previous report (Ozbek and Isaac, 1991). These graphs give some idea of the variability between individual fibres and show how the fibres in the present study compare with an "average" fibre from the same batch. The values quoted in Tables I and II for Z and L_c typically have accuracies of ±0.3°-1.0° and ±0.2-0.5nm respectively. These error estimates are based on the measurement of up to six values in each case taken from two films that were exposed simultaneously and which were analysed using two optical density ranges (0-0.5 O.D. and 0-1.0 O.D.).

The general effects of heat treatment temperature (HTT) on the microstructural parameters Z and L_c are well documented. The decrease in Z and increase in L_c with increasing HTT (in the range 2700°C to 3000°C) for the same batch of starting fibres have already been demonstrated in an earlier report (Francis et al., 1991). Thus for the present studies only the two relatively low, and hence economically more viable, temperatures of 2600°C and 2800°C have been used, and interest has been concentrated on the effects of the two process parameters load and dwell time. Nevertheless, the differences in induced properties between hot stretching at 2600°C and 2800°C is evident in most of the graphs.

The preferred orientation parameter, Z, is plotted against the process variables load and dwell time in Figures 9 and 10 respectively. These graphs indicate that Z decreases consistently with both increasing load and dwell time at 2600°C and 2800°C. The effects of increasing the temperature is also very clear from these graphs, since significantly better preferred orientation is always achieved at the higher temperature. The precise fitting of curves to these data has not been attempted here, but "best fit" straight lines have been included to indicate the trends. All the hot stretched fibres showed substantial improvements from the Z value of ~38° in the as-received fibres and, for the range of stretching conditions used in these experiments, Z values from ~21° to ~13° were found. None of the previous studies of hot stretched carbon fibres such as those by Bacon and

TABLE I *Properties measured on single carbon fibres following stretching at 2600°C with various loads and times.*

Fibre No	Stretching Load (g/filament)	Dwell Time (mins)	Young's Modulus (GPa)	Tensile Strength (GPa)	Preferred Orientation (Z°)	Crystallite Size (L_c) (nm)
1	0.10	30	378	3.56	20.6	6.6
2	0.17	30	448	3.29	19.3	6.5
3	0.17	30	400	2.36	19.7	6.8
4	0.20	30	498	3.18	18.9	7.3
5	0.20	30	476	2.49	18.6	5.5
6	0.24	30	508	2.27	18.3	7.9
7	0.24	30	544	3.29	17.9	6.9
8	0.31	30	590	4.01	17.6	8.5
9	0.31	30	559	4.68	17.7	8.0
10	0.50	30	600	4.27	16.6	8.8
11	0.50	30	582	3.12	16.7	8.9
12	0.50	30	587	5.27	17.0	
13	0.31	2	384	1.22	20.9	6.8
14	0.31	5	399	3.29	20.3	7.1
15	0.31	10	461	3.49	18.7	7.0
16	0.31	10	450	2.44	18.9	8.6
17	0.31	20	467	3.52	18.0	8.1
8	0.31	30	590	4.01	17.6	8.5
9	0.31	30	559	4.68	17.7	8.0

TABLE II *Properties measured on single carbon fibres following stretching at 2800°C with various loads and times*

Fibre No	Stretching Load (g/filament)	Dwell Time (mins)	Young's Modulus (GPa)	Tensile Strength (GPa)	Preferred Orientation (Z°)	Crystallite Size (L_c) (nm)
18	0.10	30	520	2.80	18.9	7.6
19	0.17	30	397	2.31	18.3	8.0
20	0.20	30	487	2.07	17.1	8.3
21	0.20	30	511	2.21	16.5	7.3
22	0.20	30	473	2.92	17.4	8.3
23	0.24	30	556	3.10	15.4	8.4
24	0.24	30	552	2.32	15.8	9.4
25	0.31	30	659	4.43	15.5	9.3
26	0.31	30	635	3.01	14.3	9.6
27	0.50	30	703	4.50	14.2	10.2
28	0.50	30	755	4.97	13.2	9.9
29	0.50	30	666	4.61	14.9	9.9
30	0.31	2	466	2.02	17.1	7.9
31	0.31	5	431	3.28	17.1	8.5
32	0.31	5	379	2.63	16.5	8.0
33	0.31	10	555	4.55	16.7	8.7
34	0.31	10	590	4.25	15.5	9.0
35	0.31	20	560	3.99	15.8	9.4
25	0.31	30	659	4.43	15.5	9.3
26	0.31	30	635	3.01	14.3	9.6

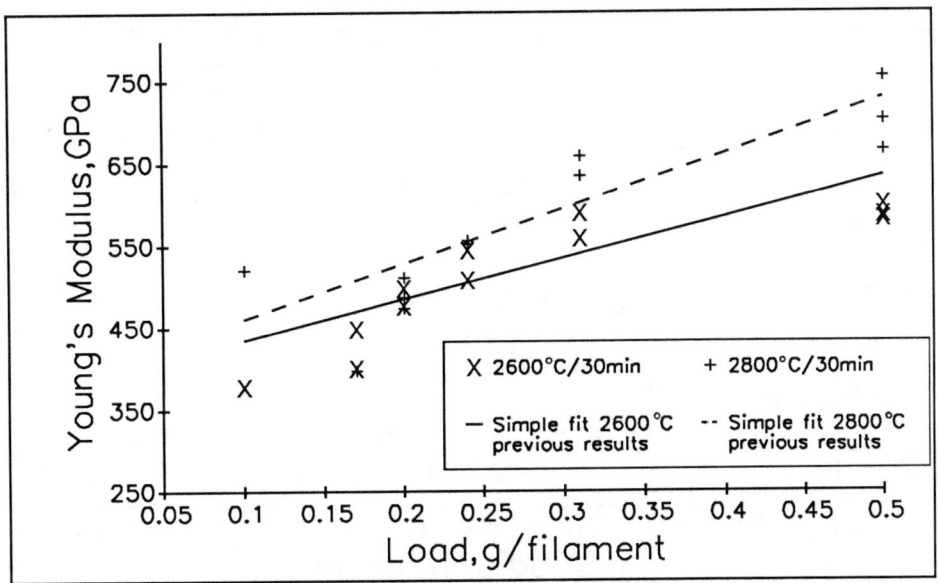

Figure 7. Young's modulus as a function of stretching load for the individual fibres considered in this report. For comparative purposes, straight lines have been included which correspond to the "best fits" obtained for larger numbers of fibres from the same stretched tow, as reported previously (Ozbek and Isaac, 1991).

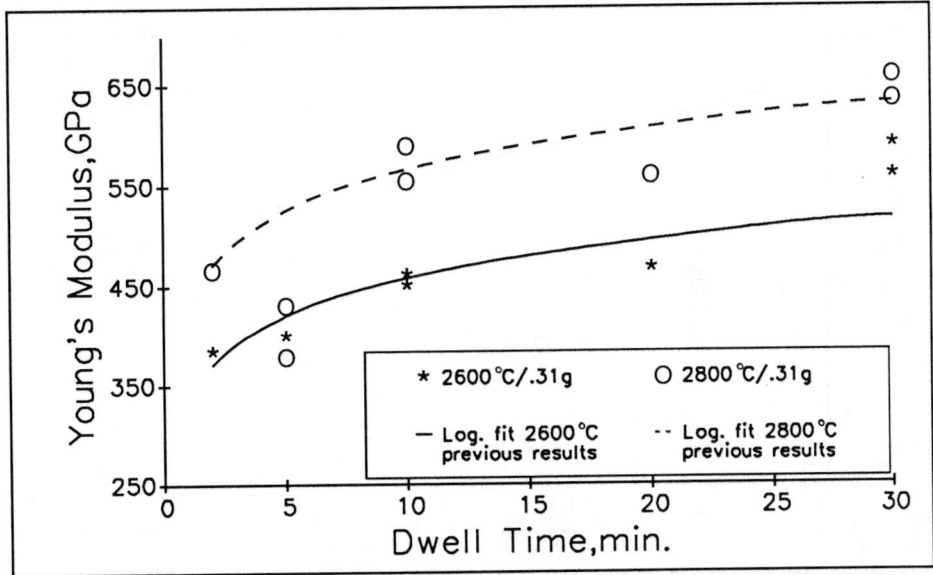

Figure 8. Young's modulus as a function of dwell time for the individual fibres considered in this report compared with the "best fit" straight lines obtained from larger numbers of fibres from the same stretched tow (Ozbek and Isaac, 1991).

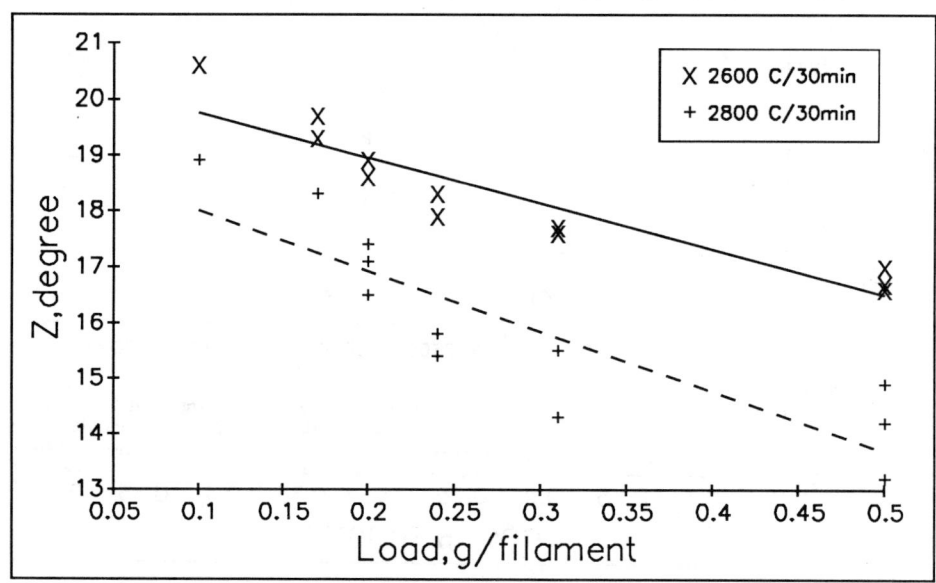

Figure 9. The preferred orientation parameter, Z, as a function of stretching load for individual fibres, together with "best fit" straight lines for each treatment temperature to illustrate the trend.

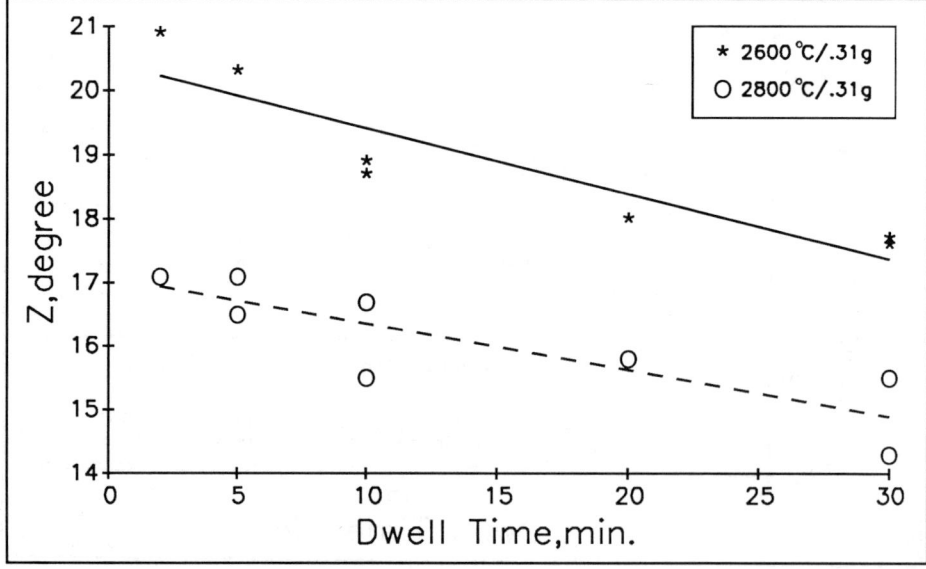

Figure 10. The preferred orientation parameter, Z, as a function of dwell time for individual fibres, together with "best fit" straight lines for each treatment temperature to illustrate the trend.

Schalamon (1969), Johnson *et al.* (1969), Johnson (1970) and Hawthorne (1971) have directly related the temperature, load and dwell time to the degree of basal plane alignment. Indeed this present report is the first to study the systematic variation in these three process parameters. Bacon and Schalamon (1969) produced fibres from a rayon precursor with a Z value of about 5°. Johnson *et al.* (1969) on the other hand claimed to have made PAN-based fibres with a considerably lower Z value. Their lowest figure ~2° is well below any other report for any form of carbon fibre (even ultra high modulus mesophase pitch-based fibres have Z ~5°) and is comparable with graphite whiskers (Reynolds and Sharp, 1974). However in the absence of details on the experimental method that was used to measure this value it is difficult to assess these results meaningfully.

Figure 11 shows the relationship between the preferred orientation parameter, Z, and Young's modulus for these PAN-based fibres. Several authors have examined this relationship in various types of carbon fibres; Bacon and Schalamon (1969) and Ruland (1969) for rayon-based fibres, Watt *et al.* (1966), Johnson *et al.* (1969), Diefendorf and Tokarsky (1975) and Francis *et al.* (1991) for PAN-based fibres, Hawthorne (1971) for pitch-based fibres and Bright and Singer (1979) for MP-based fibres. Most of these previous studies have found a non-linear relationship between the two values, the increase in modulus being more rapid at smaller values of Z. Only Diefendorf and Tokarsky (1975) and Francis *et al.* (1991) have suggested a simple linear relationship between these two parameters. Diefendorf and Tokarsky (1975) also showed the similarities between the modulus/orientation curve for rayon-based carbon fibres and for the single crystal of graphite (which is drawn from a constant stress model for modulus). On the other hand Fitzer *et al.* (1986) studied the degree of graphitisation of PAN-based carbon fibres treated at temperatures up to 2800°C and concluded that PAN-based carbon fibres can never graphitise. In these previous studies the lower end of the modulus range was generally less than 300GPa, and in most cases less than 200GPa, compared with a minimum of more than 350GPa in this work. The upper end of the modulus range was ~700GPa, roughly the same value as recorded in the present studies. Although some previous reports found a non-linear relationship between orientation and modulus over the range 200GPa to 700GPa, this relationship is approximately linear above 350GPa (Bacon and Schalamon, 1969). Thus, the linearity found in the present work may be only part of a non-linear relationship extending below 350GPa and 21°. This possibility is being considered in a series of further studies designed to extend the data to lower modulus values by stretching at lower temperatures.

Tables I and II show wide scatter in the tensile strength results, as expected, due to random nature of defects, which are thought to control the ultimate strength. Despite this scatter there is a discernible trend of increasing strength with improved orientation of the "crystallite" domains along the fibre.

It has been reported previously (Francis *et al.* 1991) that the "apparent" crystallite stacking height, L_c, after hot stretching of carbon fibres increases approximately linearly with temperature for each different loading condition. However, the effect of load on L_c was uncertain, although little variation was found for a HTT of 3000°C with loads between 0.020g/filament (for which L_c was ~10.3nm) and 0.333g/filament (L_c ~10.9nm). The values of L_c obtained in the present study are plotted against the process parameters, stretching load and dwell time in Figures 12 and 13 respectively. The best fit straight lines on these graphs indicate a trend of increasing L_c with both load and dwell time at each hot stretching temperature. These findings are in general agreement with the results of other workers (Shindo, 1961 ; Johnson, 1971) who have shown that L_c varies from ~1-2 nm at 1000°C up to ~8-11nm at 3000°C for PAN-based fibres. Fourdeux *et al.* (1969) and Perret and Ruland (1972) demonstrated that for rayon-based fibres at 2900°C and at strain rates below ~0.05sec^{-1}, L_c was not significantly affected by stretching. However at a higher strain rate (~0.16sec^{-1}), L_c increased from ~30nm to about 80nm. Johnson *et al.* (1969) also demonstrated that in PAN-based fibres, L_c was affected by stretching at high temperatures, with values following treatment at 2970°C increasing from 17nm without straining, up to 25nm after 23% strain. All of these results indicate that L_c values vary much less for PAN-based fibres than for rayon-based fibres and the variation in L_c with induced strain is relatively small compared with changes induced by temperature. This is no doubt related to the very poor graphitisation behaviour of PAN-based carbon fibres.

The relationship between L_c and Young's modulus of the hot stretched carbon fibres is shown in Figure 14 for all the conditions used in these experiments. Although Figure 14 shows the inevitable scatter associated with individual fibres, it is clear from the "best fit" straight line that there is a trend of increasing L_c with modulus. Similarly, although Tables I and II show even greater scatter for the tensile strength values, a general trend of increasing L_c with strength is evident.

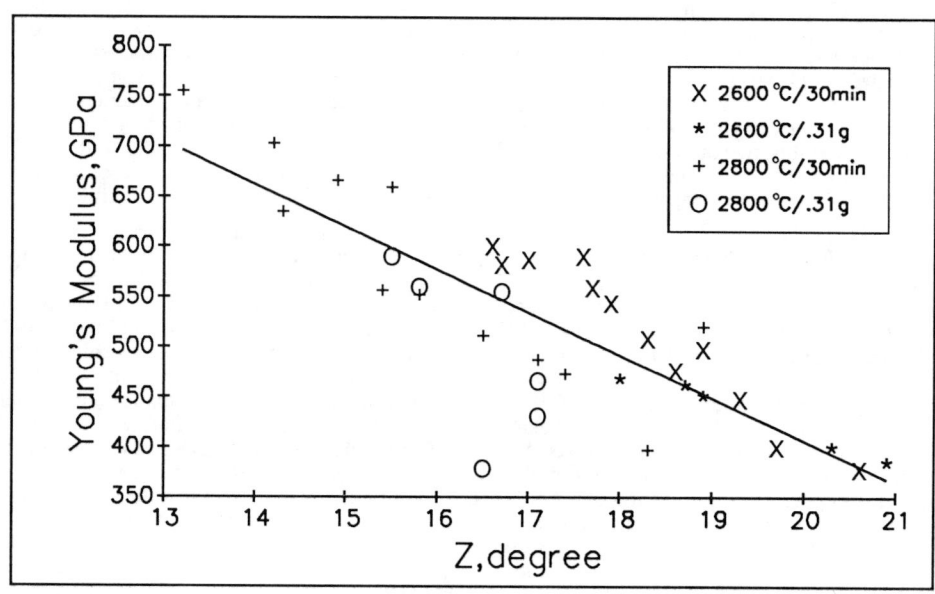

Figure 11. Young's modulus as a function of preferred orientation for all 35 individual fibres, together with a "best fit" straight line to illustrate the trend.

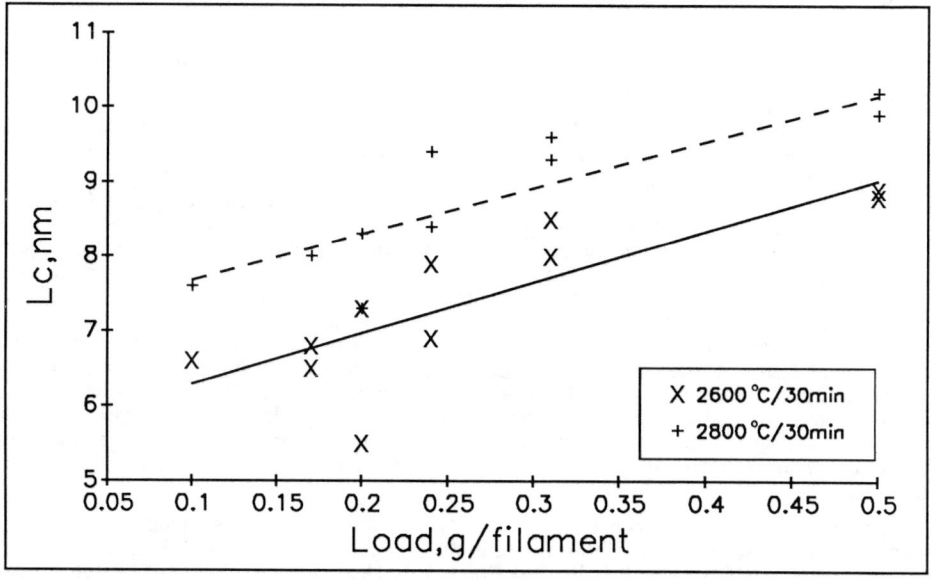

Figure 12. The "apparent" crystallite size, L_c, as a function of stretching load for individual fibres, together with "best fit" straight lines for each treatment temperature to illustrate the trend.

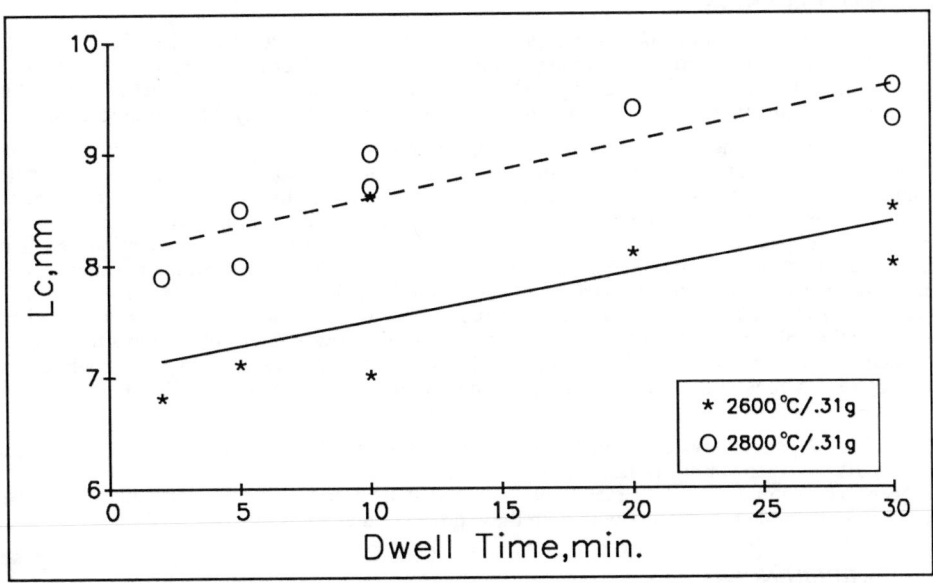

Figure 13. The "apparent" crystallite size, L_c, as a function of dwell time for individual fibres, together with "best fit" straight lines for each treatment temperature to illustrate the trend.

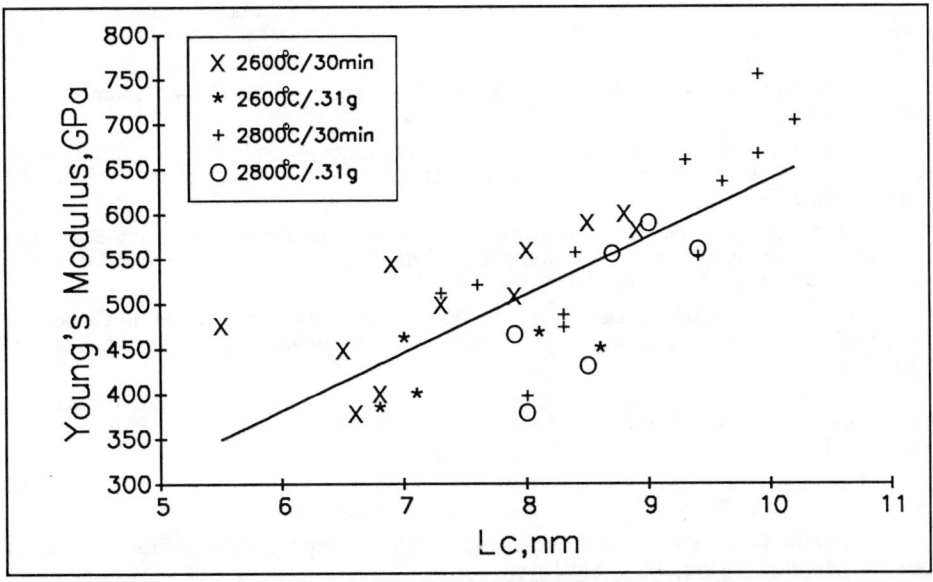

Figure 14. Young's modulus as a function of the "apparent" crystallite size for all 35 individual fibres, together with a "best fit" straight line to illustrate the trend.

4. CONCLUSIONS

It has been demonstrated that following hot stretching at 2600°C - 2800°C, PAN-based carbon fibres develop a fibrous morphology which gives rise to electron diffraction patterns with a sharp (00.2) reflection and little arcing. Although data from these electron diffraction patterns cannot easily be quantified, they are qualitatively consistent with the x-ray diffraction results.

For the first time, the development of texture and "apparent" crystallite size in carbon fibres has been monitored as a function of time, as well as applied load. The preferred orientation parameter Z decreased from 38° to between 21° and 13° depending on the specific processing conditions. The preferred orientation was found to depend on both the applied load and the dwell time, so that increasing either of these parameters improved the preferred orientation and hence decreased Z. Despite the inevitable scatter occurring for individual fibres, an approximately linear relationship between the preferred orientation parameter, Z, and Young's modulus has been noted in the range 350GPa to 750GPa. However, the small range of modulus and Z values presented here does not exclude the possibility of a non-linear relationship at lower values of modulus that has been suggested by other workers. Although scatter in tensile strength results prevents accurate curve fitting, there is a general trend of increasing strength (between ~2GPa and ~5GPa) with decreasing Z.

The "apparent" crystallite stack height L_c has been found to vary within the range ~6.5nm to ~10.2nm after hot stretching, from a value before treatment that was too small to measure. It has also been demonstrated that L_c depends on the applied load and dwell time, the number of basal planes stacked together increasing with both these process parameters.

5. REFERENCES

Bacon, R. and Schalamon, W., 1969, "Physical Properties of High Modulus Graphite Fibres Made from a Rayon Precursor", *Applied Polymer Symposia*, ed. J. Preston, American Chemical Society, **9**, 285-292.

Bright, A.A. and Singer L.S., 1979, "The Electronic and Structural Characteristics of Carbon Fibres from Mesophase Pitch", *Carbon*, **17**, 59-69.

Brydges, W.T., Badami D.V., Joiner J.C. and Jones G.A., 1969, "The Structure and Elastic Properties of Carbon Fibres", *Applied Polymer Symposia*, ed. J. Preston, American Chemical Society, **9**, 255-260.

Butler, B.L. and Diefendorf R.J., 1969, "Microstructure of Carbon Fibres", *Proceedings of the 9th Biennial Conference on Carbon, Boston College, Massachusetts*, American Carbon Society, 161-162.

Diefendorf, R.J. and Tokarsky E., 1975, "High Performance Carbon Fibres", *Polymer Engineering and Science*, **15(3)**, 150-159.

Fitzer, E., Frohs W. and Heine M., 1986, "Optimisation of Stabilization and Carbonisation Treatment of PAN-Based Fibres and Structural Characterisation of the Resulting Carbon Fibres", *Carbon*, **24(4)**, 387-395.

Fourdeux, A., Perret R. and Ruland W., 1969, "Electron Diffraction and Electron Microscopy of Carbon Fibres", *Proceedings of the 9th Biennial Conference on Carbon, Boston College, Massachusetts*, American Carbon Society, 159-160.

Francis, J.G., Ozbek. S. and Isaac D.H., 1991, "Microstructural Changes in Carbon Fibres During High Temperature Processing", in *Processing and Manufacturing of Composite Materials*, ASME, **PED-Vol.49 / MD-Vol.27**, 173-183.

Hawthorne, H.M., 1971, "Structure and Properties of Strain-Graphitized Glassy Carbon Fibres", *International Conference on Carbon Fibres and their Composites, Plastics Institute*, London, 81-93.

Johnson, D.J., 1971, "The Microstructure of Various Carbon Fibres", *1st International Conference on Carbon Fibres - Their Composites and Applications*, 52-56.

Johnson, D.J., 1987a, "Structure-Property Relationship in Carbon Fibres", *Journal of Physics D : Applied Physics*, **20**, 286-291.

Johnson, D.J., 1987b, "Structural Studies of PAN-Based Carbon Fibres", in *Chemistry and Physics of Carbon*, ed. P.A. Thrower, Dekker, New York, **20**, 1-58.

Johnson, J.W., Marjoram J.R. and Rose P.G., 1969, "Stress Graphitization of Polyacrylonitrile Based Carbon Fibre", *Nature*, **221**, 357-358.

Johnson, W., 1970, "Hot Stretching of Carbon Fibres Made From Polyacrylonitrile Fibres", *Proceedings of the 3rd Conference on Industrial Carbon and Graphite, London*, Society of Chemical Industries, London, 447-453.

Johnson, W. and Watt W., 1967, "Structure of High Modulus Carbon Fibres", *Nature*, **215**, 384-386.

Oberlin, A. and Guigon M., 1988, "The Structure of Carbon Fibres", in *Fibre Reinforcements for Composite Materials*, ed. A.R. Bunsell, Composite Materials Series, Elsevier, New York, **2**, 149-210.

Ozbek, S., and Isaac D.H., 1991, "Carbon Fibres : Effect of Processing Parameters on Mechanical Properties", in *Processing and Manufacturing of Composite Materials*, ASME, **PED-Vol.49 / MD-Vol.27**, 307-320.

Ozbek, S., Jenkins G.M. and Isaac D.H., 1991, "Mechanical Properties of Hot Stretched Carbon Fibres", *Proceedings of the 20th Biennial Conference on Carbon, Santa Barbara*, American Carbon Society, 308-309.

Perret, R. and Ruland W., 1972, "Structural Changes Induced by the Plastic Deformation of Carbon Fibres at Elevated Temperatures", *International Carbon Conference, Baden-Baden*, 318-320.

Reynolds, W.N. and Sharp J.V., 1974, "Crystal Shear Limit to Carbon Fibre Strength", *Carbon*, **12**, 103-110.

Ruland, W., 1969, "The Relationship Between Preferred Orientation and Young's Modulus of Carbon Fibres", *Applied Polymer Symposia*, ed. J. Preston, American Chemical Society, **9**, 293-301.

Shindo, A., 1961, Report No. 317 Government Industrial Research Institute, Osaka, Japan.

Watt, W., Phillips L.N. and Johnson. W., 1966, "High-Strength High-Modulus Carbon Fibres", *The Engineer*, **221**, 815-816.

FATIGUE LIFE CHARACTERISATION OF AS4/PEEK COMPOSITES

K. S. Saib, W. J. Evans, and D. H. Isaac
Department of Materials Engineering
University of Wales Swansea
Swansea, United Kingdom

ABSTRACT

Thermoplastics reinforced with continuous carbon fibres have in recent years gained increasing acceptance for applications in aerospace. In these major stress bearing roles, weight saving and strength/stiffness properties are important, but resistance to fatigue failure is also critical. This paper presents results from fatigue life (S/N) tests conducted on one of the "state of the art" materials - AS4/PEEK (continuous carbon fibres in the thermoplastic PEEK). The effect of experimental conditions such as specimen geometry (coupon or waisted), frequency (0.5, 1 and 5Hz), shape of loading waveform (sine or square), stress ratio (R=0.1 or 0.5) and test temperature (18 or -55°C) have been investigated. Further, microstructural parameters such as laminate layup ($[\pm 45]_{4s}$, $[-45,0,+45,90]_{2s}$ and $[0,90]_{4s}$) and an increase in matrix crystallinity (induced by post-forming annealing) were also considered. Regression lines were fitted to the S/N data obtained for $[\pm 45]_{4s}$ and $[-45,0,+45,90]_{2s}$ layups. The ratio of tensile strength to gradient (determined from the best fit lines) demonstrated that the response to fatigue of AS4/PEEK materials is superior to that of a wide range of composites. Fracture mechanisms are highlighted and discussed with the aid of scanning electron microscopy.

1.0 INTRODUCTION

Continuous carbon fibre reinforced composites have made significant inroads into high technology applications such as aerospace where weight saving is of paramount importance. Until recently, these materials were based almost exclusively on thermosetting epoxy resin matrices. Although strength and stiffness for these composites are generally inferior to the traditional aluminium alloys used in fuselage and secondary structures, a large improvement in specific properties is obtained. However, the brittle nature of thermoset resins is manifest as a lack of composite damage tolerance, especially at the sub-zero temperatures encountered in service at altitude. Additionally, resistance to environmental attack (fuel and oil) is low. In response to the demand for a tough and environmentally resistant polymer based composite system, ICI introduced APC (Aromatic Polymer Composite) based on their ductile and relatively high temperature capability thermoplastic, PEEK, reinforced with continuous carbon fibres. This has progressed through a number of developmental stages to APC-2/AS4 which is now a clear rival to continuous carbon fibre reinforced epoxy composites (ICI Fiberite, 1988).

PEEK based composites have excellent damage tolerance/toughness properties as a result of the polymer's ductile nature, which is utilised fully by an extremely strong fibre/matrix interface (Saib, 1991). The semi-crystalline nature of PEEK also imparts excellent chemical resistance and a temperature capability approaching that of the best thermosets. The presence of a rigid molecular chain structure in PEEK in combination with high fibre/matrix interface strength results in a high strength and stiffness for the composite.

APC materials have been successfully employed in both military and civil aircraft (Gray and Savage, 1989). In these major stress bearing applications, it is established that fatigue loading is a dominant factor in the failure of components and structures. In order to fulfil their potential, the fatigue performance of these composites must be characterised and compared with rivals such as thermoset based systems. Many studies have attempted this although the results have often proven to be equivocal (Hartness and Kim, 1983; Curtis, 1986). This can generally be attributed to non-standard testing or processing techniques.

An ongoing research programme is addressing this problem by testing a range of PEEK based and novel toughened epoxy composites using controlled experimental parameters. Both fatigue life (a measure of crack initiation) and fatigue crack propagation modes of failure are being assessed in these studies. The influence of experimental conditions such as specimen geometry, frequency of cycling, shape of loading waveform, stress ratio and sub-zero test temperature on fatigue life performance of AS4/PEEK composites (where AS4 refers to the type of carbon fibre used for reinforcement) is discussed in this paper. In addition, microstructural parameters such as laminate layup and an increase in matrix crystallinity (induced by post-forming annealing) are considered.

2.0 EXPERIMENTAL PROCEDURES

2.1 Materials and Specimens

PEEK reinforced with continuous AS4 carbon fibres has been tested in three different lay-ups:

quasi-isotropic $[-45,0,+45,90]_{2s}$ or [QI],
$[0,90]_{4s}$, and
$[\pm 45]_{4s}$.

These were supplied in the form of compression moulded plaques with dimensions of 457mm x 457mm. Width waisted specimens (Figure 1) were used to generate fatigue life data, although some parallel-sided (or coupon) samples were tested to study the effects of specimen geometry. Coupon shapes (15mm x 200mm) were cut using a diamond impregnated slitting wheel. The edges of these parallel sided specimens were dressed using wet/dry fine emery paper to ensure a smooth finish, and end-tabbed with tapered 10G/40 grade TUFNOL to prevent grip damage. Width waisted specimens were machined from coupons (25mm x 180mm) using a polycrystalline diamond tipped fly cutter. The waisted section had a minimum width of 15mm and a radius of 71mm. The purpose of the waist was to define the point of failure more reliably without inducing an excessive stress concentration. Special jigs were designed and used throughout specimen manufacture to ensure reproducability and alignment.

2.2 Fatigue Life Testing

Testing was conducted in an air conditioned laboratory (18°C) on an 100kN Instron 1341 servohydraulic testing machine and two pneumatic fatigue sites (capable of 15 and 45kN) manufactured in this Department. The number of cycles to failure (N) at a particular nominal stress level (S or σ) were recorded and the data presented in the form of traditional S/N plots for comparison of fatigue life performance. Fatigue lifetimes in the range 1 to 10^7 cycles were considered. Occasionally tests were stopped prior to failure and these are defined by the use of a horizontal arrow in the figures. Specimen surface temperatures were monitored throughout using a self adhesive Type K thermocouple.

Test frequencies of 0.5, 1 and 5 Hz with both sine and square shaped loading waveforms were investigated. The loading waveform obtained on the pneumatic fatigue sites was intermediate between sine and square shaped with a frequency of 1Hz and R=0.1 although both these variables were controllable. The effect of an increase in stress ratio (R) from 0.1 to 0.5 was also considered. Further, fatigue life behaviour at a test temperature of -55°C was compared with that at room temperature. The role of an increase in matrix crystallinity on fatigue life performance was assessed by using a post-moulding annealing treatment recommended by Chivers and Moore (1992). This involved heat treating the composite specimens at 320°C for 1 hour followed by furnace cooling, a procedure designed to maximise the matrix degree of crystallinity. To avoid warpage, the samples were first sandwiched between two metal plates.

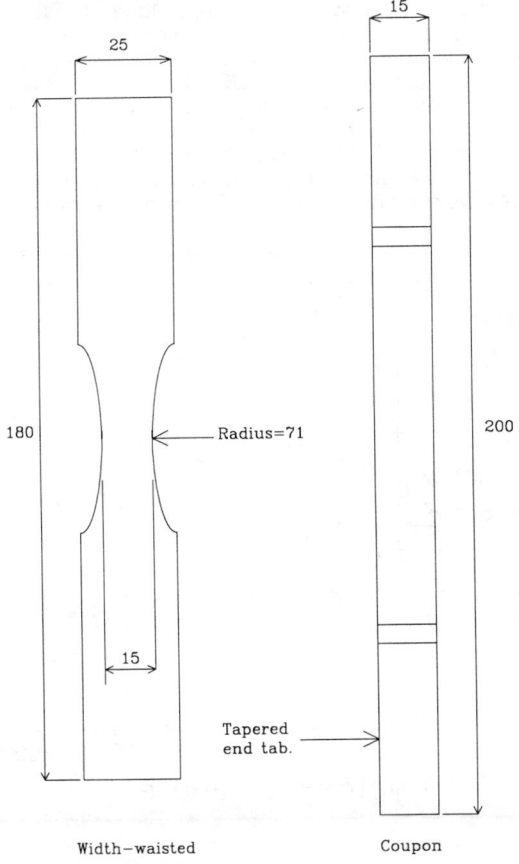

Figure 1. Fatigue specimen geometries employed (dimensions in mm).

3.0 RESULTS AND DISCUSSION

3.1 Fatigue Life Behaviour

Figures 2-8 show S/N data for various AS4/PEEK based materials and demonstrate the effects of experimental and microstructural parameters on the fatigue life behaviour.

Figure 2 includes results for both coupon (C) and width waisted (W) specimens in [0,90], [QI] and [±45] layups. It is not surprising to find that for the [0,90] layup longer fatigue lives are obtained with parallel coupon specimens. Half of the fibres in this material are in the 0° direction and these plies are normally susceptible to premature failure if waisting is employed. This effect is usually unimportant for the [QI] and [±45] layups since there are fewer 0° fibres present (none in [±45]). The [QI] material shows identical behaviour for both specimen types. However, the [±45] layup exhibits slightly lower lives for parallel sided specimens. Not withstanding this small discrepancy for the [±45] layup, the general observations are consistent with results presented by Curtis *et al.* (1988). Reduced strengths with angle-ply composites are normally associated with stress concentration effects resulting from misalignment of end-tabs. In all the cases presented for end-tabbed specimens, however, failure occurred within the gauge length and well away from the gripping regions. Further tests are being performed to investigate the significance of the small difference in results between the [±45] coupons and waisted samples and to explain the behaviour. Nevertheless, the differences are small enough to give confidence in the results obtained for both layups, and all subsequent tests reported are from waisted samples.

In common with the findings of Curtis *et al.* (1988), a change in loading waveform shape from square to sine is seen to have negligible effect on fatigue life response for both [QI] and [±45] layups (Figure 3). This is perhaps a surprising result since a square waveform is normally considered to be the most severe loading excursion as a result of the extremely high strain rates and times at maximum load involved. However, AS4/PEEK composites are known to have significant creep resistance (Carlile *et al.*, 1989) at room temperature even in matrix dominated layups. Additionally, the effects of strain rate on monotonic mechanical properties of continuous carbon fibre/PEEK composites have been found to be small (Dickson *et al.*, 1985). Hence, even though a sine waveform imposes a lower strain rate and time at maximum load, the effects of this difference are minimal and cannot be resolved on fatigue life plots. The practical advantage of this result is that

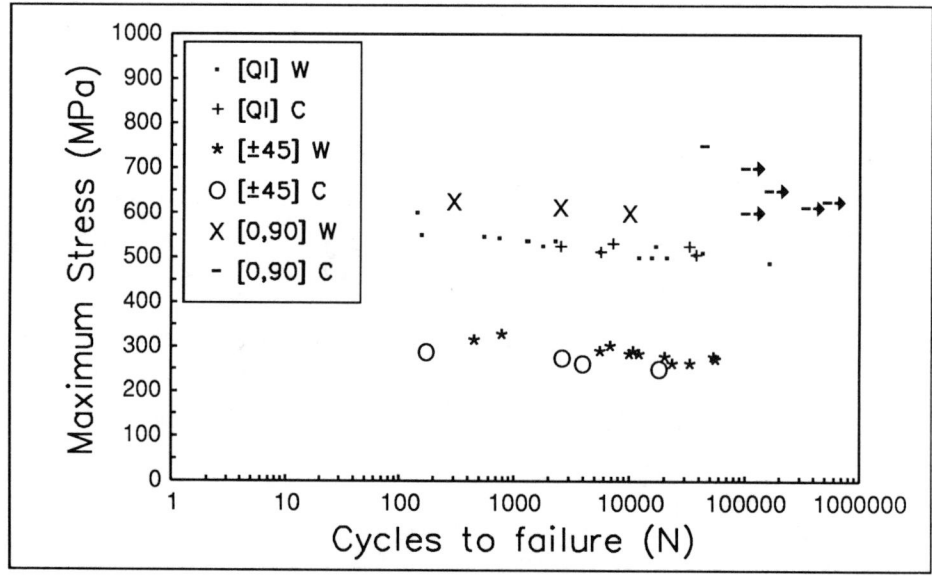

Figure 2. Effect of specimen geometry on fatigue life behaviour: (W = Width-waisted, C = Coupon).

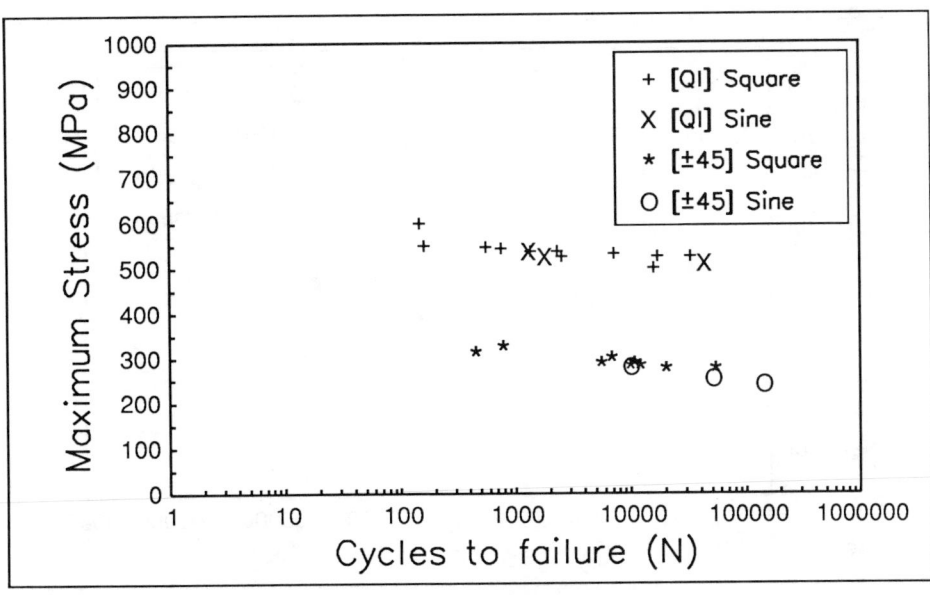

Figure 3. Effect of loading waveform shape on fatigue life behaviour of [QI] and [±45].

it allows accurate and consistent data acquisition from simple fatigue machines, such as pneumatic sites which provide less control over parameters such as frequency and waveform than servohydraulic facilities. This is especially useful for low stress/high lifetime tests in which equipment can be occupied for extremely long periods.

Typical fatigue life data obtained on both servohydraulic and pneumatic fatigue machines are presented in Figure 4 for [QI] and [±45] materials. Although the loading waveform shape on the pneumatic sites was non-standard, it is encouraging to note that the data from both types of machine clearly superimpose, confirming that the fatigue lives are not waveform dependant.

An increase in frequency of cycling from 0.5Hz to 1Hz is seen to have negligible effect on fatigue life performance for both [QI] and [±45] materials (Figure 5). Measurements during testing confirmed that, for both materials, temperature rises were not excessive at either of these frequencies and although rises were slightly higher at 1Hz the temperature always stabilised. Thus testing times could be halved without the risk of thermal failures occurring. However, the limited data obtained for the [±45] material with a 5Hz square wave demonstrates the process of thermal fatigue failure. As a result of significant hysteretic heating (temperature rises of the order of 50°C were evident at this frequency), the softening of the matrix manifests itself as substantially reduced fatigue strength. The behaviour is more significant in the [±45] layup since its mechanical properties are matrix dominated. Additionally, it was noted that when samples were tested using a 5Hz sine waveform at equivalent stresses to those used with a 5Hz square waveform, rupture of specimens was not observed. This contrasts with the effect of waveform at low frequencies. Clearly, at a frequency of 5Hz, hysteretic heating effects associated with the higher strain rates encountered when using a square waveform are substantially more severe than at 0.5 or 1Hz.

The role of stress ratio (a variable which affects the mean stress) during fatigue life testing of [QI] and [±45] layups can be seen in Figure 6. In both cases, a significant increase in cycles to failure at the higher R-value is evident for equivalent maximum stress. This is a common response with many materials at higher stress ratios and results from the reduction in stress amplitude during the fatigue cycle. The implication is that these materials are more sensitive to stress amplitude than to the creep component of loading, which increases for a higher R value. This is consistent with the high creep resistance of AS4/PEEK found by Carlile et al. (1989). Further work is being carried out to assess the effects of a compressive component in the fatigue loading cycle during fully reversed testing (R=-1).

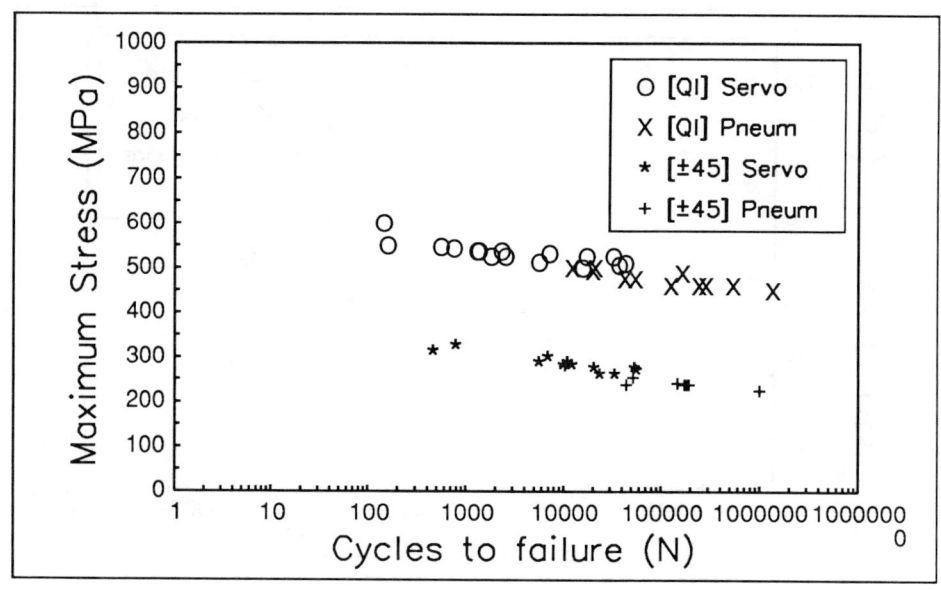

Figure 4. Effect of testing site on fatigue life behaviour of [QI] and [±45].

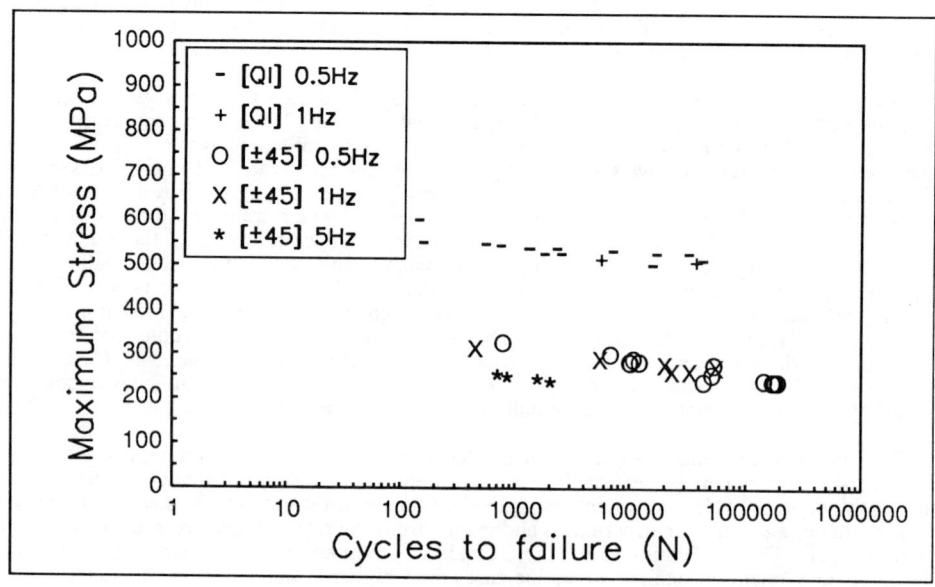

Figure 5. Effect of test frequency on fatigue life behaviour of [QI] and [±45].

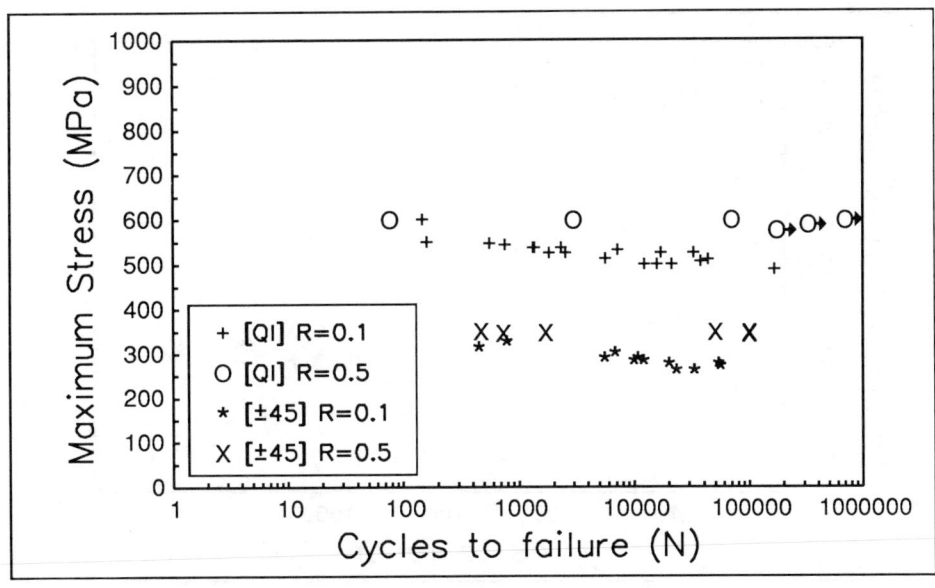

Figure 6. Effect of stress ratio (R) on fatigue life behaviour of [QI] and [±45].

Figure 7 demonstrates the effect of reducing the test temperature to -55°C on the fatigue life behaviour of [±45] AS4/PEEK. This is a matrix dominated layup and so should be most sensitive to temperature. Certainly, a clear increase in fatigue strength is evident at the lower temperature. This effect can be attributed directly to an increase in the matrix strength and stiffness. This behaviour is unusual in that a lower temperature (especially sub-zero) often reduces toughness, and fatigue behaviour is affected correspondingly. The mechanisms of failure are discussed in the following section.

A similar though less distinct increase in fatigue strength can be induced by post-annealing the [±45] layup, as seen in Figure 8. The post-annealing treatment has negligible effect on spherulitic morphology but results in some lamellar thickening and local ordering of crystallites. The consequence of this process is an increase in degree of crystallinity and correspondingly, strength and stiffness levels. However, the role that this effect plays during fatigue failure is not clear cut. It has been shown that an increase in matrix crystallinity can reduce fracture toughness but simultaneously improve the fatigue performance of short fibre reinforced PEEK composites (Saib, 1991). This behaviour was related to the relative damage zone sizes in cyclic and static loading. With S/N testing, stress levels are relatively high and so the bulk response of the material is more significant. Hence, although stiffness has risen, the ductility and toughness of the polymer matrix are reduced. This may explain why Curtis et al. (1991) found that a slow cooling rate during moulding, which increased the matrix crystallinity, had no effect on fatigue life behaviour for [±45] carbon fibre/PEEK composites. In the present work, the small improvement in fatigue life performance for [±45] AS4/PEEK after post-annealing may be a consequence of the slightly different processes occurring during annealing and slow cooling from the melt, or may arise from the relief of internal stresses during annealing.

The role of laminate layup may also be determined from Figure 2. It is clear that if more 0° fibres are present in the laminate, then the fatigue strength is increased. Hence, [0,90] is far superior to [QI] in terms of absolute fatigue strength. Similarly, [QI] has significantly higher fatigue strength than the [±45] layup. However, the mechanisms controlling the progress of damage are complex and different in each layup, and this aspect is discussed in the following section.

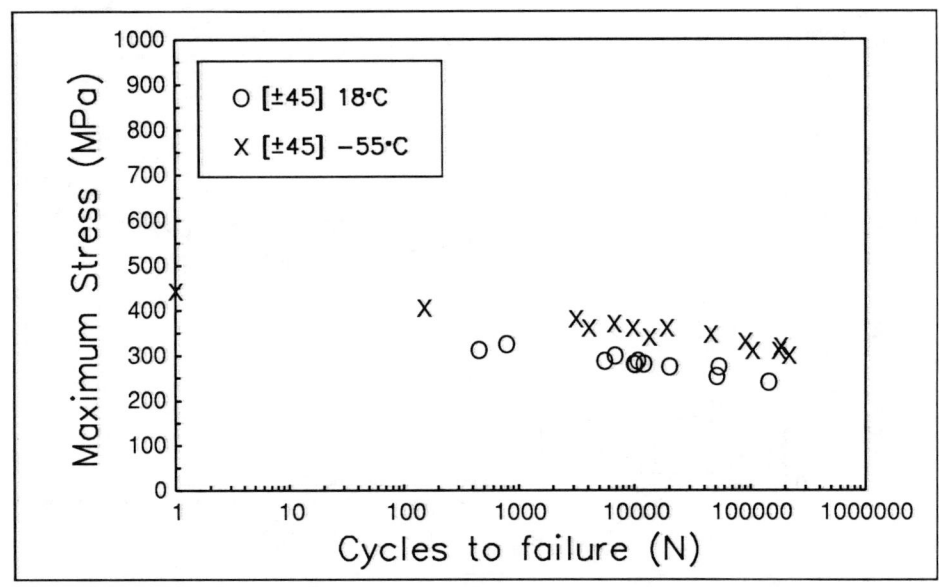

Figure 7. Effect of test temperature on fatigue life behaviour of [±45].

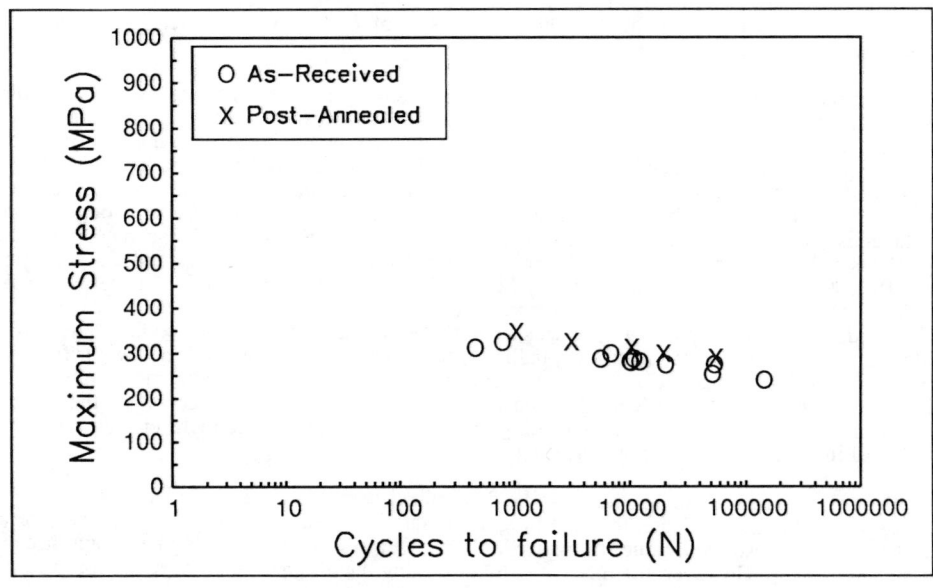

Figure 8. Effect of post-annealing on fatigue life behaviour of [±45].

3.2 Fatigue Mechanisms

It is well known that fatigue failure in continuous fibre composites involves a number of mechanisms occurring separately or in unison. Differences in S/N data are usually a consequence of varying strengths and stiffnesses (which can be allowed for by dividing fatigue stress by tensile strength) and changes in fracture mechanisms are normally associated more with changes in slope. For a wide range of glass fibre reinforced composites (both short and continuous fibres with a variety of matrices), Mandell (1982) suggested that fatigue failure was always fibre dominated. It was found that the ratio of $\sigma_U/B \approx 10$ (where σ_U is the composite's tensile strength and B is the absolute gradient of the regression line fitted to the S/log N data). Thus the reciprocal of σ_U/B is the fractional loss in strength per decade of fatigue cycles and hence a measure of the materials resistance to fatigue damage.

In this study, values of σ_U/B obtained for [QI] and [±45] were 21.5 and 13.9 respectively (using tensile strength values predicted by the regression equations - see Table 1). These values suggest that differing mechanisms (or contributions of mechanisms) are operating in the two materials. The higher value of σ_U/B emphasises the fact that damage progresses more slowly through the [QI] layup. It also is clear that AS4/PEEK in both [QI] and [±45] layups is substantially more fatigue resistant than the majority of composites tested by Mandell (1982). Dickson et al. (1985) have attributed this to the superior fatigue properties of the PEEK matrix and the strength of the carbon fibre/PEEK matrix interface.

It has been established that the mechanisms occurring in the three relevant layups are (Beaumont, 1987; Ogin et al. 1985):

[0,90]- matrix cracking in the 90° plies leading to final tensile fibre failure in the 0° plies,

[QI]- matrix cracking in the 90° plies followed by matrix cracking in the 45° plies. Final failure occurs via delamination of the plies.

[±45]- matrix cracking in the 45° plies followed by delamination of the plies.

It is clear that the values of σ_U/B determined for [QI] and [±45] layups are consistent with a change in fracture mechanism of the type outlined above.

Work carried out by Curtis et al. (1991) has shown an increase in fatigue strength for [QI] carbon fibre/PEEK when the temperature is lowered to -55°C. However, it has been suggested that the fatigue mechanisms occurring in this layup are unaffected by the drop in temperature although their proportionate contributions to the overall fatigue process may have altered.

A similar analysis to that described above yields a σ_U/B value of 15.0 for [±45] AS4/PEEK at -55°C. This small increase suggests that a change in mechanism or contribution of mechanisms may have occurred at -55°C. Evidence for this can be seen in the fractographs of [±45] AS4/PEEK at 18°C and -55°C (Figure 9 and 10 respectively).

These four micrographs demonstrate the intralaminar and interlaminar (delamination) fracture surfaces of [±45] AS4/PEEK S/N specimens. From the earlier listing of mechanisms that can occur during fatigue failure in this layup, it is clear that intralaminar fracture is the precursor to delamination cracking.

Fibre failure does not play an important role in the [±45] layup and so an examination of the matrix fracture morphology can help to elucidate any change in the contribution of mechanisms that may occur. Although there is evidence of ductility in the matrix fracture morphology at -55°C (Figure 10) this is somewhat reduced from that at room temperature (Figure 9). The photographs of Figures 9 and 10 are not conclusive, but they suggest that both intra and inter laminar cracking of the matrix is less ductile at -55°C than at room temperature. Although there is a rise in fatigue strength at -55°C, this does not translate into a significant rise in σ_U/B: damage progression in the material is relatively unchanged. This can probably be attributed to the reduction in matrix ductility at -55°C. However, there is still ample matrix ductility remaining for the material (unlike many polymer based materials) to retain a large proportion of its toughness even at sub-zero temperatures.

It must be pointed out however, that matrix fracture morphology during fatigue in short fibre reinforced PEEK has been seen to change with stress intensity levels/fatigue crack growth rate (Saib, 1991). This type of information is not available from simple S/N tests although crack growth rates are assumed to be high since propagation does not constitute a large proportion of the lifetime. A more complete picture of the processes would require fatigue crack growth testing to be carried out where crack growth kinetics are accurately known, and such investigations are continuing.

Layup	σ_U	B	σ_U/B
[QI]	634.2	29.5	21.5
[±45]	406.3	29.2	13.9
[±45], -55°C	484.1	32.2	15.0

Table 1. Statistical parameters obtained from S/N data

(a)

(b)

Figure 9. Fracture surface of sample of [±45], fatigue tested at room temperature: (a) intralaminar and (b) interlaminar.

(a)

(b)

Figure 10. Fracture surface of sample of [±45], fatigue tested at -55°C: (a) intralaminar and (b) interlaminar.

4.0 CONCLUSIONS

1. The fatigue life behaviour of three AS4/PEEK layups has been characterised. A quantitative analysis of S/N data for [±45] and [QI] layups has demonstrated that these materials have excellent fatigue resistance.

2. Experimental parameters such as machine type (servohydraulic and pneumatic), two low test frequencies (0.5 and 1Hz) and waveform shape (square and sine) were found to have negligible effect on fatigue strength.

3. A higher frequency of 5Hz (resulting in thermal failures for [±45]) and stress ratio of 0.5 were found to affect fatigue strength significantly (for [±45] and [QI]).

4. Both a reduced test temperature of -55°C and a post-annealing heat treatment increased the fatigue strength of [±45] as a result of the modification of matrix fracture properties.

5.0 SYMBOLS USED

R = stress ratio (minimum stress/maximum stress).

σ_U = tensile strength of the composite (in this instance obtained from the regression line).

B = gradient of the regression line fitted to the S versus log N data.

6.0 REFERENCES

Beaumont, P.W.R., 1987, "The fatigue damage mechanisms of composite laminates", *CUED/C/MATS/TR* **139**.

Carlile, D.R., Leach, D.C., Moore, D.R. and Zahlan, N., 1989, "Mechanical properties of the carbon fibre/PEEK composite APC2/AS4 for structural applications", *ASTM STP*, **1044**, 199.

Chivers, R.A. and Moore, D.R., 1992, "The effect of molecular weight and crystallinity on the mechanical properties of injection moulded PEEK", *Polymer*, (in press).

Curtis, D.C., Davies, M., Moore, D.R. and Slater, B., 1991, "The fatigue behaviour of continuous carbon fibre reinforced PEEK", *ASTM STP*, **1110**, 581-595.

Curtis, D.C., Moore, D.R., Slater, B. and Zahlan, N., 1988, "Fatigue testing of multi-angle laminates of CF/PEEK", *Composites*, **19**, 446-452.

Curtis, P.T., 1986, "An investigation of the mechanical properties of improved carbon fibre composite materials", *RAE Technical Report*, **86021**.

Dickson, R.F., Jones, C.J., Harris, B., Leach, D.C. and Moore, D.R., 1985, "The environmental fatigue behaviour of carbon fibre reinforced poly ether ether ketone", *J.Mat.Sci.*, **20**, 60-70.

Gray, G. and Savage, G.M., 1989, "Advanced thermoplastic composite materials", *Metals and Materials*, **5**, 513-517.

Hartness, J.T. and Kim, R.Y., 1983, "A comparative study on fatigue behaviour of PEEK and Epoxy with reinforced graphite cloth", *28th SAMPE Symposium*, April 1983, 535.

ICI Fiberite Data Sheet, 1988, "APC-2, The product of high technology".

Mandell, J.F., 1982, in "Developments in reinforced plastics - 2", Ed. G. Pritchard, App.Sci.Pub., London.

Ogin, S.L., Smith, P.A. and Beaumont, P.W.R., 1985, "Matrix cracking and stiffness reduction during fatigue of a (0/90), GFRP laminate", *Comp. Sci. Tech.*, **22**, 23-31.

Saib, K.S., 1991, "Effects of microstructure on fatigue behaviour in short fibre reinforced PEEK composites", PhD Thesis, University of Wales.

7.0 ACKNOWLEDGEMENTS

The authors would like to thank Dr. D.R Moore of ICI Wilton Research Centre for valuable discussions throughout the work. Financial support was provided by SERC and ICI through a Co-operative Award. All materials were kindly supplied by ICI Materials.

THERMOPLASTIC POWDER COMPOSITE MANUFACTURING USING A WET SLURRY METHOD

Karthik Ramani, Michail Tryfonidis, and Chris Hoyle
Composites Manufacturing Laboratory
School of Mechanical Engineering
Purdue University
West Lafayette, Indiana

John Gentry
Delco Products
Indianapolis, Indiana

ABSTRACT

Thermoplastic matrix components are predominantly manufactured using preimpregnated materials. This increases the manufacturing costs significantly. A critical need exists to develop economical methods of in-situ impregnation for thermoplastic composites. This project is directed towards development and evaluation of an economical impregnation procedure using a slurry based powder technology. Such a process can be used for economical prototyping purposes.

The approach presented is to develop a process where the resin to fiber ratio is controlled accurately before consolidation. The influence of the process parameters on the impregnation, preheating and consolidation are studied. The velocity of fibers couples the impregnation and preheating stages. A thermal preheating model is used to determine the preheater temperature and the velocity of the fibers. A systematic procedure for obtaining the slurry composition to give the required resin to fiber ratio for the maximum fiber velocity is presented.

Introduction

The eventual success of structural composites depends upon developing economical manufacturing processes. Structural and semi-structural composites are finding wide applications in the industries where the mission can justify the high processing costs. To date, thermosetting matrix composites have been used more than thermoplastics. Thermoplastic matrix composites have several advantages over thermosets, some of them being reprocessability, repairability, recyclability, impact resistance, environmental and temperature resistance. Thermoplastics have not gained significant market in spite of potential advantages they offer due to the high processing costs. One of the difficulties in processing of thermoplastic matrix continuous fiber composites is the resin impregnation procedure. In this investigation, an economical prototyping process for impregnating thermoplastic resin powder using a wet slurry method is explored.

Background

Mixing high performance thermoplastics with continuous fiber tows is a difficult task. These thermoplastics are difficult to impregnate because of the high melt viscosity and this makes reasonable process times impossible. If the melt temperature of the thermoplastic is close to the decomposition temperature, reduction in the viscosity by increasing the temperature is not possible. Much of the previous work in thermoplastic composites used intermediate material forms such as prepregs, towpregs, comingled fibers etc. This increases material costs, reduces design and manufacturing flexibility, and increases the manufacturing cycle time. Solution prepregging is not a viable alternative due to limited solubility in solvents. Solvent removal poses added problems; they can be environmentally hazardous and solvents can be expensive. Drawbacks of hot melt and solution prepregging for thermoplastic matrices has lead to the development of powder impregnation methods.

Powder impregnation methods can be classified as dry powder impregnation methods and wet powder impregnation methods. All of the dry powder impregnation methods produce towpreg and typically involve two key stages, a deposition chamber and an oven for fusing the powder on the tow. In all the dry processes the tow is contacted with the powder suspended in air. Thermoplastic prepregs have been developed by electrostatic deposition of charged and fluidized polymer powder by Muzzy et al. [4]. Subsequent to deposition the powder particles are fused in a radiant oven. At the NASA Langley Research Center, the powder contacts the tow in a fluidised bed [3]. The process developed at Michigan State [2] the fiber surface is treated and then contacted with powder in a chamber. Electrostatic attraction or a binder on the surfae is used as a means for adhering the powder. The present investigation explores the use of a wet slurry method as a means for economical prototyping [1].

Process description

The fibers (E-glass) are spread on rollers in a bath of the slurry. The slurry is a water suspension of the thermoplastic powder. The constituents of the slurry are the resin powder, thickener and the wetting agent. The thickener keeps the powder in suspension in the water. The wetting agent increases the surface tension between water and fibers. Next the fibers go through a close tolerance orifice (CTO) that controls the amount of resin and water that adhere to the fibers. The fibers then go through a ceramic eye that traverses back and forth. The fibers are now wound on a frame. During the initial experiments the water was driven off by hang drying in an oven. The charge was subsequently consolidated in a press. See Figure 1 for a schematic of the process. The thermoplastics considered were PEEK, PPS and Polycarbonate. The resin were ground into a fine powder of 200 mesh size (particle size of 74 μ) for the initial experiments.

Resin impregnation:

The experiments were conducted in a test bed and important variables that affect the resin impregnation and wet-out were identified. The important variables controlling the resin pickup are outlined below:

1. **Fiber yield**[1]: Higher fiber yield promotes better pickup of resin, since smaller tows are easier to spread, and easier to work outwards in the resin bath. The surface area of the fibers exposed to the resin will in general be greater for larger strands of higher yield glass than a equivalent lower yield glass. For example, three rovings of 360 yield would have greater exposed area than a single roving of 120 yield. Figure 2 shows the fiber bundle on a roller, comparing the exposed

---

[1] Number of yards per pound weight of the fiber

areas of many small bundles (N), to an equivalent large bundle. The ratio of increased surface area ($S2$) to the original surface area ($S1$) of a single bundle can be shown to be $\sqrt{N}$.

2. **CTO size**: The effect of the CTO size on the amount of resin pickup was also studied. A smaller CTO picks up lesser amount of slurry than a larger CTO, and hence the glass content with a smaller CTO is greater than a larger CTO. This effect becomes pronounced at higher velocities of the tow in the slurry. Figure 3 compares the resin pickup with an orifice of 0.097 inches diameter with the resin pickup of an orifice of 0.103 inches in diameter.

3. **Velocity of roving in slurry bath**: As we found out in this investigation, the velocity of roving in the slurry is one of the important control variables. Long residence time of the fiber in the bath results in large pickup of the powder resin. We cannot control the velocity profile in filament winding for a constant rotation rate of the part, since this will be dictated by the part geometry. Figure 4 shows the influence of the velocity on the glass content for a particular size of the CTO. The influence of the slurry composition on the resin pickup is reduced at low tow velocities.

4. **Slurry composition**: The effect of slurry composition is pronounced at higher velocities. At lower velocities the resin pickup is high due to longer residence times.

The resin pick up depends on the powder concentration, water concentration, pulling velocities, CTO sizes and slurry compositions. As determined in the preliminary work, the ability to impregnate and wet-out the fibers depends on many parameters including the concentration of the powder in the slurry (P_c), additive wetting agent concentration (A_c), work done on the fibers in the bath (W_b), and the velocity of the fibers in the bath (V_f). Based on the trends noticed in the preliminary work, we have

$$\text{wet-out} \propto \frac{P_c A_c W_b}{V_f} \quad (1)$$

Preheating: The fiber velocity couples the impregnation and preheating stages. The maximum velocity of the fibers in this process is determined by the preheating stage. Excessive temperatures of the preheater will cause the surface of the tow to overheat and degrade the polymer. The preheating takes place in an infrared tunnel heater. The function of this stage is to heat the resin to the desired temperature and remove the water. In the preheating station we can assume that the predominant direction of heat transfer is perpendicular to the direction of travel of the fibers. Hence a simple one-dimensional heat transfer analysis with convective boundary condition was used. The dependance of the exit temperature distribution on the velocity of the fibers (residence time), was found from this model. The analysis may then be used to predict the optimum velocity of the fibers through the preheater and the appropriate temperature of the preheater so that the resin reaches the desired temperature at the exit of the preheater.

The tow is assumed to have a circular cross-section of radius R and a uniform temperature T_T before entering the preheater. The temperature of the preheater is constant at T_H. A thin control volume is fixed to the tow and the temperature distribution in this control volume is simulated at various time instances. The fiber bundle moves at a constant velocity V. The optimum velocity is the ratio of the length of the preheater (L) to the time (t_{min}) it takes for the resin to reach a nearly uniform temperature at the exit of the preheater.

The equation of conservation of energy for this problem is:

$$\frac{1}{r}\frac{d}{dr}\left(k_e r \frac{dT}{dr}\right) + \dot{Q}_{cr} = \rho_e c_{pe} \frac{dT}{dt} \qquad (2)$$

where r represents the coordinates of a point in the control volume in the radial direction; t is the time; ρ_e, c_{pe} and k_e are the equivalent density, specific heat and the transverse thermal conductivity of the composite bundle respectively. The composite cylinder assemblage model ([8],[9]) was used to estimate the effective transverse thermal conductivity of the slurry and fibers. The surface energy reduction due to sintering of the particles is neglected in this analysis since negligible sintering was noticed in the preliminary experiments. This is possibly due to the high melt viscosity of the polymers studied and temperatures that were lower than the melting point during a major portion of the residence of the polymer in the tunnel oven.

The heat absorbed in reducing the crystallinity for semi-crystalline thermoplastics is [7]

$$\dot{Q}_{cr} = -\rho_r v_r H_T \frac{dc_r}{dt} \qquad (3)$$

where ρ_r and v_r are the matrix density and matrix volume fraction, H_T is the ultimate heat of crystallization and $\frac{dc_r}{dt}$ is the rate of change of the crystallinity of the matrix.

This term will be zero for amorphous thermoplastics. The rate of change of crystallinity can be shown to be

$$\frac{dc_r}{dt} = \exp\left(\frac{-\phi(T)}{\left(\frac{dT}{dt}\right)^n}\right) \left\{ \frac{\frac{d\phi(T)}{dT}}{\left(\frac{dT}{dt}\right)}^{(n-1)} - n\frac{\frac{d^2T}{dt^2}\phi(T)}{\left(\frac{dT}{dt}\right)^{(n+1)}} \right\} \quad (4)$$

where $\phi(t)$ and n were calculated by Lee and Springer [10] by performing DSC experiments and fitting the expression proposed by Ozawa [7]. The initial temperature of the tow is uniform. The finite difference scheme was used to convert the energy equation 2 to its numerical analog. If T_2 and T_4 are the temperatures of nodes adjacent to node with temperature T_0 as shown in Figure 5, then

$$T_2|_{t+\Delta t} > T_0|_{t+\Delta t} > T_0|_t$$

Hence a bisection scheme can be used to solve for $T_0|_{t+\Delta t}$.

At the outer surface layer of the bundle convective boundary conditions were imposed. Specifically the temperature of the surface node is determined by an energy balance on the control volume around the radiative surface of the heater. The energy radiated into the bundle by the heater is,

$$P = h_r(T_H - T_T)A \quad (5)$$

where h_r is the convective heat transfer coefficient on the surface of the bundle,

$$h_r = F\epsilon\sigma(T_H^2 + T_T^2)(T_H + T_T) \quad (6)$$

F is the view factor; ϵ is the gray factor of the radiant heater; σ is the Boltzman's constant and A is the area of the heater. The heat transfer coefficient h_r, on the surface of the bundle is a function of the position of the control volume along the length of the preheater since it depends on both T_T and T_H.

The resin, fiber and the water content were assumed to be homogeneous at the start of the simulation. At each time step the surface temperature is determined by substituting equation 6 and equation 5.The heat absorbed due to reduction in crystallinity ($\dot{Q}_{cr}$) from equations 4 and 3 is substituted in equation 2 and the finite difference form of the resulting equation is solved for the temperature at each time step for the rest of the nodes. The amount of heat transfered into a specific layer after $100°C$ was used to evaporate the water and the mass of water in that layer is recalculated. The water is replaced by air and as the simulation proceeded, effective density, specific heat and thermal conductivities of each layer were also recalculated. The graph in Figure 7 shows the preheating of PEEK slurry. A heater surface temperature of 680 K was used to avoid overheating of the fibers and a residence time of 60 seconds was required to heat the tow to a nearly uniform temperature.

The velocity of the fibers will also determine the resin content and the water content in the fibers. The slurry composition can now be adjusted to suit this velocity to give the desired fiber loading. The submodel from the previous step allows us to fine tune the resin content.

Consolidation: Preliminary consolidation experiments were performed in a press. The charge of fibers wound on a frame was consolidated. Due to the high melt viscosities of the high performance thermoplastic resins it is very difficult to get any resin squeeze off during consolidation. Hence the fibers have to be loaded uniformly with the correct amount of resin prior to consolidation or forming. Hence the design of the process to control the resin to fiber ratio is fundamental to the processing of thermoplastics. Fiber wander occured in the center of the mold where the resin carried the fibers and prevented the platens from closing. This problem was solved by increasing the tension during the winding of the fibers. The cooling rate of the plaques controls the crystallinity and hence the resin controlled properties such as toughness and resistance to hostile solvents [11]. Several rectangular plaques of Polycarbonate, PPS and PEEK were made with glass content close to the targeted glass content ($70\% \pm 1\%$). Detailed models for consolidation of powder impregnated tows are being developed.

Appendix

Effective values of density, specific heat and transverse thermal conductivity: The effective density is given by

$$\rho = \frac{1}{\frac{f_r}{\rho_r} + \frac{f_f}{\rho_f} + \frac{f_w}{\rho_w}}$$

where,

f_r is the weight fraction of the resin

f_f is the weight fraction of the fiber

f_w is the weight fraction of the water

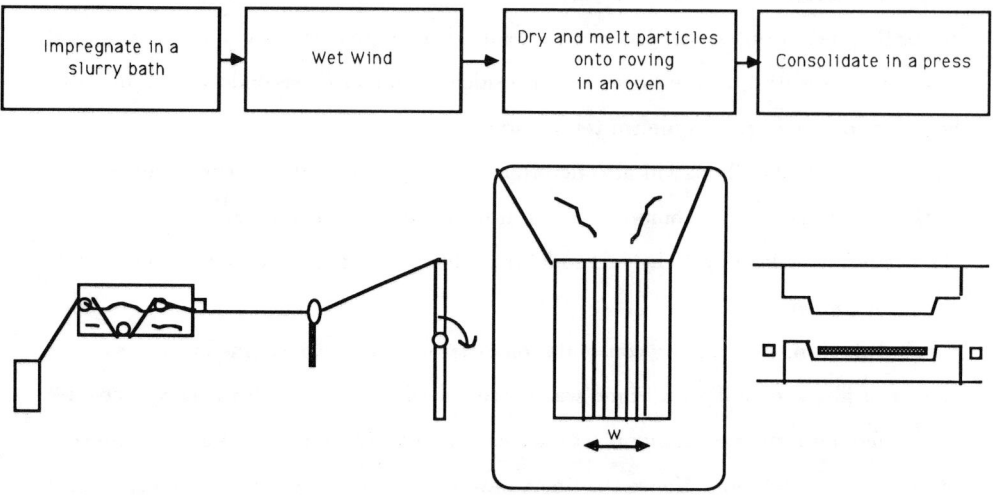

FIGURE 1: Experimental setup used for preliminary work

Higher yield
    Better Pickup
    Easier to work outwards, better wetting
    Better seating of roving

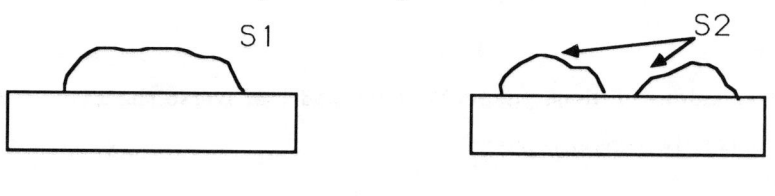

$$S2/S1 \;=\; rN/R \;=\; \sqrt{N}$$

FIGURE 2: Greater exposed area of fibers promotes wetting

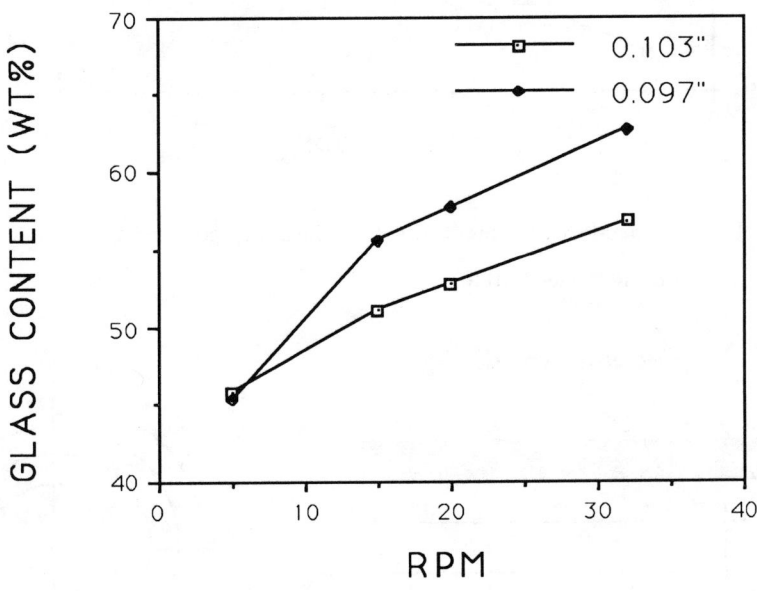

FIGURE 3: Dependence of glass content on orifice size

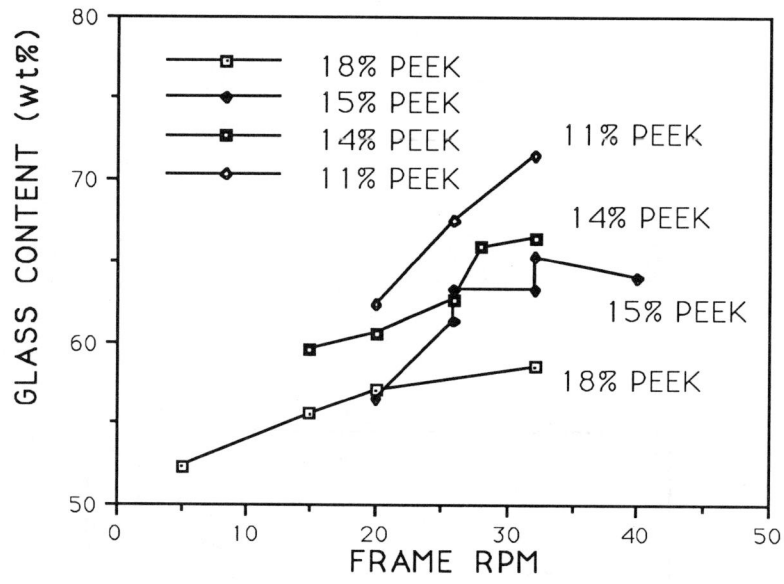

FIGURE 4: Dependence of glass content on the slurry composition

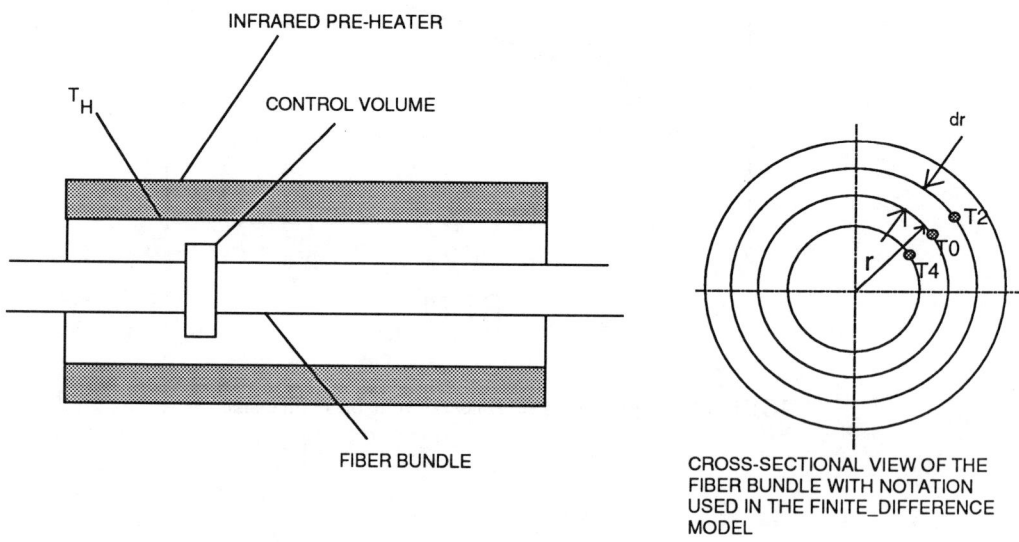

FIGURE 5: Cross-section of the fiber bundle used in the simulation

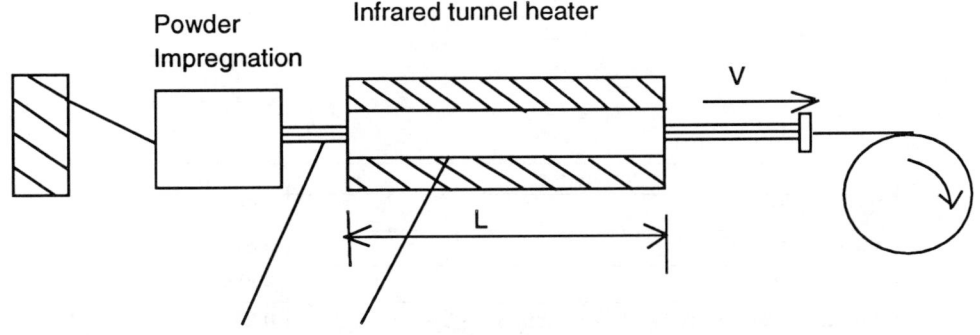

FIGURE 6: Impregnation and preheating in a tunnel oven

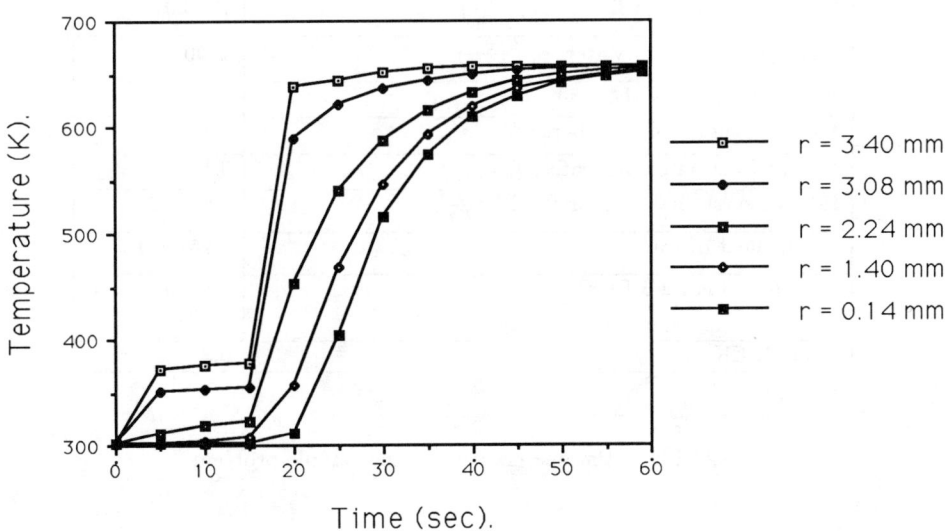

FIGURE 7: Sample simulation of the tow preheating stage showing the temperature at different radii of the tow along the preheater

Parameter	Value
Radius, r (m)	0.0035
View Factor, F	0.8
Power, P (W)	200.0
Thermal Conductivity of the resin, $k_r \left(\frac{W}{m^\circ K}\right)$	0.25
Transverse thermal conductivity of the fiber, $k_f \left(\frac{W}{m^\circ K}\right)$	15.0
Thermal conductivity of the water, $k_w \left(\frac{W}{m^\circ K}\right)$	0.65
Density of the resin, $\rho_r \left(\frac{kg}{m^3}\right)$	1400.0
Density of the fiber, $\rho_f \left(\frac{kg}{m^3}\right)$	1900.0
Density of the water, $\rho_w \left(\frac{kg}{m^3}\right)$	1000.0
Heat capacitance of resin, $c_r \left(\frac{J}{KgK}\right)$	1675.0
Heat capacitance of fibers, $c_f \left(\frac{J}{KgK}\right)$	7353.0
Heat capacitance of water, $c_w \left(\frac{J}{KgK}\right)$	4290.0
Weight fraction of the resin, f_r	0.2
Weight fraction of the fibers, f_f	0.7
Weight fraction of the water, f_w	0.1
Heat of crystallization for PEEK $\left(\frac{J}{Kg}\right)$	120,000.0
Latent heat of water $\left(\frac{J}{Kg}\right)$	2.257×10^6
$\phi(T)$ function for PEEK	$\ln(-.037T + 11.3)$
n for PEEK	0.8

TABLE 1: Parameters used in the sample problem

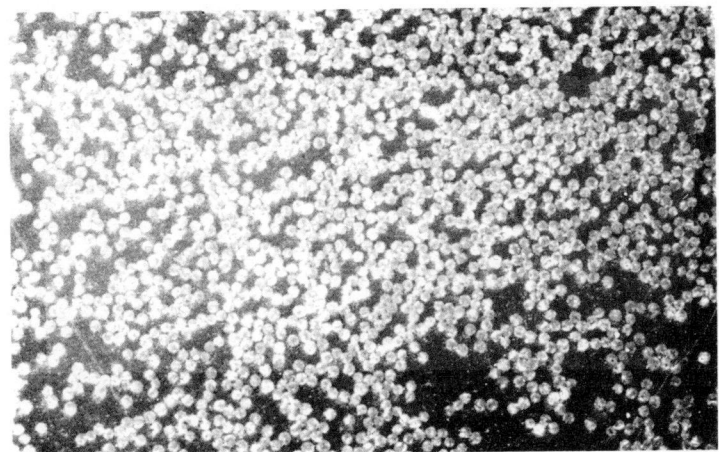

FIGURE 8: Cross-section of the component showing excellent impregnation with low void content

ρ_r is the density of the resin

ρ_f is the density of the fiber

ρ_w is the density of the water.

The volume fraction of each component is given by: $v_r = \frac{f_r \rho}{\rho_r}$, $v_f = \frac{f_f \rho}{\rho_f}$, $v_w = \frac{f_w \rho}{\rho_w}$.

The problem of the transverse thermal conductivity is analogous to the problem of longitudinal shearing. The effective value of the transverse thermal conductivity of the fiber bundle was determined using the composite-cylinder-assemblage model [8] modified in the form described below.

If k is the effective transverse thermal conductivity of the composite bundle, then

$$k = k_{rw} \left(\frac{k_{rw}(v_r + v_w) + k_f(1 + v_f)}{k_{rw}v_f + k_f(v_r + v_w)} \right)$$

where, k_{rw} is the effective thermal conductivity of the resin-water slurry and air that replaces the water evaporated.

Using thermal circuits we can model the mixture of the resin-water slurry and air as thermal resistances connected in series or in parallel. Since both the models are not exact, an average value of the thermal conductivity predicted by these models is used for the thermal conductivity of the slurry (k_{rw}).

For a parallel model of thermal resistances,

$$\left. \frac{1}{k_{rw}} \right|_{parallel} = \frac{1}{k_r + k_w + k_a}$$

For a series model of thermal resistances,

$$\frac{1}{k_{rw}}\bigg|_{series} = \frac{k_r k_w + k_w k_a + k_r k_a}{k_r k_w k_a}$$

The effective thermal conductivity of the resin-water slurry was found using,

$$k_{rw} = \frac{k_{rw}|_{parallel} + k_{rw}|_{series}}{2}$$

The effective value for the specific heat is given by,

$$c_p = f_r c_{pr} + f_f c_{pf} + f_w c_{pw} + f_a c_{pa}$$

Conclusions and recommendations

Several plaques have been made using this process with good wet-out and low void content. The control of the resin impregnation process and the resin fiber ratio is difficult in wet processes. However, the equipment needs for this process are minimal and an existing winding setup can be modified to impregnate thermoplastics. This process can serve as an economical method of impregnating thermoplastic resin for prototyping purposes. Removal of water in this process places an upper limit on the velocity of the fibers and hence the process throughput. Powder impregnation procedures that are fast, economical, and provide good control of the resin content have to be developed for thermoplastic powder impregnation technologies to be successful.

Acknowledgements

The early portion of this work was explored jointly by Delco Products and Wright Patterson Air Force Base. The authors would like to thank Harry Couch for his support that made this work possible.

REFERENCES

[1] K. Ramani, J. Gentry *"Filament winding of thermoplastic composites,"* Technical Report, General Motors Inland Division, Dayton, OH, June 1989.

[2] S.R.Iyer, L.T.Drzal *"Manufacture of Powder Impregnated Thermoplastic Composites,"* Journal of Thermoplastic Composite Materials, Vol.3-October 1990.

[3] M.K.Hugh, J.M.Marchello, R.M.Baucom, N.J.Johnston *"Composites from Powder coated towpreg: Studies with variable tow sizes,"* 37th International SAMPE symposium, Vol.37, pp.1040-1051, 1992.

[4] J.Muzzy, B.Varughese, V.Thammongkol, W.Tincher *"Electrostatic prepregging of Thermoplastic Matrices,"* 24th International SAMPE Symposium, May 8-11, 1989.

[5] J.L.Throne, M.K.Sohn *"Electrostatic dry powder pregfing of carbon fiber,"* 35th International SAMPE Symposium, April 2-5, 1990.

[6] A.C.Loos and M.C. Li *"Heat Transfer Analysis of Compression Molded Thermoplastic Composites,"* Advanced Materials: The Challenge for the Next Decade, Society for Advancement of Material and Process Engineering, (1990), pp. 557-570.

[7] T. Ozawa *"Kinetics of Non-Isothermal Crystallization,"* Polymer, Vol.12(1971), pp.150-158.

[8] Z. Hashin *"Theory of fiber reinforced materials,"* NASA CR-1974, National Aeronautics and Space Administration, 1972.

[9] Z. Hashin *"Analysis of properties of fiber reinforced composites with anisotropic constituents,"* J. Appl. Mech., Vol. 46, 1979, pp. 543-550.

[10] W.I.Lee, and Springer, G.S *"A Model of the Manufacturing Process of Thermoplastic Matrix Composites,"* Journal of Composite Materials, Vol.21, 1987, pp. 1017-1055.

[11] M.F.Talbott, G.S.Springer, and L.A.Berglund *"The effects of Crystallinity on the Mechanical Properties of PEEK Polymer and Graphite Fiber Reinforced PEEK,"* Journal of Composite Materials, Vol. 21, (1987), pp. 1056-1081.

[12] J.J.Scobbo, Jr. *"Radiative heating of thermoplastic/graphite fiber composites,"* 22nd International SAMPE technical conference, November 6-8, 1990.

[13] S.C.Mantell *"Manufacturing Process Models for Thermoplastic Matrix Composites,"* Doctoral Dissertation, Stanford University, June 1991.

[14] I.Brewster and Cattanach, J.B., *"Engineering with long Fiber Thermoplastic Composites,"* Sampe Meeting in Europe, London, March 1983.

DUCTILE FRACTURE PROCESS OF DISCONTINUOUS FIBER REINFORCED COMPOSITES

S. B. Biner
Ames Laboratory
Iowa State University
Ames, Iowa

ABSTRACT

In this study the role of fiber morphology, interface failure and void nucleation mechanisms within the matrix on the deformation and fracture behavior of discontinuous fiber reinforced composites was numerically investigated. The matrix was modelled using a constitutive relationship that accounts for strength degradation resulting from the nucleation and growth of voids, and fibers are assumed to be elastic. The debonding behavior at the fiber interfaces was simulated in terms of a cohesive zone model which describes the decohesion by both normal and tangential separation. The results indicate that in the absence of interface failure, for a given matrix failure mechanism the fiber morphology strongly affects the strength and ductility of discontinuous fiber reinforced composites. The weak interfacial behavior can significantly increase the ductility without sacrificing the strength for certain fiber morphology and for certain void nucleation mechanisms.

INTRODUCTION

The ductile failure of the matrix by nucleation, growth and coalescence of voids is reported to be the dominant failure mode in many metal matrix composite systems which are reinforced with particulate and whisker or discontinuous fibers (Christman and Suresh, 1988; You, et al., 1987; Nair, et al., 1985; Davidson, 1987; Gungor and Laiw, 1991). Several studies on the mechanism of ductile fracture indicate that void nucleation can occur in two different modes (stress-controlled nucleation and strain-controlled nucleation) that depend on local stress-strain state, the most important being the size, shape and distribution of void nucleating intermetallic inclusions and/or dispersoids (Argon, et al., 1975;

Goods and Brown, 1979; McClintock, 1968). The nucleated voids grow in size as a function of applied strain. On the other hand, the rate of void growth is considerably influenced by the hydrostatic stress state (McClintock, 1968; Rice and Tracey, 1969; Spitzig, 1990).

Recent numerical analyses of discontinuous fiber reinforced composites have shown that significant levels of tensile hydrostatic stresses develop in the matrix as a consequence of constrained deformation (Tvergaard, 1990; Llorca, et al., 1991; Papazian and Adler, 1990; Biner, in press). Of course, the magnitude of this hydrostatic stress state depends upon the geometrical parameters of the reinforcement (e.g. aspect ratio and shape) and the distribution of reinforcement within the matrix. The hydrostatic stress level within the matrix increases monotonically during the far field loading, thereby causing the apparent flow strength and strain hardening exponents to rise with increasing deformation. However, this build-up can be relieved by one or a combination of micromechanical mechanisms such as: a) interface separation between the reinforcement and the matrix, b) void nucleation and growth within the matrix, and c) formation of intense shear bands. Therefore, apart from the geometrical parameters of the reinforcement and distribution parameters within the matrix, the interface behavior and ductile fracture characteristics of the matrix can be controlling parameters of the strength and ductility of the composite system.

In this study, in order to elucidate the role of fiber morphology, interface failure and void nucleation mechanisms within the matrix, the deformation and fracture behavior of discontinuous fiber reinforced composites was numerically investigated. The matrix was modelled using a constitutive relationship that accounts for strength degradation resulting from the nucleation and growth of voids, and fibers are assumed to be elastic. The debonding behavior at the fiber interfaces was simulated in terms of a cohesive zone model which describes the decohesion by both normal and tangential separation.

NUMERICAL ANALYSIS

MATERIAL MODEL

Matrix Material
The matrix was modelled using a constitutive relationship that accounts for strength degradation resulting from the nucleation and growth of micro-voids. The basis for the constitutive model is a flow potential introduced by Gurson (1975,1977), in which voids are represented in terms of a single internal variable, f, the void volume fraction:

$$\phi = \frac{\sigma_e^2}{\sigma_m^2} + 2f^*q_1 \cosh\left(\frac{q_2\sigma_h}{2\sigma_m}\right) - 1 - q_1^2 \, f^{*2} = 0 \qquad (1)$$

where

$$\sigma_e = \frac{3}{2}\sigma':\sigma' \qquad \sigma_h = I:\sigma \qquad \sigma' = \sigma - \frac{1}{3}\sigma_h I \qquad (2)$$

and σ_e is the equivalent stress, σ_h is the hydrostatic stress, I is the Identity matrix and σ_m is the flow strength of the matrix. The parameters q_1 and q_2 were introduced by Tvergaard (1981, 1982) in order to provide a better relationship between unit-cell analysis and eq.1; the case $q_1 = q_2 = 1$ corresponds to Gurson's original formulation.

The function f^* was proposed by Tvergaard and Needleman (1988) to account for the effect of rapid void coalescence at failure. Initially $f^* = f$ as originally proposed by Gurson, but at some critical void fraction, f_c, the dependence of f^* on f is changed. This function is expressed by

$$f^* = \begin{cases} f & f \leq f_c \\ f_c + \dfrac{f_u^* - f_c}{f_f - f_c}(f - f_c) & f \geq f_c \end{cases} \qquad (3)$$

The constant f_u^* is the value of f^* at zero stress in eq. 1 (i.e. $f_u^* = 1/q_1$) and f_f is the void fraction at fracture. As $f \to f_f$, $f^* \to f_u^*$ and the material loses all stress carrying capacity.

The increase in void volume fraction $\dot{f}$ arises from the growth of existing voids and from the nucleation of new voids. Thus

$$\dot{f} = (\dot{f})_{growth} + (\dot{f})_{nucleation} \qquad (4)$$

The growth rate is related to the macroscopic dilation rate by

$$(\dot{f})_{growth} = (1-f)\delta^{ij}\dot{\eta}_{ij}{}^P \qquad (5)$$

where $\dot{\eta}_{ij}{}^P$ is the plastic part of the rate of deformation.

The increase in the void volume fraction due to the nucleation process is assumed to occur in two different modes;
i) Strain Controlled Nucleation:

$$(\dot{f})_{nucleation} = D\dot{\varepsilon}_m^P \qquad (6)$$

where ε_m^P is the effective plastic strain in the matrix and void nucleation is assumed to follow a normal distribution as suggested by Chu et al. (1980). Thus, plastic strain controlled nucleation is specified by

$$D = \frac{f_N}{s_N\sqrt{2\pi}} \exp\left[-\frac{1}{2}\left(\frac{\varepsilon_m^P - \varepsilon_N}{s_N}\right)^2\right] \qquad (7)$$

where f_N is the volume fraction of void nucleating particles, ε_N is the mean strain for nucleation, and s_N is the corresponding standard deviation. Similarly;
ii) Stress-Controlled Nucleation:

$$(\dot{f})_{nucleation} = B(\sigma_{max} + \sigma_k^k/3) \tag{8}$$

with

$$B = \frac{f_N}{s_n\sqrt{2\pi}} \exp\left[-\frac{1}{2}\left(\frac{(\sigma_{max} + \sigma_k^k/3) - \sigma_N}{S_n}\right)^2\right] \tag{9}$$

where σ_N is the mean stress for nucleation and f_N and s_N have the same meaning as in eq. 7.

In the present investigation a rate-sensitive version of Gurson's model was employed. In the matrix, the microscopic effective plastic strain rate ε_m^P is represented by the power law relation

$$\dot{\varepsilon}_m^P = \dot{\varepsilon}_o \left[\frac{\sigma_m}{g(\varepsilon_m^P)}\right]^{1/m} \tag{10}$$

where m is the strain rate hardening exponent, $\dot{\varepsilon}_o$, is a reference strain rate and ε_m^P is the current value of the effective plastic strain representing the actual microscopic strain state in the matrix. The function $g(\varepsilon_m^P)$ represents the effective tensile flow stress in the matrix material in a tensile test carried out at strain rate that is equal to reference strain rate, $\dot{\varepsilon}_o$. For a power hardening matrix material the function $g(\varepsilon_m^P)$ is taken to be

$$g(\varepsilon_m^P) = \sigma_o \left(\frac{E_m \varepsilon_m^P}{\sigma_o} + 1\right)^N \qquad g(0) = \sigma_o \tag{11}$$

with strain hardening exponent N, Young's modulus E_m and reference stress σ_o.

Using $\dot{\Phi}=0$ as the plastic potential together with the consistency condition, the values f and σ_m can be determined from the known strain rates and macroscopic stress-rates; full formulation can be found in refs. (Llorca, et al., 1991; Tvergaard & Needleman, 1988; Peirce, et al., 1984).

The material parameters of the matrix appearing in Eqs. 10 and 11 were chosen as $E_m=500\sigma_o$, $\nu=0.3$, $N=0.1$, $m=0.01$ and the reference strain rate $\dot{\varepsilon}_o=2\times10^{-3}$. Initially, zero volume fraction of voids is assumed in the matrix. The parameters appearing in Eqs. 6-9 for void nucleation were taken as $f_N=0.04$, $s_N=0.1$, $\sigma_N=2.2\sigma_o$, and $\varepsilon_N=0.3$. For accelerated void growth, the parameters appearing in Eq.3 were chosen as $f_f=0.25$, $f_c=0.10$ and $f_u^*=1/1.25$. Also $q_1=1.25$ and $q_2=q_1^2$ were selected for Eq.1.

Fiber Material

The fiber material is assumed to be linear-elastic. Young's modulus and Possion's ratio of the fibers were taken as $E_f=10E_m$ and $\nu=0.3$ respectively.

INTERFACE MODEL

To study the effects of fiber debonding and subsequent fiber pull-out during the deformation of fiber reinforced composites, it is necessary to simulate the interface failure by normal and tangential separation. A debonding model has been developed by Needleman (1987), in terms of a potential that specifies the dependence of interface traction on the displacement differences at the interface. The potential used in Needleman (1987), which defines the nonlinear variation of interface traction as a function of interface displacements, also contains three parameters σ_i, δ_i (δ_N, δ_T) and α where σ_i is the interfacial strength; complete separation is assumed to occur at $u_N = \delta_N$; and α specifies the ratio of shear to normal stiffness of the interface. These parameters are assumed to be intrinsic material properties. However, as discussed in (Tvergaard, 1990) the interface constitutive relationship given in (Needleman, 1987) describes the debonding only by normal separation. Therefore it is not suitable for tangential separation and fiber pull-out that occur under significant normal compression. An alternative model introduced in Tvergaard (1990) is utilized in this study. The normal and tangential traction between the fiber and the matrix are given by:

$$T_N = \frac{u_N}{\delta_N} F(\lambda) \tag{12}$$

$$T_T = \alpha \frac{u_T}{\delta_T} F(\lambda) \tag{13}$$

where $F(\lambda)$ is chosen as

$$F(\lambda) = \frac{27}{4} \sigma_i (1-2\lambda+\lambda^2) \qquad \text{for } 0 \leq \lambda \leq 1 \tag{14}$$

and

$$\lambda = \sqrt{\left(\frac{u_N}{\delta_N}\right)^2 + \left(\frac{u_T}{\delta_T}\right)^2} \tag{15}$$

Eqs. 12 and 13 are valid as long as λ is monotonically increasing. In purely normal separation ($u_T \equiv 0$) the variation in the normal traction with interface separation distance is shown in Fig.1. The maximum traction is σ_i, total separation occurs at $u_N = \delta_i$ and the work of separation per unit interface area is $9\sigma_i\delta_i/16$. In purely tangential separation ($u_N \equiv 0$), the maximum traction is $\alpha\sigma_i$, total separation occurs $u_T = \delta_i$ and the work of separation per unit area is $9\alpha\sigma_i\delta_i/16$.

For decreasing λ, a type of elastic unloading is imposed to simulate the partly damaged interface:

For $\lambda \leq \lambda_{max}$ or $\dot{\lambda} \leq 0$

$$T_N = \frac{u_N}{\delta_N} F(\lambda_{max}) \qquad (16)$$

$$T_T = \alpha \frac{u_T}{\delta_T} F(\lambda_{max}) \qquad (17)$$

If Eqs.12 and 13 were used instead of Eqs. 16 and 17 this would mean that partly damaged material repaired itself when the loading was reversed. Under normal compression, however, elastic springs with a large stiffness are used to approximately represent the contact instead of Eqs.16 and 17.

During the analysis, interface parameters were chosen as $\sigma_i=2.5\sigma_o$, $\delta_N=\delta_T=5\times10^{-4}r_f$ and $\alpha=1$, and r_f is the fiber radius.

UNIT-CELL MODEL

In the analysis of a deforming two-phase material, it is often necessary to make simplifying assumptions about the shape and distribution of the phases in order to make the problem tractable. A unit-cell around the periodic array of aligned discontinuous fibers as shown in Fig.2a was approximated by an axisymmetric model as shown in Fig.2b. Initial cell and fiber geometry were specified by cell half length l_c and cell radius r_c; fiber half length l_f and fiber radius r_f. The fiber volume fraction is

$$V_f = \frac{r_f^2 l_f}{r_c^2 l_c} \qquad (18)$$

The initial fiber aspect ratio, a_f, and cell aspect ratio, a_c, are defined by

$$a_f = \frac{l_f}{r_f} \qquad a_c = \frac{l_c}{r_c} \qquad (19)$$

The deformation of the unit-cell must be constrained to maintain the compatibility and equilibrium with the adjacent material. This constraint requires that the cell boundaries remain straight and orthogonal, and free of shear traction. Several methods for imposing these requirements on FEM models have been suggested (Tvergaard, 1981; Needleman, 1972; Smelser & Becker, 1990); the procedure outlined in (Smelser & Becker, 1990) is utilized in this study.

A typical mesh used during the analysis is shown in Fig.2c. The elements used are quadrilaterals each built up of four triangular axisymmetric linear displacement elements. The ductile fibers constitute 20% of the total cell volume (i.e. $V_f=0.20$) and cell aspect ratio was chosen as $a_c=5$. Fiber aspect ratio a_f was varied between 2.5 and 10. The axial deformation rate of the unit cell was the same as the reference strain rate in Eq.10. During the analysis the uniform stress and strain values were computed from the resulting reaction forces and axial displacements. The integration of the stress rate requires small time steps for stable numerical integration. The tangent modulus method of Peirce et al. (1984) is used to increase the stable time step size. The tangent modulus provides a forward gradient estimate of the deformation rate based on a Taylor series expansion about the current deformation rate.

The material failure in the matrix is implemented by element vanish technique. When the failure condition is met within an element, i.e. $f=f_f$, that element no longer contributes to the virtual work. To avoid numerical instabilities, the nodal force arising form the remaining stresses in failed elements are redistributed in several iterations.

RESULTS AND DISCUSSION

The strengthening behavior of the composite system in the absence of any interface failure between the fibers and the matrix (i.e. rigid interface) and without any damage development in the form of nucleation and growth of voids within the matrix is considered first. For a constant volume fraction of reinforcement $V_f=0.20$, the effect of fiber aspect ratio a_f on the strengthening behavior is given in Fig.3 together with the strength data of unreinforced matrix material. As can be seen, the most significant increases occur in the modulus of elasticity, in the strain hardening characteristics and in the ultimate strength of the composite with increasing aspect ratio of the discontinuous fibers. The increase in the proportional limit (i.e. transition from elastic to plastic behavior) occurs to a lesser extent. For these cases, the distribution of the axial stress component within the unit-cell and the distribution of hydrostatic stress only within the matrix are shown in Figs.4 and 5. The axial stress carried by the fibers increases with increasing aspect ratio of the fibers. Near the fiber ends within the matrix the development of large axial stress component caused by the stress concentration effect of sharp corners of the fibers can also be seen in Fig.4. The increase in the proportional limit of the composite to a lesser extent than other strength parameters is associated with this early stress elevation in the area surrounding the fiber ends in the matrix. The magnitude of the hydrostatic stress within the matrix also increases with increasing fiber aspect ratio, Fig.5. The increase in the strain hardening and flow strength seen in Fig.3 are the result of this build-up in the hydrostatic stress which lowers equivalent stress level, rather than the stress portioning between the fibers and the matrix. Therefore, the predictions based solely on the volume fraction without taking into account the geometrical parameters of the reinforcement and their distribution, such as simple "rule of mixtures," usually fail to accurately estimate the strength characteristics of this type of composite systems.

In the next set of calculations, the interface between the fibers and matrix was assumed to be rigid and the nucleation and growth of the voids within the matrix was the sole damage mechanism. Two modes of void nucleation mechanisms, namely stress-controlled and strain-controlled, were separately investigated.

For each mechanism, the volume fraction of void nucleating particles f_N, their standard deviation s_N appears in Eqs.7 and 8; and strain hardening characteristics of the matrix were assumed to be the same. These analyses were carried out for three fiber aspect ratios a_f of 2.5, 5 and 10. The resulting stress-strain curves together with the strength data in the absence of any damage development in the composite are given in Figs.6-8. When no damage mechanism is operative, the composite flow strength increases continuously with increasing deformation. Due to the damage development in the matrix in the form of nucleation and growth of voids, the strength data exhibit a maximum, and with void coalescence (i.e. by the attainment of critical void fraction $f \geq f_f$ a rapid drop occurs in the strength data. This large drop in the stress carrying capacity is assumed to be the failure strain and is indicative of the ductility of the composite. As can be seen from Figs.6-8, the strain-controlled nucleation mechanism gave higher strength and ductility in comparison to stress-controlled nucleation. For both nucleation mechanisms, a slight increase in strength accompanied by considerable decrease in the ductility can be seen with increasing aspect ratio of the fibers. Particularly for the largest aspect ratio $a_f=10$ very little of matrix strain hardening capacity was utilized before the failure. Of course, this behavior is associated with the larger build-up in hydrostatic stress with increasing fiber aspect ratio as shown earlier in Fig.5. As soon as the voids nucleate, the grow more rapidly to coalescence, and a faster drop in stress capacity occurs as the regions over which $f \geq f_f$ expand more rapidly.

In following simulations, the role of interface separation between the fibers and the matrix is investigated. For these cases, again the damage development in the matrix is assumed to occur separately by two different nucleation processes. In these analyses, since the fiber diameter varies with the aspect ratio, the work of separation per unit interface area is also affected due to the changes in the separation distance via the $\delta_N = \delta_T = 5 \times 10^{-4} r_f$ relationship. Therefore, an increase in the fiber aspect ratio yields a smaller separation distance δ_N and δ_T, hence a lower value of the work separation for the constant value of $\sigma_i = 2.5 \sigma_o$ and $\alpha = 1$. For all cases investigated the interface failure started first at the top surface of the fibers; after complete separation of the top surface, with continued deformation the fiber pull-out progressively occurred by the debonding of side surface. For comparison purposes, the resulting stress-strain data from these simulations are also summarized in Figs.6-8. The interface separation had little effect on the strength behavior of composite failing with stress-controlled nucleation mechanism. For this failure mechanism the ductility of the composite is significantly increased particularly for small fiber aspect ratios. For the strain-controlled nucleation, interface failure resulted in much larger reductions in strength than its stress-controlled counterpart. Also, larger increases in the ductility of the composite were observed for increasing fiber aspect ratios.

For the cases investigated the damage evolution within the matrix are given in Figs.9-11 for three different fiber aspect ratios. Either in the presence or absence of interface failure, the area fraction of matrix material undergoing damage formation was much smaller in strain-controlled than in stress-controlled nucleation due to the localization of the plastic flow. For both nucleation mechanisms, the role of interface separation can be clearly seen. In stress-controlled nucleation, the areas which exhibit higher amount of void growth shifted away from the fiber ends, and also the damage accumulation occurred at a slower rate in comparison to rigid interface. Similarly in strain-controlled nucleation, reduction in the damage levels and the spread of the void nucleation to the regions along the fiber length due to fiber pull-out can be observed. These changes in the damage morphology, with resulting variations in the strength and ductility (Figs.6-8) are the direct result of the reduction and redistribution of hydrostatic stress within the matrix due to the interface separation.

In the case of rigidly bonded fibers, the void nucleation appears to be the key controlling parameter of the composite strength and ductility, hence, of the fracture toughness. In this case, a very rapid void growth and void coalescence is followed immediately the void nucleation. By fiber debonding and fiber pull-out mechanisms, the build-up in the hydrostatic stress state can be relaxed considerably. As a result of this, it can be seen that the damage tolerance of the composite increases significantly (Figs.6-8). The ductility in this case appears to be controlled by the void growth. If the key issue is to increase the toughness without extensively sacrificing the strength, the results indicate that this can be achieved by controlling the interface behavior. However, it also appears that the weakening of the interface does not have the same effectiveness for all fiber geometries and matrix characteristics (Figs.6-8). As can be seen this can only be effectively utilized for certain fiber morphologies and for certain void nucleation mechanisms of the matrix.

CONCLUSIONS

The role of fiber morphology, interface failure and void nucleation mechanisms within the matrix on the deformation and fracture behavior of discontinuous fiber reinforced composites was investigated. The results indicate that:
1. In the absence of interface failure, for a given matrix failure mechanism the fiber morphology strongly affects the strength and ductility of the discontinuous fiber reinforced composites.
2. The weak interfacial behavior can significantly increase the ductility without sacrificing strength for certain fiber morphology and for certain matrix void nucleation mechanisms.

ACKNOWLEDGMENTS

This work was performed for the United States Department of Energy by Iowa State University under contract No. W-7405-Eng-82. This research was supported by the Director of Energy Research, Office of Basic Energy Sciences.

REFERENCES

Argon, A.S., Im, J., and Safoglu, R., 1975, *Metallurgical Transactions*, Vol. 6A, pp. 825-837.

Biner, S.B., *Materials Science and Engineering A* (in press).

Christman, T. and Suresh, S., 1988, *Materials Science and Engineering*, Vol. 102A, pp. 211-216.

Chu, C.C. and Needleman, A., 1980, *Journal of Engineering Materials Technology*, Vol. 102, pp. 249-256.

Davidson, D.L., 1987, *Metallurgical Transactions*, Vol. 18A, pp. 2115-2128.

Goods, S.H. and Brown, L.M., 1979, *Acta Metallurgica*, Vol. 27, pp. 1-15.

Gungor, M.N. and Liaw, P.K. (eds), 1991, *Fundamental Relationship Between Microstructure and Mechanical Properties of Metal Matrix Composites*, TMS, Warrendale, PA.

Gurson, A.L., 1975, "Plastic Flow and Fracture Behavior of Ductile Materials Incorporating Void Nucleation, Growth and Interaction," Ph.D. Thesis, Brown University, Providence, RI.

Gurson, A.L., 1977, *Journal of Engineering Materials Technology*, Vol. 99, pp. 2-15.

Llorca, J., Needleman, A., and Suresh, S., 1991, *Acta Metallurgica*, Vol. 39, pp. 2317-2335.

McClintock, F.A., 1968, *Journal of Applied Mechanics*, Vol. 35, pp. 363-371.

Nair, S.V., Tien, J.K., and Bates, B.C., 1985, *International Metallurgical Review*, Vol. 30, pp. 275-290.

Needleman, 1987, *Journal of Applied Mechanics*, Vol. 54, pp. 525-531.

Needleman, A., 1972, *Journal of Applied Mechanics*, Vol. 39, pp. 964-970.

Papazian, J.M. and Adler, P.N., 1990, *Metallurgical Transactions*, Vol. 21A, pp. 411-420.

Peirce, D., Shih, C.F., and Needleman, A., 1984, *Computers and Structures*, Vol. 18, pp. 875-887.

Rice, J.R. and Tracey, D.M., 1969, *Journal of Mechanics and Physics of Solids*, Vol. 17, pp. 201-217.

Smelser, R.E. and Becker, R., 1990, *Abaqus User Conference Proceedings*, pp. 207-218.

Spitzig, W.A., 1990, *Acta Metallurgica*, Vol. 38, pp. 1445-1453.

Tvergaard, V., 1990, *Materials Science and Engineering*, Vol. A125, pp. 203-213.

Tvergaard, V., 1990, *Acta Metallurgica*, Vol. 38, pp. 185-194.

Tvergaard, V., 1982, *International Journal of Fracture*, Vol. 18, pp. 237-252.

Tvergaard, 1981, *International Journal of Fracture*, Vol. 17, pp. 389-408.

Tvergaard, V. and Needleman, A., 1988, *Acta Metallurgica Materiala*, Vol. 32, pp. 157-169.

You, C.P., Thompson, A.W., and Bernstein, I.M., 1987, *Scripta Metallurgica*, Vol. 21, pp. 181-185.

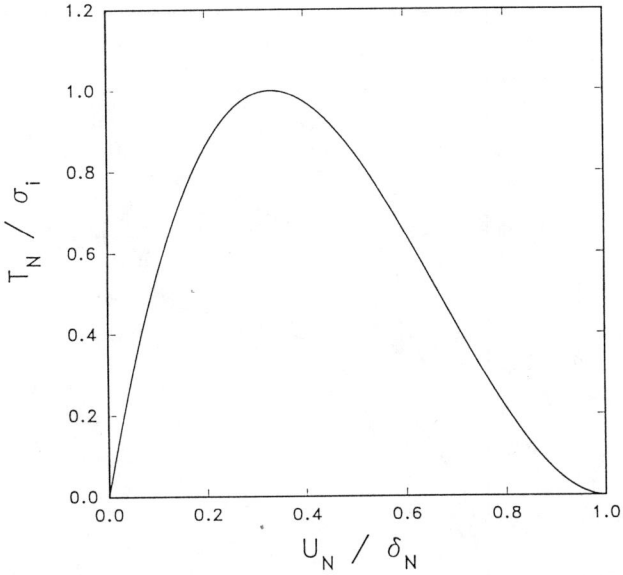

Fig. 1. Schematic representation of variation in interface strength with interface separation.

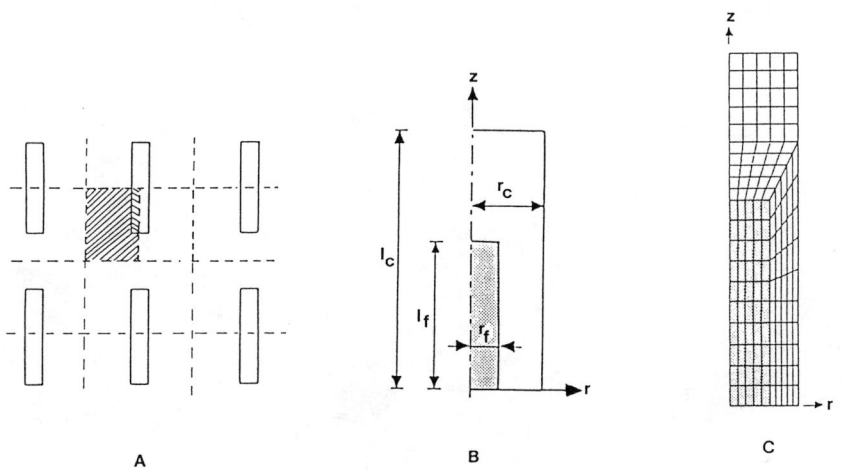

Fig. 2. a) Unit-cell.
b) The parameters of the unit-cell.
c) The typical FEM mesh used during the analysis.

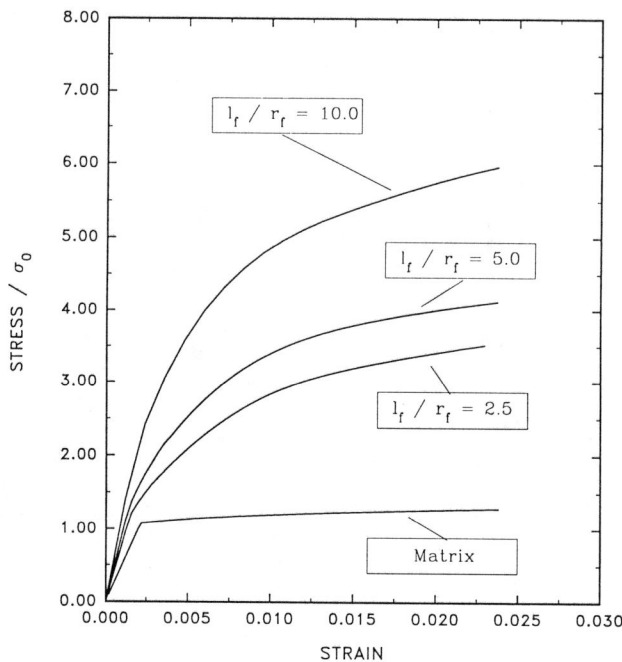

Fig. 3. In the absence of any operating damage mechanism, the variation of stress-strain behavior of the composite containing 20 vol.% reinforcement with fiber aspect ratio.

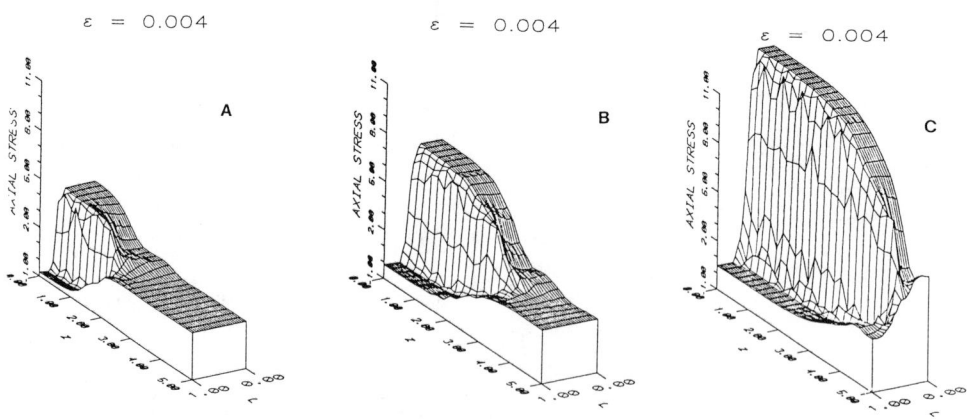

Fig. 4. In the absence of any damage development, variation of axial stress within the unit-cell with fiber aspect ratios of a) 2.5, b) 5, and c) 10, respectively.

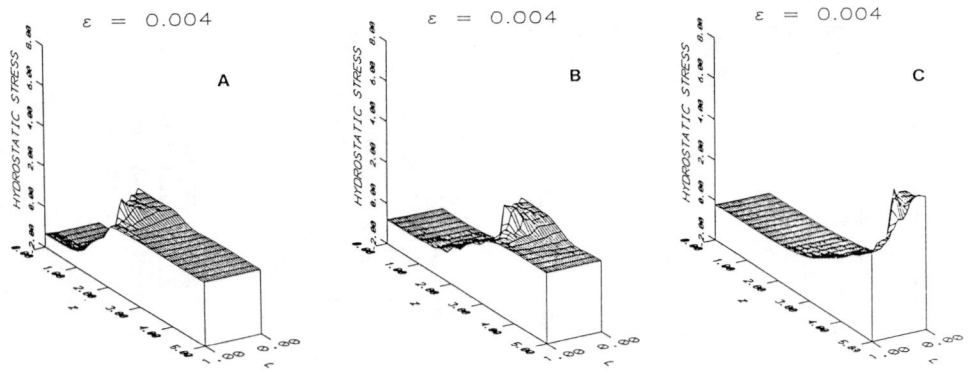

Fig. 5. In the absence of any damage development, variation of hydrostatic stress in the matrix with fiber aspect ratios of a) 2.5, b) 5, and c) 10, respectively.

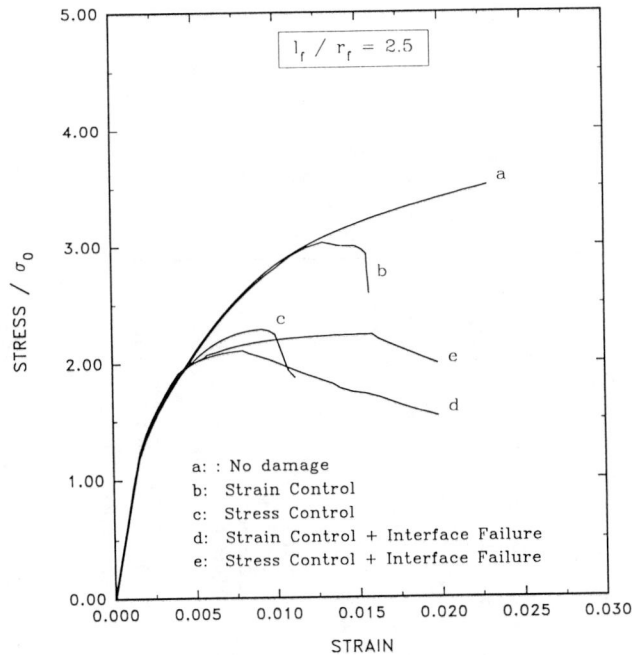

Fig. 6. Stress-strain behavior of composite in the presence of matrix and interface failure. Data for fiber aspect ratio of 2.5 and 20 vol.% reinforcement.

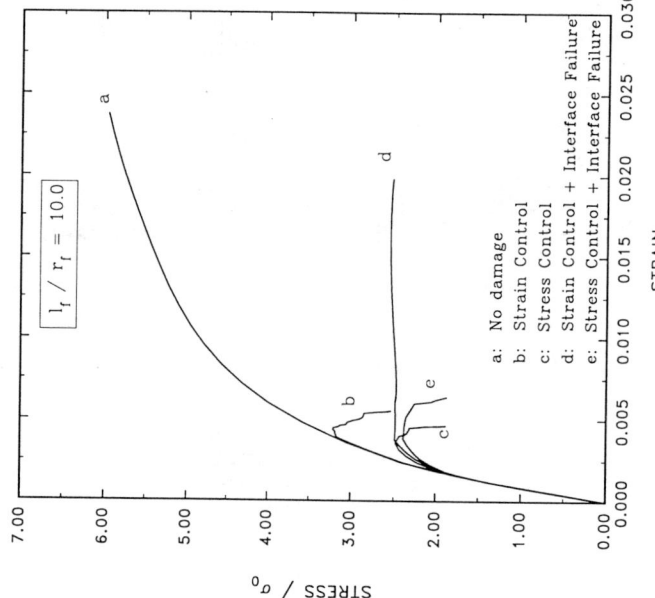

Fig. 8. Stress-strain behavior of composite in the presence of matrix and interface failure. Data for fiber aspect ratio of 10 and 20 vol.% reinforcement.

a: No damage
b: Strain Control
c: Stress Control
d: Strain Control + Interface Failure
e: Stress Control + Interface Failure

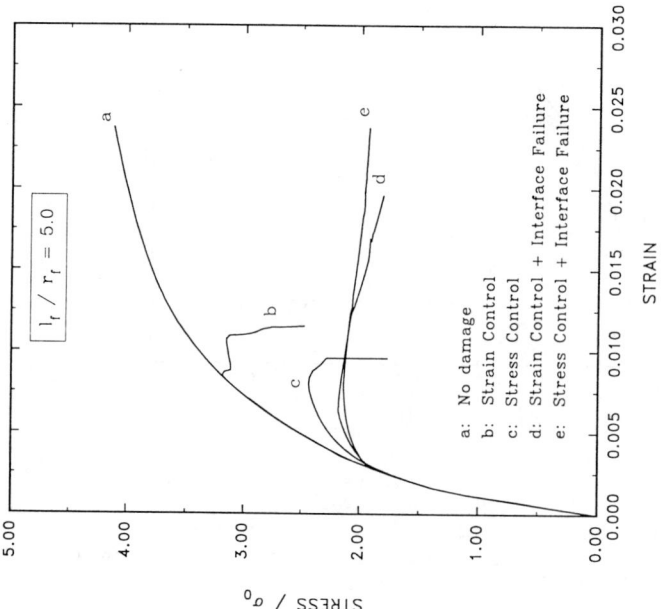

Fig. 7. Stress-strain behavior of composite in the presence of matrix and interface failure. Data for fiber aspect ratio of 5 and 20 vol.% reinforcement.

a: No damage
b: Strain Control
c: Stress Control
d: Strain Control + Interface Failure
e: Stress Control + Interface Failure

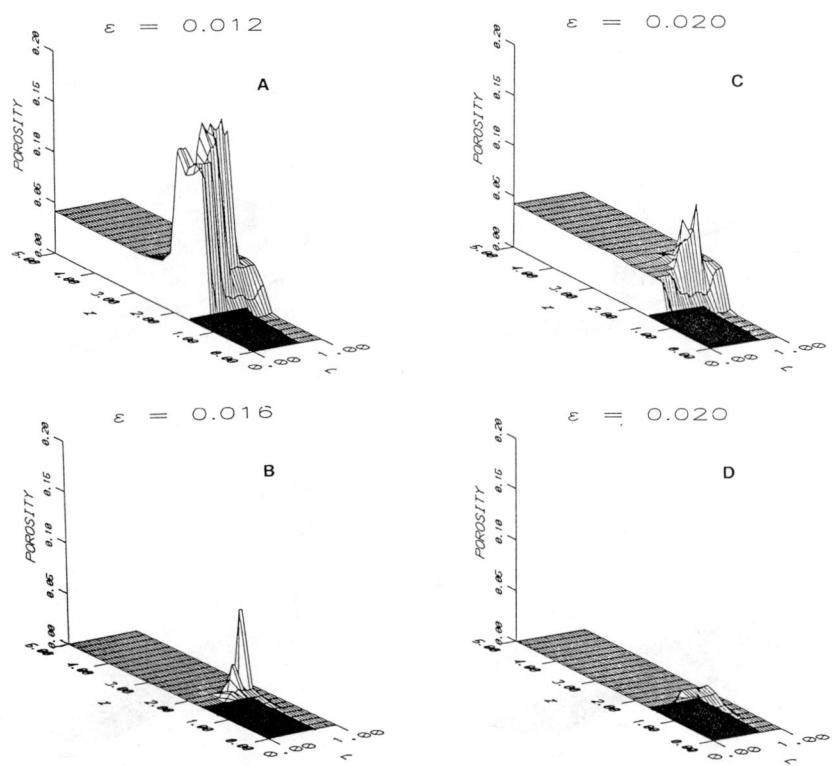

Fig. 9. For different failure mechanisms, the damage evolution within the matrix. a) Stress-Controlled void nucleation. b) Strain-Controlled void nucleation. c) Stress-Controlled void nucleation plus interface failure and d) Strain-Controlled void nucleation plus interface failure. Data is for fiber aspect ratio of 2.5. The solid dark areas are the fibers.

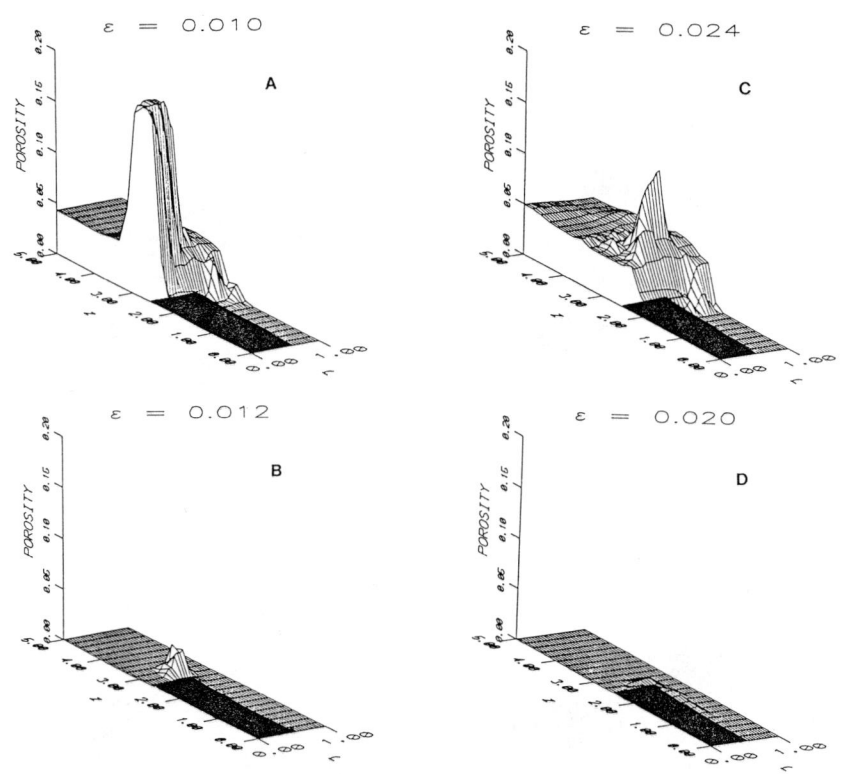

Fig. 10. For different failure mechanisms, the damage evolution within the matrix. a) Stress-Controlled void nucleation. b) Strain-Controlled void nucleation. c) Stress-Controlled void nucleation plus interface failure and d) Strain-Controlled void nucleation plus interface failure. Data is for fiber aspect ratio of 5. The solid dark areas are the fibers.

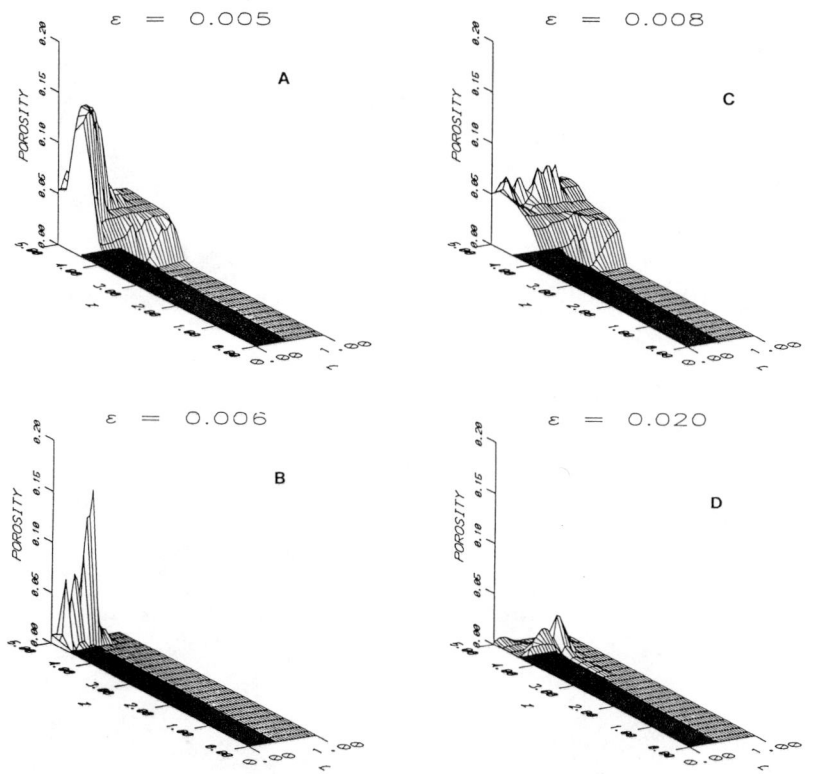

Fig. 11. For different failure mechanisms, the damage evolution within the matrix. a) Stress-Controlled void nucleation. b) Strain-Controlled void nucleation. c) Stress-Controlled void nucleation plus interface failure. d) Strain-Controlled void nucleation plus interface failure. Data is for fiber aspect ratio of 10. The solid dark areas are the fibers.

INFLUENCE OF SPRAY ATOMIZATION AND DEPOSITION ON THE MICROSTRUCTURE AND MECHANICAL BEHAVIOR OF ALUMINUM-COPPER BASED METAL MATRIX COMPOSITES

Manoj Gupta
Department of Mechanical and
Aerospace Engineering
Materials Science and Engineering
University of California
Irvine, California

T. S. Srivatsan
Department of Mechanical Engineering
University of Akron
Akron, Ohio

Farghalli A. Mohamed and Enrique J. Lavernia
Department of Mechanical and Aerospace Engineering
Materials Science and Engineering
University of California
Irvine, California

ABSTRACT

In this study, aluminum-copper based metal matrix composites were synthesized utilizing the spray atomization and co-deposition technique. Microstructural characterization studies were carried out with an emphasis on understanding the effects associated with the co-injection of silicon carbide and aluminum oxide particulates. The results demonstrate the aging kinetics of the spray deposited and hot extruded metal matrix composites to be the same as those of the monolithic aluminum-copper material. Results of ambient temperature mechanical tests demonstrate that the presence of particulate reinforcement in the metal matrix does little to improve strength, and degrades the ductility of the matrix material. A model is formulated to compute the critical volume fraction of reinforcement. The results obtained using this model suggest that an optimum volume fraction of silicon carbide is essential in order to realize a strength improvement in the metal matrix composite, relative to the monolithic counterpart.

1. INTRODUCTION

Discontinuously-reinforced aluminum (DRA) composites have over the years been the subject of intensive study due to their innate ability to combine superior strength, high stiffness, low density and fracture resistance [1]. Commonly utilized matrix materials include aluminum alloys based on the 2XXX (Al-Cu), 6XXX (Al-Mg-Si) and more recently the 7XXX (Al-Zn-Mg-Cu) series. The discontinuous reinforcements include particulates, chopped fibers and whiskers. The primary advantage of using discontinuous reinforcements is that the metal-matrix composites are relatively easy to fabricate by both powder metallurgy (PM) and ingot metallurgy (IM) techniques and the resulting products exhibit near isotropic behavior. Commonly used reinforcements for aluminum alloy matrices include silicon carbide (SiC) and aluminum oxide (Al_2O_3).

In recent studies, investigators have succeeded in tailoring the properties of existing aluminum alloys to specific applications through compositional modifications. One such example is alloy AA 2519 (Al-Cu-Mg), which has demonstrated excellent strength and ballistic performance coupled with good stress corrosion cracking resistance [2]. In an effort to achieve additional improvements in the properties of alloys, such as AA 2519, a variety of synthesis techniques are actively being studied. One such technique is spray atomization and deposition which has received considerable attention for the synthesis of aluminum base alloys [3-5] and metal-matrix composites (MMCs) [6-14]. This novel technique involves processing in a regime of the phase diagram where the alloy is a mixture of solid and liquid phases. Such an approach would inherently avoid the extreme thermal excursions with concomitant degradation in interfacial properties and extensive macrosegregation normally associated with conventional casting processes [15, 16]. Furthermore, this approach also eliminates the need to handle fine reactive particulates, as is necessary with powder metallurgical (PM) processes [17, 18].

The objective of the present study was to provide an insight into the effects of co-injection of SiC and Al_2O_3 particulates during spray atomization and co-deposition processing, henceforth referred to as spray processing, on the microstructure and mechanical behavior of aluminum alloy AA 2519. The effects of SiC and Al_2O_3 particulates on the microstructure during solid state cooling, that is after the matrix/particulate mixture has arrived on the substrate, were investigated with particular emphasis on the rate of grain growth. To identify the role of particulate reinforcement on grain boundary migration, the reinforced matrix and unreinforced material were exposed to various isothermal heat treatments. The intrinsic microstructural features of the spray processed metal-matrix composite are characterized and discussed in light of alloy composition and processing variables. The ambient temperature mechanical properties of the composite are correlated with microstructural features.

2. EXPERIMENTAL PROCEDURE

2.1 Processing

The experimental studies were conducted on an Al - (5.0-7.0)Cu - (0.1-0.3)Mg - (0.0-0.8)Mn - (0.0-0.1)Ti - (0.0-0.25)V - (0.0-0.25)Zr - (0.0-0.5)Fe - (0.0-0.5)Si - (0.0-0.12)Zn (in weight percent)(designated as alloy AA 2519) alloy. The alloy was provided, in the form of rolled plates, by the Army Materials Technology Laboratory (AMTL: Watertown, Massachusetts, U.S.A.). The matrix alloy will be henceforth referred to as Al-Cu. The size distributions of the SiC (α phase) and Al_2O_3 (α phase) particulates were Gaussian and exhibited an average size of 3 µm (d_{50}).

The MMCs were synthesized according to the following procedure. The Al-Cu matrix material was superheated to the temperature of interest (see Table I), and disintegrated into a fine dispersion of micrometer-sized droplets using atomizing gas at a pre-selected pressure. Simultaneously, two jets containing one type of ceramic particulate reinforcement (either SiC or Al_2O_3) and positioned at 180° with respect to each other were injected into the atomized matrix material at a previously selected flight distance. The flight distance was determined on the basis of a numerical analysis of the temperature and fraction solid contained in the atomized matrix material. The reinforcement injection distance for Expt. 3, Expt. 4 and Expt. 5 was 0.21 m, and the substrate position used was 0.41 m for all of the experiments. The selection of this injection distance was made on the basis of a study by Gupta et al. [19]. Their results showed that at a particular flight distance, the atomized aluminum alloy droplets have lost approximately 40-50% of their original enthalpy. Following co-injection, the mixture of rapidly quenched, partially solidified droplets with interdispersed ceramic particulates was deposited on a water cooled deposition surface, eventually collecting as a coherent preform. The microstructure of the preform is dictated by the solidification conditions during impact. In order to avoid extensive oxidation of the aluminum alloy matrix during processing, the experiments were conducted in an environmental chamber. The latter was evacuated to a negative pressure of 0.020 MPa, and backfilled with inert gas to a positive pressure of 0.014 MPa, prior to melting and atomization. A schematic diagram of the experimental arrangement used in this study is shown in Figure 1. A total of five experiments were conducted. The primary experimental variables used for each experiment are summarized in Table I.

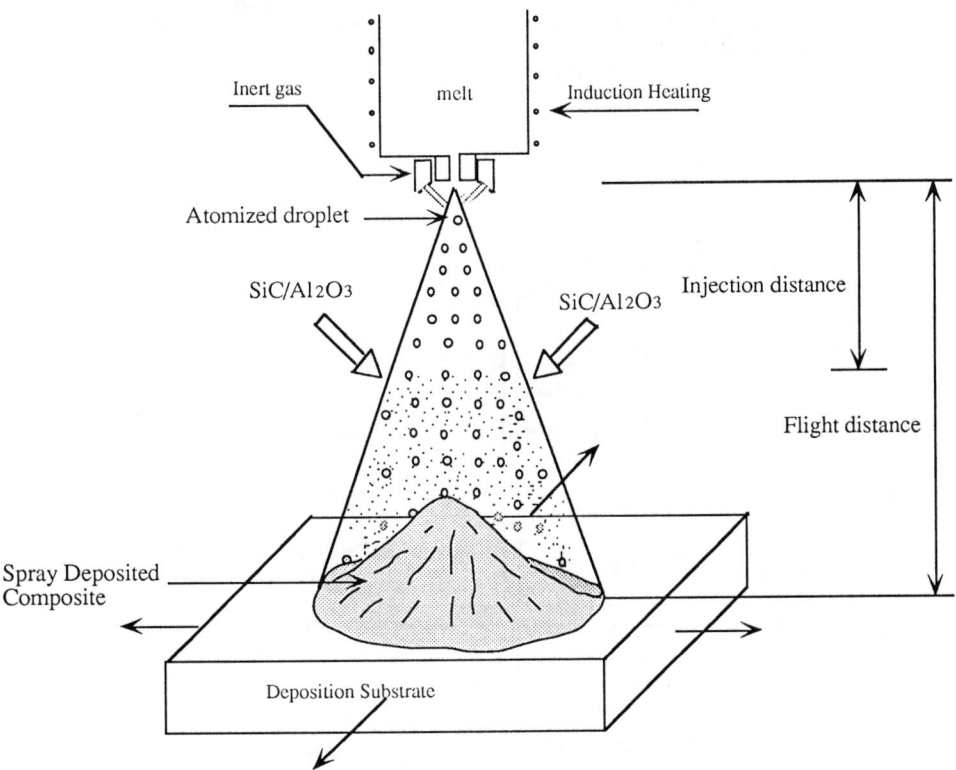

Figure 1. Schematic diagram of experimental arrangement used for spray processing.

Table I. Experimental Parameters.

Variable	Experiment #					Units
	1	2	3	4	5	
Matrix alloy	Al-Cu	Al-Cu	Al-Cu	Al-Cu	Al-Cu	wt. %
Reinforcement	--	--	SiC	SiC	Al_2O_3	--
Atomization pressure	1.21	1.21	1.21	1.21	1.21	MPa
Atomization gas	Ar	N_2	N_2	N_2	N_2	--
Superheat Temp.	1023	1073	1073	1073	1073	K
SiC carrier gas	--	--	N_2	N_2	N_2	--
Pressure of carrier gas	--	--	0.17	0.17	0.17	MPa
Metal flow rate	0.034	0.039	0.039	0.039	0.039	kg/s
Gas flow rate	0.013	0.018	0.018	0.018	0.018	kg/s

The SiC and Al_2O_3 particulates were introduced into the atomized Al-Cu spray using an injector (see Figure 2). The injector consisted of a coaxial tube that entrained the ceramic particulates as the gas flowed from the inlet to the outlet orifices. The injection of the ceramic particulates was carried out at ambient temperature. A comprehensive discussion of the experimental details are provided elsewhere [6] and will not be reiterated here.

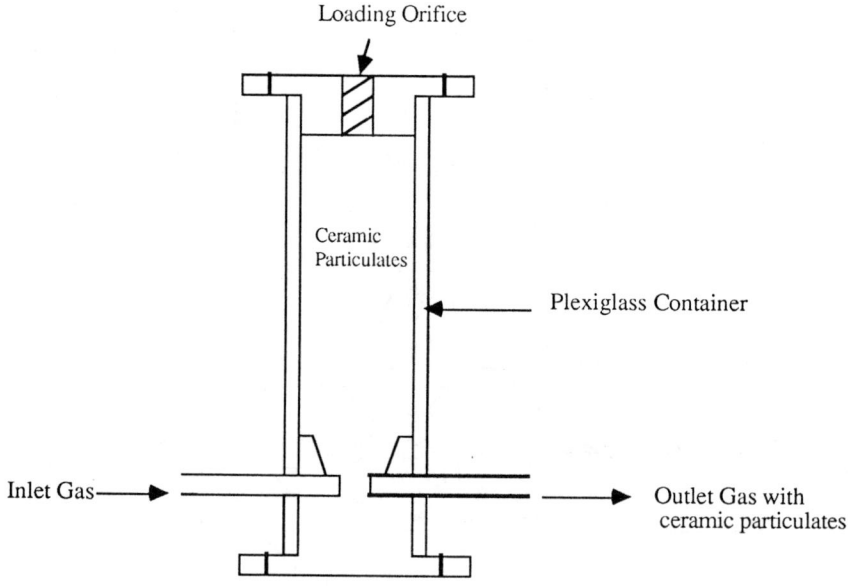

Figure 2. Schematic diagram showing coaxial tube injector.

2.2 Microstructure

Microstructural characterization studies were conducted on both the unreinforced and reinforced matrix materials in order to determine: (a) the grain size, (b) volume fraction of SiC and Al_2O_3 particulates, and (c) the presence of secondary phases in the spray-atomized and deposited samples. In addition, density measurements were carried out on spray-deposited and extruded samples in order to ensure the closure of micrometer sized porosity present in the as-spray deposited samples.

Optical microscopy was conducted on both polished and etched as-deposited samples using conventional and Differential Interference Contrast (DIC) techniques. The use of DIC microscopy facilitated identification of the ceramic particulates in the matrix. The samples were sectioned to a thickness of 0.5 cm, polished using conventional techniques, and etched using Keller's reagent (0.5 HF-1.5 HCl-2.5 HNO_3-95.5 H_2O). The grain size was measured using the linear intercept method, as described in ASTM E 112-84.

Density measurements were conducted on the polished, extruded samples utilizing Archimedes' principle. The weight of each sample was determined, using a Fisher Scientific A-250 Balance, to an accuracy of $\pm$ 0.0001 g. Ethylene glycol was used as the fluid.

The volume fractions of ceramic particulate (SiC or Al_2O_3) was determined using a chemical dissolution method. This method involved: (i) measuring the mass of composite samples, (ii) dissolving the samples in dilute hydrochloric acid (38.0% max.), followed by (iii) filtering to separate the ceramic particulates. The particulates were then dried and the weight fraction determined. The weight fraction was converted to volume fraction, $V_{f(R)}$, using the equation:

$$V_{f(R)} = \frac{\text{wt. \% R}/\rho_R}{\text{wt. \% R}/\rho_R + \text{wt. \% Matrix}/\rho_{matrix}} \quad (1)$$

where, ρ_R and ρ_{matrix} represent the densities of the reinforcing particulates and the matrix, respectively.

Scanning electron microscopy (SEM) studies were conducted using a HITACHI S-500 microscope. Samples taken from the: (i) as-received material, (ii) as-spray deposited material, and (iii) spray-deposited and hot-extruded materials from different experiments, were sectioned to a thickness of 0.5 cm and polished using conventional techniques. The polished samples were then examined in secondary electron mode for intrinsic microstructural features. In addition, SEM studies were conducted on fractured samples taken from Expt. 1, Expt. 4 and Expt. 5 in order to provide an insight into the quasi-static fracture behavior.

X-ray diffraction analysis (XRD) was conducted on the spray deposited samples from Expt. 2, Expt. 3, and Expt. 5 using a Philips Norelco vertical diffractometer. Thin samples were exposed to Cu Kα radiation (λ = 1.5418A°) using a scanning speed of 0.24 deg/min. A plot of intensity versus 2θ was obtained, illustrating peaks at different Bragg angles. The Bragg angle corresponding to each of the different peaks was noted and the value of interplanar spacing, 'd', was calculated using Bragg's law (λ = 2dsinθ). The values of 'd' obtained were matched with standard values for aluminum and other phases.

2.3 Thermomechanical Treatment

Following spray processing, samples from Expt. 2, Expt. 3 and Expt. 5 were isochronally annealed at 673 K and 773 K in order to study the influence of temperature and annealing time on grain growth. In order to assess the mechanical behavior, both the unreinforced and reinforced spray atomized materials were hot extruded. The extrusion step was accomplished using a press (capacity: 82.7 MPa) at a temperature of 673 K for the unreinforced matrix material and at 723 K for the reinforced material. An area reduction ratio of 16:1 was used for all samples. The extrusion step was used in this study in order to close the micrometer size porosity that is normally associated with spray atomized and deposited materials [6, 7, 13]. In order to provide an insight into the aging behavior, Rockwell hardness measurements were made on the as-spray deposited and hot extruded samples from Expt. 1, Expt. 4 and Expt. 5, solutionized at 802 K for 2 hours and isochronally annealed at 436 K for different intervals of time.

Smooth bar tensile properties were determined in accordance with ASTM E8-81. Tensile specimens were prepared from materials taken from Expt. 1, Expt. 4 and Expt. 5. Tensile tests were conducted using a servohydraulic structural testing machine. The specimens were deformed at a constant crosshead speed of 0.0254 cm per minute.

3. RESULTS

3.1 Macrostructure

The overall dimensions of the spray processed preforms from the five experiments were approximately 36 cm in length, and 18 cm in width. The thickness of the preforms decreased from 5.0-7.5 cm in the central portion to approximately 0.5 cm in the thickness dimension. All of the structural characterization studies were performed on material removed from the central portion (80-90 %) of the preforms. The remaining portion (10-20 %) of the preforms was considered too porous for detailed analysis.

3.2 Microstructure

Optical microscopy conducted on coupons of the unreinforced matrix material taken from experiments 1-5 revealed the presence of an equiaxed grain morphology. The results of grain size measurements are summarized in Table II. An example taken from Expt. 2 is shown in Figure 3.

The results of density measurements revealed: (a) a density value of 2.82 g/cm^3 for the as-received AA 2519 alloy and the extruded samples taken from Expt. 1, and (b) density values for the as-extruded materials taken from Expt. 4 and Expt. 5 to be 2.85 g/cm^3 and 3.00 g/cm^3, respectively.

The results of the acid dissolution experiments are summarized in Table II. The volume percentages of SiC particulates present in the as-spray deposited material were estimated to be approximately 8.4 for Expt. 3 and 11.1 for Expt. 4, respectively. The volume percent of Al_2O_3 for Expt. 5 was approximately 16.8 for the as-spray deposited sample. The interparticulate spacings were calculated using the formula suggested by Nardone and Prewo, for discontinuous reinforcement [20]:

$$\lambda = (l\, t\, /\, V_f)^{1/2} \tag{2}$$

where: λ is the interparticulate spacing; and t, l and V_f are the thickness, length, and volume fraction of the ceramic particulates, respectively. The results of the computed interparticulate spacings, λ, are also shown in Table II.

Table II. Microstructural Characterization of As-Spray Deposited MMCs.

Experiment #	Reinforcement size (d_{50}, μm)	Grain size (μm)	Reinforcement V_f (%)	Porosity V_f (%)	[2]Interparticulate spacing (λ, μm)
[1]I.M	–	44.0 ± 3.0	--	0.0	--
1	–	25.0 ± 2.1	--	[3]N.D	--
2	–	35.5 ± 3.0	--	1.1	--
3	3	28.3 ± 2.2	8.4	[3]ND	10.4
4	3	23.0 ± 0.5	11.1	6.9	9.0
5	3	27.7 ± 2.5	16.8	6.4	7.3

[1] The material was obtained in the form of rolled plates.
[2] These values were computed using Equation 2.
[3] ND: Not determined.

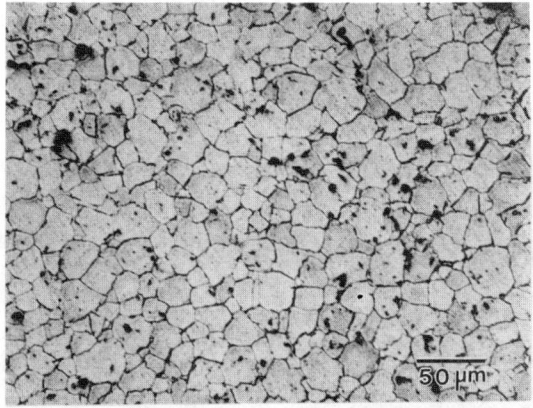

Figure 3.   Optical micrograph showing equiaxed grain morphology of a spray deposited Al-Cu sample taken from Expt. 2.

Scanning electron microscopy of samples taken from Expt. 1 revealed the presence of a finite amount of unconnected porosity and the presence of needle shaped particles (Figure 4). The presence of a dark region surrounding the needle-like particles is evident in the micrographs. A scanning electron micrograph taken from the as-received plates shows the plate-like morphology of these particles with a low aspect ratio (Figure 5). The distribution of reinforcing particulates in the Al-Cu matrix is shown in Figures 6 and 7.

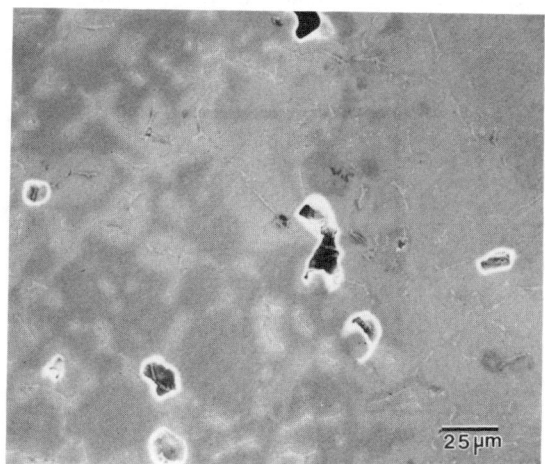

Figure 4. Scanning electron micrograph showing the presence of unconnected porosity and needle shaped particles in the as-spray deposited material from Expt. 1.

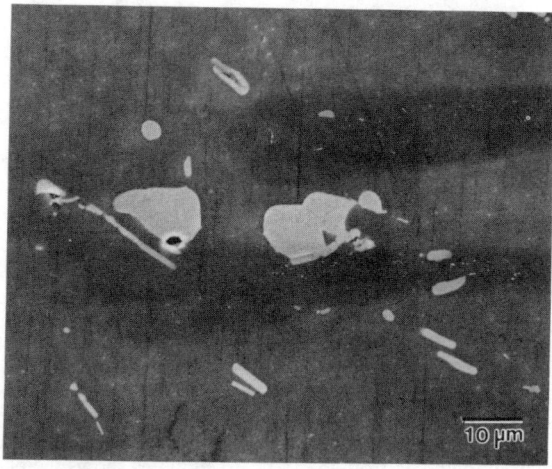

Figure 5. Scanning electron micrograph showing the presence of plate-shaped particles in the as-received Al-Cu plates.

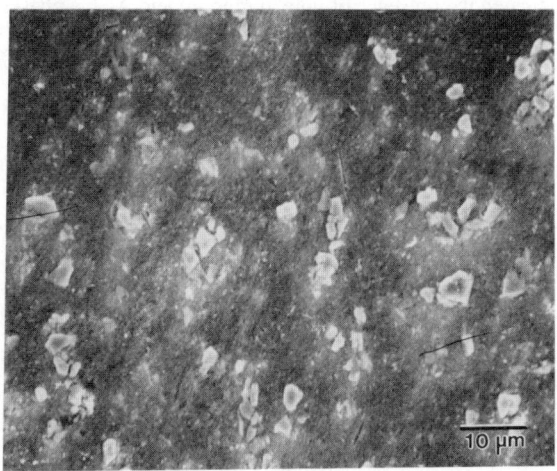

Figure 6. Scanning electron micrograph showing the distribution of SiC particulates in the as-spray deposited and extruded sample taken from Expt. 4.

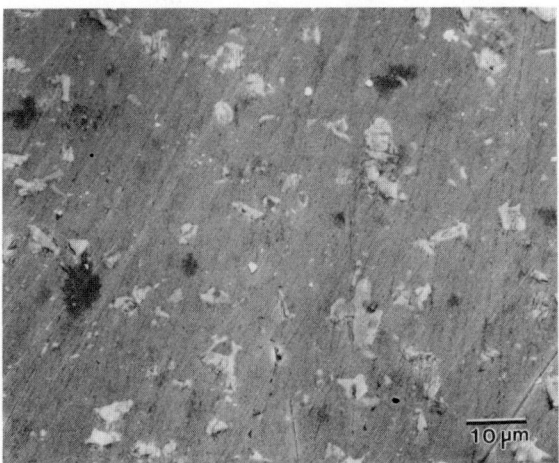

Figure 7. Scanning electron micrograph showing the distribution of Al_2O_3 particulates in the as-spray deposited and extruded sample taken from Expt. 5.

The X-ray diffraction spectra corresponding to as-spray deposited samples from Expt. 2, Expt. 3, and Expt. 5 were analyzed. The lattice spacings (d) corresponding to the observed Bragg angles are shown in Table III. The XRD spectrum corresponding to the as-spray deposited Al-Cu material (Expt. 2) indicated the presence of pure aluminum and Al_2Cu phases. The XRD spectrum corresponding to the as-spray deposited Al-Cu/SiC material (Expt. 3) indicates the presence of pure aluminum, Al_2Cu and α-SiC phases, while the XRD spectrum corresponding to the Al-Cu/Al_2O_3 composite material (Expt. 5) revealed the presence of pure aluminum and α-Al_2O_3 phases.

Table III. Results of X-ray Diffractometry Studies.

Experiment 2 (Al-Cu)								
Angle	38.38	42.58	44.67	47.09	47.74	65.16	78.38	
Calculated d values	2.34	2.12	2.03	1.93	1.90	1.43	1.22	
Experiment 3 (Al-Cu/SiC)								
Angle	34.16	35.68	38.56	42.60	42.68	44.76	47.36	60.08
Calculated d values	2.62	2.51	2.33	2.12	2.11	2.02	1.91	1.54
Experiment 5 (Al-Cu/Al_2O_3)								
Angle	35.16	38.56	43.40	44.80	57.48			
Calculated d values	2.55	2.33	2.08	2.02	1.60			
Standard 'd' values								
Al	2.338	2.025	1.432	1.221	1.169			
α-Al_2O_3	2.55	2.09	1.60					
Al_2Cu	1.91	2.12	4.29					
α-SiC	2.51	2.63	1.54					

3.3 Grain Growth Behavior

In order to provide an insight into the effects of the SiC and Al_2O_3 particulates on the microstructure of the composite matrix during solid state cooling, grain growth studies were conducted on the samples taken from Expt. 2, Expt. 3, and Expt. 5. The grain sizes were determined using the linear intercept method, after isochronal anneals at 673 K and 773 K (Tables IV and V and Figures 8 and 9). The results of grain size measurements shown in Figures 8 and 9 reveal a logarithmic progression of grain growth with time for the unreinforced Al-Cu, and reinforced Al-Cu/SiC and Al-Cu/Al_2O_3 materials. Not unexpectedly, it can be seen that the final grain sizes of the Al-Cu/SiC and Al-Cu/Al_2O_3 composite materials, at 673 K and 773 K, are lower than that of the unreinforced matrix material. The microstructure of the as-spray deposited monolithic alloy (Al-Cu) and the SiC and Al_2O_3 reinforced Al-Cu matrices consisted of equiaxed grains, both before and after the isochronal heat treatments.

Table IV. Results of the Grain Size Measurements for Different Intervals of Time at 673 K for As-Spray Processed Samples of Al-Cu, Al-Cu/SiC and Al-Cu/Al$_2$O$_3$ Materials.

Experiment #	2	3	5
Time (min)	Grain size (μm)	Grain size (μm)	Grain size (μm)
[1]0	35.5 ± 3.0	28.3 ± 2.2	27.7 ± 2.5
1	36.6 ± 3.8	28.7 ± 1.1	28.3 ± 0.9
10	39.5 ± 1.8	31.5 ± 1.0	31.1 ± 0.7
50	43.0 ± 1.3	34.2 ± 0.91	35.0 ± 0.8
100	44.8 ± 2.1	38.8 ± 0.6	35.1 ± 2.2

[1] Time 0 refers to as-spray deposited grain size.

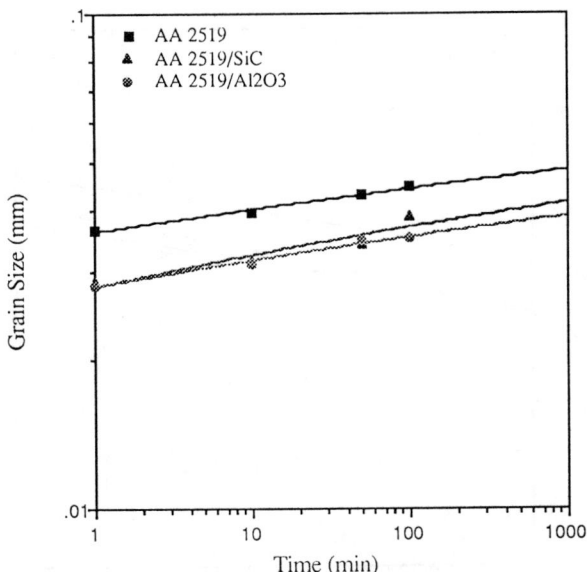

Figure 8. Graphical representation of the logarithmic grain growth relationship at 673 K observed in as-spray processed Al-Cu, Al-Cu/SiC and Al-Cu/Al$_2$O$_3$ materials.

Table V. Results of the Grain Size Measurements for Different Intervals of Time at 773 K for As-Spray Processed Samples of Al-Cu, Al-Cu/SiC and Al-Cu/Al$_2$O$_3$ Materials.

Experiment #	2	3	5
Time (min)	Grain size (μm)	Grain size (μm)	Grain size (μm)
[1]0	35.5 ± 3.0	28.3 ± 2.2	27.7 ± 2.5
1	37.8 ± 8.0	29.1 ± 1.8	29.3 ± 1.6
10	39.9 ± 2.6	35.2 ± 4.7	34.7 ± 2.3
50	46.2 ± 2.3	36.2 ± 1.4	37.5 ± 1.0
100	48.8 ± 2.7	39.1 ± 1.2	42.2 ± 2.8

[1] Time 0 refers to as-spray deposited grain size.

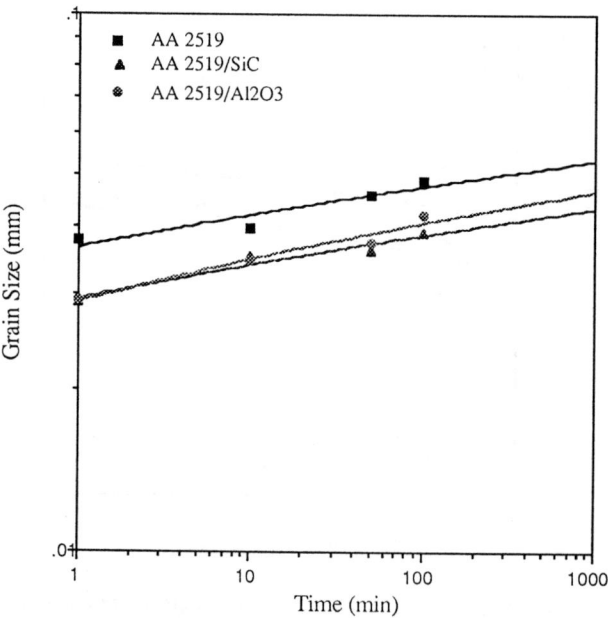

Figure 9. Graphical representation of the logarithmic grain growth relationship at 773 K observed in as-spray processed Al-Cu, Al-Cu/SiC and Al-Cu/Al$_2$O$_3$ materials.

3.4 Aging studies

The results of the aging studies conducted on samples removed from Expt. 1, Expt. 4 and Expt. 5 are shown in Figure 10. The results exhibit the presence of a well defined peak at 12.0 hrs for the samples taken from Expt. 1, Expt. 4 and Expt. 5. The results also reveal the maximum peak hardness is achieved in samples taken from Expt. 4 followed by samples taken from Expt. 5 and Expt. 1, respectively. In addition, the results also show the as-quenched hardness of the composites samples to be higher than the monolithic counterpart.

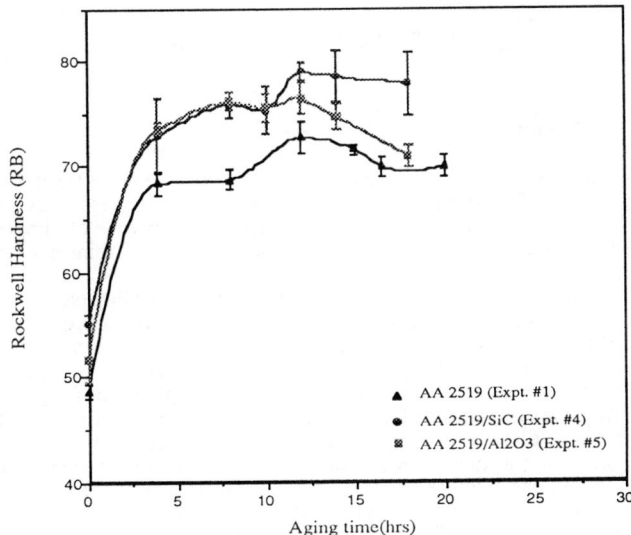

Figure 10. Graphical representation of the aging studies conducted on as-spray processed and extruded Al-Cu, Al-Cu/SiC and Al-Cu/Al$_2$O$_3$ materials. All materials were solution treated at 802 K for 2 hrs.

3.5 Mechanical Behavior

The results of ambient temperature testing on the spray-deposited and hot extruded unreinforced and reinforced matrices (Al-Cu), aged to peak hardness, are summarized in Table VI. Also shown in this table are the properties of equivalent material prepared by the ingot metallurgy route. The results in Table VI reveal that the ambient temperature mechanical properties of the unreinforced spray deposited materials accords well with the ingot metallurgical material [21]. The results also show that the presence of particulate reinforcement (SiC or Al$_2$O$_3$) in the Al-Cu matrix does not help in improving strength, and, in fact, degrades the ductility of the composite matrices.

Table VI. Results of the Room Temperature Mechanical Properties.

Experiment #	Y.S (MPa)	U.T.S (MPa)	Ductility (%)
1	318.7 ± 0.0	410.3 ± 10.7	14.7 ± 0.7
4	301.2 ± 5.3	422.3 ± 8.5	10.8 ± 0.5
5	299.6 ± 4.6	411.8 ± 5.3	6.5 ± 1.3
IM	[1]332.3	[2]--	[2]--

[1]Cast samples solutionized at 802 K and aged at 436 K for 16 hrs [21].
[2]Not reported.

3.6 Fracture Behavior

Macroscopic investigation of the fracture surfaces of samples taken from Expt. 1 reveal the presence of a cup and cone type of fracture. Figure 11 is a representative scanning electron micrograph showing morphology of the fracture surface. The presence of dimples of uniform size is indicative of ductile failure. Fracture surfaces samples of the SiC reinforced composites (Expt. 4) reveal matching fracture surfaces tilted at an angle of ~ 8°, with respect to the horizontal plane, suggestive of brittle failure. Figure 12a is a low magnification scanning electron micrograph of the fractured sample. Cracks can be seen emanating and randomly distributed throughout the fracture surface. Figure 12b shows regions of highly localized plastic deformation and that of brittle failure on the fracture surface of a sample taken from Expt. 4. In addition, interfacial debonding was observed between SiC particulates and the matrix, in the sample taken from Expt. 4 (Figure 13). Finally, macroscopic examination of the fractured samples taken from Expt. 5 revealed the fractured surfaces to be tilted by ~ 7° to the horizontal plane, suggestive of brittle failure. Figure 14a is a low magnification scanning electron micrograph revealing the presence of cracks randomly distributed throughout the fracture surface. Figure 14b is a representative scanning electron micrograph revealing a mixed mode type of fracture. The regions corresponding to highly localized plastic deformation and features reminiscent of brittle failure can easily be discerned in this figure.

Finally, the presence of cavities of ~5 μm to ~15 μm were observed on the fractured surfaces of the samples taken from Expt. 4 and Expt. 5. These cavities were noted to be associated with either the large ceramic particulates or clusters of small particulates.

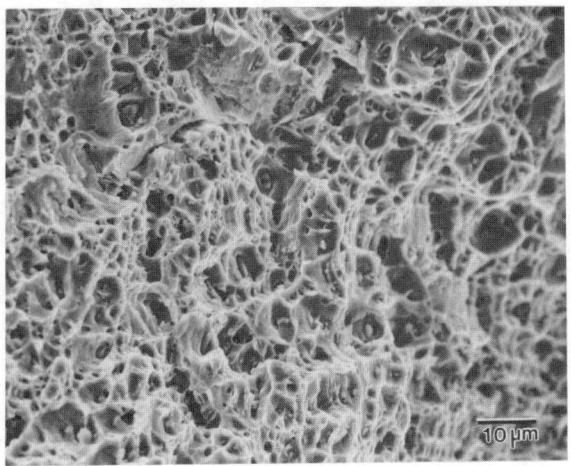

Figure 11.   Scanning electron micrograph showing fracture surface morphology of the sample taken from Expt. 1.

Figure 12.   Representative scanning electron micrograph showing: a) cracks distribution on the fracture surface, and

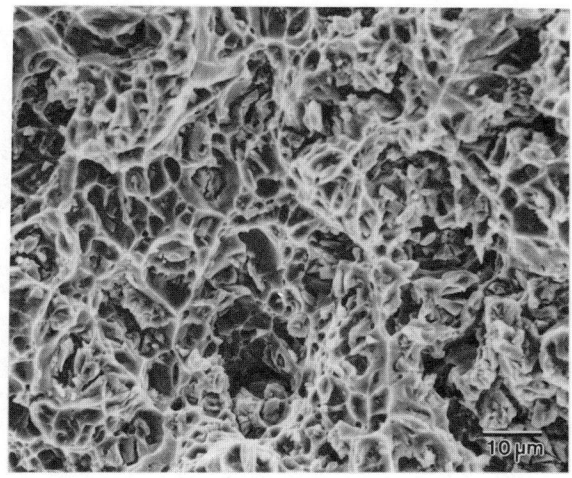

Figure 12. (cont'd)   b) mixed mode type of failure observed in the fractured samples taken from Expt. 4.

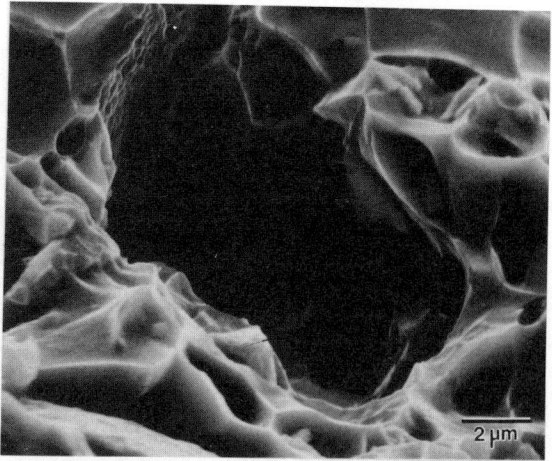

Figure 13.   Scanning electron micrograph showing the interfacial debonding between matrix and SiC reinforcement observed on the fractured surface of the sample taken from Expt. 4.

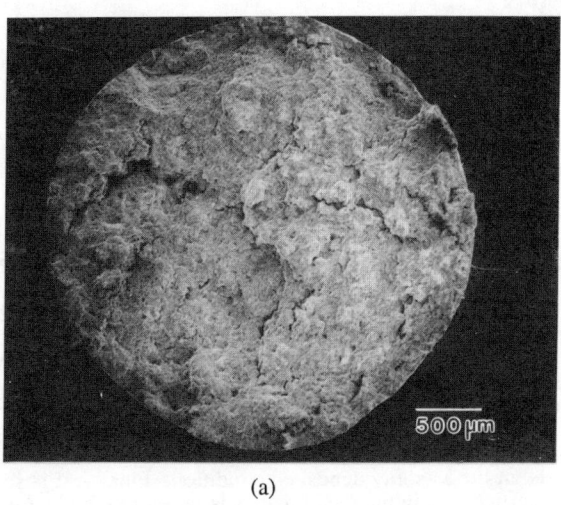

(a)

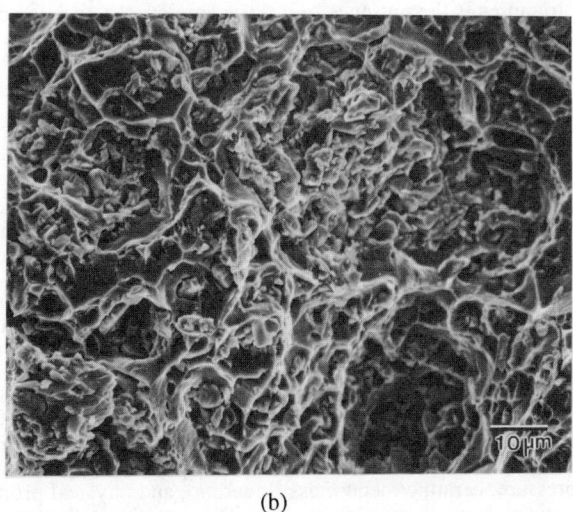

(b)

Figure 14. Representative scanning electron micrograph showing: a) cracks distribution on the fracture surface, and b) mixed mode type of failure observed in the fractured samples taken from Expt. 5.

4. DISCUSSION

4.1 Microstructure

Three salient features are associated with the microstructure of spray processed ceramic particulate reinforced metal matrix composites:

(a) the grain structure,
(b) the presence of micrometer sized pores, and
(c) the amount and distribution of reinforcing particulates.

The grain morphology of the spray processed MMCs was equiaxed, in agreement with the results obtained by other investigators [22-30]. The mechanisms associated with the formation of equiaxed grains during spray processing are addressed elsewhere [6, 22-26, 30], and will not be reiterated here. The results from Table II reveal the experimentally observed grain sizes for the unreinforced and reinforced spray deposited materials to range from 23.0 to 35.5 µm. The relatively finer grained microstructure noted for Expt. 1 (25.0 µm), relative to that obtained in Expt. 2 (35.5 µm), is consistent with the lower superheat temperature used in Expt. 1 (1023 K), relative to that used in Expt. 2 (1073 K) [31]. The results also show that increasing the volume fraction of reinforcement effectively decreases the grain size of the as-spray deposited material as observed in Expt. 2, Expt. 3, and Expt. 4 (see Table II). Moreover, the results accordingly show that the grain size of the as-received plates is higher than that observed for unreinforced and reinforced materials, in the as-spray deposited condition. Finally, it is seen that the presence of 11.1% (V_f) of SiC particulates in the metal matrix is far more effective in refining the grain size when compared to 16.8% (V_f) of Al_2O_3, under identical processing conditions (Table II). A thorough discussion of the effects of the processing variables on the resulting microstructure obtained during spray processing is provided elsewhere [6, 7, 24].

A second important microstructural characteristic frequently associated with spray processed microstructures is the presence of a finite amount of non-interconnected porosity [22-29]. The overall amount of porosity present in spray processed materials depends on the following:

(a) the thermodynamic properties of the material,
(b) the thermodynamic properties of the gas, and
(c) the processing parameters.

Under conditions typical for aluminum alloys, the amount of porosity present in spray processed materials has been reported to be in the 1-10% range [6, 22, 25] (Figure 4). The intrinsic mechanisms governing the formation of pores during spray processing can be found elsewhere [32-34, 35]. Finally, the results of this study reveal the presence of ceramic particulates (SiC/Al_2O_3) increases the volume fraction of porosity in the as-spray deposited materials when compared to that of the unreinforced matrix material, processed under identical conditions (Table II). This is attributed to enhanced heat transfer and the concurrent increase in fraction solid of the droplets prior to deposition [13].

The resultant size, amount and distribution of reinforcing particulates is of interest since the mechanical behavior of the ceramic particulates reinforced metal matrix composites is linked with the presence of these particulates in the matrix. The volume fraction of ceramic particulates present in spray processed materials has been correlated with processing parameters such as injection angle, injection pressure, ceramic/metal mass flow ratio, and physical properties such as surface tension of the atomized droplets [23, 25, 36, 37]. The ceramic particulates may be incorporated into the aluminum alloy matrix by two possible mechanisms:

a) the ceramic particulates penetrate the atomized droplets during co-injection and remain entrapped in the matrix during subsequent impact with the deposition surface [37], or
b) the SiC particulates remain on the surface of the atomized droplets and are entrapped by the matrix after impact with the deposition surface.

Gupta et al. [6, 24] proposed that the extent of particulate entrapment after impact will depend on the conjunct influence of magnitude of the impact and repulsive forces present at the metal/ceramic interface. If entrapment fails to take place either during co-injection, or subsequently, during deposition, the microstructure of the spray processed materials will be characterized by a high concentration of ceramic particulates at the prior droplet boundaries. Such a situation has been reported by Ibrahim et al. [38] for a spray processed 6061 Al/SiC metal matrix composite.

A comparison of the observed grain sizes with the measured interparticulate spacing provides an insight into the extent of entrapment that occurred during the experiments (Table II). The results show that the interparticulate spacings in materials from Expt. 3, Expt. 4, and Expt. 5 were substantially smaller than the measured grain sizes, suggesting entrapment of the ceramic particulates by a large proportion of the droplet population. However, on the basis of the present data, it was not possible to discern as to whether the entrapment of ceramic particulates occurred either during atomization, or during subsequent deposition. Furthermore, it is worth noting that the SiC particulates were more homogeneously distributed in the Al-Cu matrix when compared to the Al_2O_3 particulates. The Al_2O_3 particulates showed a tendency to agglomerate into clusters (Figure 7). Further work continues in this area.

On the basis of the experimental findings, it appears that the co-injection of ceramic particulates during spray processing influences the precipitation kinetics. Phase equilibria, as inferred from the Al-Cu phase diagram [39], is consistent with the XRD results obtained for Expt. 2 and Expt. 3 which confirmed the presence of Al_2Cu precipitates. However, failure to detect the presence of Al_2Cu-type particles using SEM in samples taken from Expt. 3, suggests that the precipitates were refined by the presence of reinforcing phases. This phenomenon is consistent with the results of Salvo et al. [40] who suggested that the finer precipitate size that is commonly observed in MMCs may be attributed to a greater number of nucleation sites associated with a higher density of dislocations that is present in the reinforced metal matrices. Finally, failure to detect Al_2Cu diffraction lines in samples taken from Expt. 5 is attributed to:

a) volume fraction of Al_2Cu is below the XRD detection limit, and/or
b) precipitation of Al_2Cu is completely suppressed.

The latter phenomenon is unlikely as a result of the high quench rates that are necessary to completely suppress this reaction. In the related studies it has been suggested that the formation of microsegregation free solidification is governed by the absolute stability criterion or solute trapping [41-43]. In order to achieve considerable solid solubility extension the amount of solute, C, has to be less than a critical amount, C_{cr}. If $C < C_{cr}$, the formation of microsegregation free solidification will be governed by the absolute stability criterion and substantial extension of solid solubility can be realized. However, if $C > C_{cr}$, the formation of microsegregation free solidification will be governed by solute trapping and the material system will exhibit a relatively high resistance to solid solubility extension. In order to analyze the present system, C_{cr}, was computed considering the present alloy as a simple binary Al-Cu system, on the basis of the following equation [41-43]:

$$C_{cr} = \frac{k^2 \Gamma}{m_L (1-k) a_o} \qquad (3)$$

where, m_L is the liquidus slope, k is the partition coefficient, Γ is the Gibbs-Thompson coefficient, and a_o is the interatomic distance. The values of k (0.16) and m_L (3.4 K/wt. %) were taken from Ref. [44], while the value of $\Gamma = 1.08 \times 10^{-7}$ Km was taken from Ref. [45]. Substitution of these values in Equation 3 predicts a value of C_{cr} to be 2.39 wt. % Cu. Hence, for the alloy composition used in $C > C_{cr}$, the formation of microsegregation free solidification will be governed by solute trapping. Moreover, in view of results provided and discussed elsewhere [7, 46], which suggest that the solidification front velocity during spray processing is on the order of 1-2 mm/s., it is unlikely that complete suppression of copper precipitation as Al_2Cu occurred in this study. In either case a reduction/absence in volume fraction of Al_2Cu is consistent with an enhancement in heat transfer and an appreciable refinement of Al_2Cu resulting from the presence of ceramic particulates during deposition [26].

4.2 Solid State Cooling Effects

Once the mixture of solid, liquid, and mushy droplets impact the deposition substrate, the newly formed grains will continue to grow during solid state cooling. In order to gain an insight into the growth of grains in both the reinforced matix and unreinforced matrix, kinetic analysis of the data, given in Tables IV and V, was used to calculate the grain growth exponent in the as-spray deposited materials. The grain growth exponent, n, represents the slope of the line when the grain size (mm) is plotted as a function of time (min) on a bilogarithmic plot [47-49]. The empirical relationship correlating grain size with annealing time and grain growth exponent can be expressed as:

$$D = C (t)^n \qquad (4)$$

where D is the average grain diameter; t is the annealing time; and C and n are constants. The numerical values of C and n depend on both alloy composition and annealing temperature (Table VII). The values of n have been reported to range from 0.05 to 0.50 [50]. A comprehensive discussion of the significance of grain growth exponent, n, can be found elsewhere [50, 51]. The main assumptions involved in the development of Equation 4 are:

a) the grains have an equiaxed morphology,
b) there is no prior deformation, and
c) the grain growth is normal [47-49].

Regarding the grain morphology, the presence of equiaxed grains has been established in the preceding section. Finally, the linear relationship observed between grain growth and annealing time (Figures 8 and 9) provides an experimental basis for the assumption of normal growth.

Table VII. Results of grain growth exponents.

Experiment #	Temperature	Value of C	Value of n
2	673 K	0.036	0.044
	773 K	0.037	0.056
3	673 K	0.028	0.059
	773 K	0.029	0.060
5	673 K	0.028	0.050
	773 K	0.029	0.074

A few comments are in order regarding the values of grain growth exponent (n) obtained in this study. It is observed (Table VII) that the value of grain growth exponent (n), for a constant temperature, is lower for the Al-Cu material (0.044-0.056) relative to those of the Al-Cu/SiC (0.059-0.060) and Al-Cu/Al_2O_3 (0.050-0.074) composite materials. Furthermore, the results also reveal the value of n to increase with an increase in temperature for the unreinforced and reinforced materials, consistent with the results obtained by other investigators [50, 51]. This behavior is rationalized by considering the extrinsic and intrinsic effects associated with the co-injection of ceramic particulates.

The co-injection of the ceramic particulates during spray atomization and deposition increases the rate of heat transfer from the atomized spray and, thereby, promotes the retention of alloying elements in solid solution in the matrix [26]. In related studies, Gupta et al. [52] used X-ray diffractometry to show that a spray processed Al - 4.0 Ti material retained 0.88 wt. % Ti in solid solution, whereas, the corresponding as-spray processed Al - 2.3 wt. % Ti / SiC material retained up to 1.13 wt. % Ti in solution. An excess solid solubility of the alloying elements in the matrix will tend to reduce the volume fraction of precipitates in the matrix and, in effect, allow for easier grain boundary mobility. This rationale is supported by the results of X-ray diffraction studies which failed to reveal the presence of Al_2Cu-type precipitates in the as-spray processed Al-Cu/Al_2O_3 composite samples. Hence, on the basis of the results obtained, it appears that the presence of Al_2Cu-type precipitates in the matrix inhibits grain growth more effectively than the reinforcing ceramic particulates.

4.3 Aging Studies

The results of this study reveal that the as-quenched hardness of the metal matrix composites is higher than that of the monolithic counterpart. This is attributed in part to the high dislocation density present in the composite matrix due to CTE mismatch between the metal matrix and the ceramic reinforcement [53]. The results also show that the aging time for peak hardness was the same for the unreinforced Al-Cu, and reinforced Al-Cu/SiC and Al-Cu/Al_2O_3 materials. These results are consistent with the findings and observations of Chawla et al. [54] and Salvo et al. [40]. These researchers showed a negligible difference in aging kinetics between unreinforced and reinforced materials when they were aged at relatively low temperatures (~423 K). Moreover, in related studies conducted on aluminum alloy 6061, Rack and Krenzer [55] showed that in some instances the aging kinetics of unreinforced material may even be faster than the reinforced materials due to changes in the precipitation sequence, caused by the presence of a large number of matrix dislocations. This behavior has also been reported for spray processed 6061 composites during a time interval of $0.6 > t > 0.0$ hr [56]. In addition, the higher hardness observed for the SiC reinforced composites as compared to the Al_2O_3 reinforced composites in both the as-quenched and peak aged conditions, can be attributed to the higher hardness of SiC (Vickers hardness: 3000-3500 kg/mm^2) as compared to Al_2O_3 (Vickers hardness: 2600 kg/mm^2) [57]. These results are consistent with the work of Salvo et al. [40] who studied 6061 aluminum alloy reinforced with SiC and Al_2O_3 particulates, and processed using a compocasting technique.

4.4 Mechanical Behavior

The results of the mechanical behavior studies indicate that the co-injection of SiC or Al_2O_3 particulates in the Al-Cu matrix reduces the yield strength (YS) and ductility of the matrix material and does not significantly improves the ultimate tensile strength (UTS). The reduction in UTS values for the metal matrix composites is not unusual and has also been reported by several other investigators [27, 38, 58]. In related studies, Ibrahim et al. [38] showed that strength of a metal matrix composite increased appreciably over that of the monolithic counterpart only when the volume fraction of the reinforcing particulates was increased over 28 %. Friend [59] suggested that unless there is a critical volume fraction of reinforcing phase in the matrix, the load transfer between the matrix and the reinforcement will not be effective and a concomitant strength improvement may not be realized.

To provide an insight into the strengthening behavior of the MMCs used in this study, a simple numerical formulation was developed. The objective of this formulation was to calculate the critical volume fraction of reinforcing particulates required above which strengthening is to be expected in discontinuously reinforced metal matrices. The model formulation incorporates the following important assumptions:

(a) the reinforcing particulates are equiaxed (e.g., aspect ratio = 1),
(b) the reinforcing particulates acts as load carriers until the onset of debonding, and
(c) the reinforcing particulates are uniformly distributed in the matrix.

The strength of MMCs may be estimated on the basis of Rule of Mixture theory for uniaxial continuous fiber composites as [59]:

$$\sigma_c = \sigma_{uf} V_f + \sigma_m^* (1-V_f) \qquad (5)$$

where, σ_c is the composite strength, σ_{uf} is the strength of reinforcing phase and σ_m^* is the matrix stress at the reinforcement failure strain.

Equation 5 needs to be modified in order to account for the presence of surface flaws that are typically associated with the reinforcing ceramic phase. For example, the presence of the sharp edges on ceramic particulates with concomitant stress concentration effects, are known to promote early void nucleation [60]. Moreover, the occurrence of interfacial debonding has also been shown to play a critical role in the deformation behavior of metal matrix composites [61]. In view of these findings Equation 5 was modified to account for the fact that composite strength is intimately linked to strength of the interface. Hence, replacing σ_{uf} by σ_i:

$$\sigma_c = \sigma_i V_f + \sigma_m^* (1-V_f) \qquad (6)$$

where, σ_i is the interfacial bond strength between the soft and ductile matrix and the hard and brittle reinforcement.

Now in the limit, when the volume fraction of reinforcement (V_f) equals the critical volume fraction (V_{CRIT}), the strength of the composite will be equal to that of the unreinforced matrix material. Hence we can write Equation 6 as follows:

$$\sigma_{mu} = \sigma_i V_f + \sigma_m^* (1-V_f) \qquad (7)$$

where, σ_{mu} is the ultimate strength of the unreinforced matrix.

Rearranging Equation 7 we obtain:

$$V_{CRIT} = (\sigma_{mu} - \sigma_m^*)/(\sigma_i - \sigma_m^*) \qquad (8)$$

On the basis of work discussed elsewhere [11, 62, 63], an interfacial bond strength value of σ_i = 1690 MPa, was used to calculate V_{CRIT} for the SiC particulate reinforced metal matrix composites used in this study. A similar calculation was not attempted for the Al_2O_3 particulate reinforced metal matrix composites due to the unavailability of a σ_i value for this material. The results of this calculation, summarized in Table VIII suggest that V_{CRIT} = 8.2 % for the SiC particulate reinforced MMCs. The small difference between the calculated V_{CRIT} (8.2 %) and that of the actual V_f (11.1 %) is consistent with the marginal improvement in strength that was experimentally observed. The computed value of V_{CRIT}, however, should be considered as a lower bound estimate since, in practice, a completely uniform distribution of reinforcing particulates in the metal matrix is difficult to achieve, and hence, there are always clusters or agglomeration sites present in the matrix which promote early crack nucleation. This is consistent with the presence of cavities (5 μm - 15 μm) that were noted on the fracture surfaces of some of the MMC samples. These cavities are thought to originate as a result of the interfacial debonding originating at clusters of reinforcing ceramic particulates. The calculated value of V_{CRIT}, however provides an insight into the minimum volume fraction of reinforcement that is required to realize an improvement in strength in discontinuously reinforced MMCs. Moreover inspection of Equation 8 reveals some interesting trends. First, Equation 8 suggests that higher the bond strength σ_i, lower the volume fraction of ceramic reinforcement that is required for an improvement in strength. Secondly, an increase in strength of the matrix, σ_{mu}, is accompanied by an increase in V_{CRIT}. Regarding the first observation, a high value of σ_i is indicative of a more effective matrix-reinforcement load transfer. Hence, the MMC will necessitate the need for a smaller volume fraction of reinforcement relative to one with a poorly bonded reinforcement (e.g., low σ_i). The second observation is consistent with the results reported by McDanels for 20 vol. % SiC_w/Al [64]. In this study, he noted the strength improvement that was realized in a 6061 MMC to be higher than that noted for 2124 and 7075 MMCs, consistent with the lower matrix strength of the 6061 alloy relative to alloys 2124 and 7075.

Table VIII. Input parameters and the results of numerical model.

Variable	Value	Units
σ_i	1690.0	MPa
$^1\sigma_m^*$	361.4	MPa
σ_{mu}	470.6	MPa
V_{CRIT}	8.2	%

[1]assuming 0.67 % strain to failure.

4.5 Fracture Behavior

The extent of brittle fracture, as determined from fractographic studies, ranged (in ascending order): from unreinforced matrix to SiC reinforced MMC to Al_2O_3 reinforced MMC. These results are consistent with the mechanical properties (Table VI) which show a maximum ductility of 14.7 % for the unreinforced matrix material and a minimum ductility of 6.5 % for Al_2O_3 reinforced matrix. In addition, the relatively uniform and finer dimple size of the unreinforced material as compared to the reinforced counterparts (Expt. 4 and Expt. 5), are indicative of ductile failure [64]. The extent of particulate breakage and interfacial debonding noted on the fracture surface of samples taken from Expt. 4 is consistent with results obtained in earlier studies on SiC reinforced Al-7 Si [61]. The relatively low difference in atomic contrast between the Al_2O_3 particulates and the matrix made it difficult to analyze the fracture surface of samples taken from Expt. 5 with respect to interfacial debonding and particulate breakage.

5. CONCLUSIONS

1. The results of grain size measurements and grain growth studies conducted on the as-spray processed materials indicate that silicon carbide (SiC) particulates are more effective in refining grain size than the aluminum oxide (Al_2O_3) particulates.
2. The aging kinetics of the spray processed and hot extruded MMCs remain the same as those of the monolithic material.
3. The results of the present study also show that the presence of particulate reinforcement (SiC or Al_2O_3) in the aluminum alloy matrix (AA 2519) does not help in improving strength, and, in fact, reduces the ductility of the composite material.

4. Regarding strengthening behavior, preliminary results obtained, on the basis of a simple numerical formulation, suggest that a minimum of 8.2 volume percent of SiC particulates are required in order to realize a strength improvement for the Al-Cu matrix material used in this study.

ACKNOWLEDGMENTS

The authors wish to acknowledge the Army Research Office (grant # DAALO3-89-K-0027), and the National Science Foundation (Grant No. MSS 8957449) for their financial support. In addition, the authors would like to thank Mr. Irwin Sauer for his assistance with the experimental part of the study.

REFERENCES

1. P.S. Gilman: **Journal of Metals**, 43 (8), 1991, p. 7.

2. J.H. Devletian, Sandra M. Devincent, and Steven A. Gedeon: Report # MTL TR 88-47, Dec. 1988.

3. E.J. Lavernia, G. Rai, and N.J. Grant: **International Journal of Powder Metallurgy**, 22, 1986, p. 9.

4. J. Megusar, E.J. Lavernia, P. Domalavage, O.K. Harling, and N.J. Grant: **Journal of Nuclear Materials**, 122/123, 1984, p. 789.

5. R.W. Evans, A.G. Leatham, and R.G. Brooks: **Powder Metallurgy**, 28, 1985, p. 13.

6. M. Gupta, F.A. Mohamed, and E. Lavernia: **International Journal of Rapid Solidification**, 6, 1991, p. 247.

7. E.J. Lavernia: **International Journal of Rapid Solidification**, 5, 1989, p. 47.

8. T.C. Willis: **Metals and Materials**, 4, 1988, p. 485.

9. C.L. Buhrmaster, D.E. Clark, and H.O. Smart: **Journal of Metals**, 40, 1988, p. 44.

10. A.R.E. Singer: **Annals of the CIRP**, 32, 1983, p. 145.

11. I.A. Ibrahim, F.A. Mohamed, and E.J. Lavernia: **Journal of Materials Science**, 26, 1991, p. 1137.

12. J. White, I.G. Palmer, I.R. Hughes, and S.A. Court: in Aluminum-Lithium Alloys V, vol. 3, T.H. Sanders, Jr. and E.A. Starke, Jr., editors, March 27-31, 1989, Williamsburg, Virginia, p. 1635.

13. M. Gupta, F.A. Mohamed, and E.J. Lavernia: **Metallurgical Transactions**, 23A, 1992, p. 845.

14. K.A. Kojima, R.E. Lewis, and M.J. Kaufman: in Aluminum-Lithium Alloys V, Vol. 1, T. H. Sanders, Jr. and E. A. Starke, Jr., editors, March 27-31, 1989, Williamsburg, Virginia, p. 85.

15. A.P. Divecha, S.G. Fishman, and S.D. Kumar: **Journal of Metals**, 3 (9), 1981, p. 12.

16. S. Ochiai, and K. Osamura: **Metallurgical Transactions**, 18A, 1987, p. 673.

17. D.L. Erich: **International Journal of Powder Metallurgy**, 23 (1), 1987, p. 45.

18. J. Papazian: **Metallurgical Transactions**, 19A, 1988, p. 2845.

19. M. Gupta, F.A. Mohamed, and E.J. Lavernia: **Materials Science and Engineering**, 144A, 1991, p. 99.

20. V.C. Nardone, and K.W. Prewo: **Scripta Metallurgica**, 20, 1986, p. 43.

21. R.E. Sanders Jr., and I. Jocelyn: U.S Patent number 4610733, Sept. 9, 1986.

22. T. Chanda, W.E. Frazier, F.A. Mohamed, and E.J. Lavernia: in Metal and Ceramic Matrix Composites: Processing, Modelling & Mechanical Behavior, R. B. Bhagat, A. H. Clauer, P. Kumar, A. M. Ritter, eds., The Metalurgical Society, Warrendale, PA, 1990, p. 47.

23. T. Chanda, W.E. Frazier, F.A. Mohamed, and E.J. Lavernia: in Low Density, High Temperature Powder Metallurgy Alloys, W. E. Frazier, M. J. Koczak, P. W. Lee, eds., The Metallurgical Society, Warrendale, PA, 1990, p. 83.

24. M. Gupta, F.A. Mohamed, and E.J. Lavernia: **Materials and Manufacturing Processes**, 5, 1990, p. 165.

25. M. Gupta, I.A. Ibrahim, F.A. Mohamed, and E.J. Lavernia: **Journal of Materials Science**, 26 (24), 1991, p. 6673.

26. M. Gupta, F.A. Mohamed, and E.J. Lavernia: **Metallurgical Transactions**, 23A, 1992, p. 831.

26. M. Gupta, F.A. Mohamed, and E. Lavernia: **International Journal of Rapid Solidification**, 6, 1991, p. 247.

27. J. White, I.G. Palmer, I.R. Hughes, and S.A. Court: Aluminum-Lithium Alloys V, vol. 3, T. H. Sanders, Jr. and E. A. Starke, Jr., eds., March 27-31, 1989, Williamsburg, VA, 1984, p. 1635.

28. K. A. Kojima, R. E. Lewis, and M. J. Kaufman: Aluminum-Lithium Alloys V, vol. 1, T. H. Sanders, Jr. and E. A. Starke, Jr., eds., March 27-31, 1989, Williamsburg, VA, 1984, p. 85.

29. R.H. Bricknell: **Metallurgical Transactions**, 17A, 1986, p. 583.

30. X. Liang, and E.J. Lavernia: **Scripta Metallurgica**, 25, 1991, p. 1199.

31. E. Lavernia, R.H. Rangel, and T.S. Srivatsan: **International Journal of Atomization and Sprays**, 1992, in press.

32. K. Ogata, E.J. Lavernia, G. Rai, and N.J. Grant: **International Journal of Rapid Solidification**, 2, 1986, p. 21.

33. P. Bewlay, and B. Cantor: Rapidly Solidified Materials, P. Lee, R. Carbonara, eds., American Society for Metals, Metals Park, OH, 1986, p. 15.

34. E.J. Lavernia, T. Ando, and N.J. Grant: Rapidly Solidified Materials, P. Lee, R. Carbonara, eds., American Society of Metals, Metals Park, Ohio, 1986, p. 29.

35. V.G. McDonnell, E.J. Lavernia, and G.S. Samuelson: Synthesis and Analysis in Materials Processing: Advances in Characterization and Diagnosis of Ceramic and Metal Particulate Processing, E.J. Lavernia, H. Henein and I. Anderson, eds., The Metallurgical Society, Warrendale, PA, 1989, p. 13.

36. E.J. Lavernia: SAMPE Quarterly, 22 (2), 1991, p. 2.

37. Y. Wu: Ph.D dissertation, University of California, Irvine, 1991.

38. I.A. Ibrahim, F.A. Mohamed, and E.J. Lavernia: Advanced Aluminum and Magnesium Alloys, T. Khan, G. Effenberg, eds., ASM International, Amsterdam, 1989, p. 745.

39. M. Hansen, and K. Anderko: Constitution of Binary Alloys, McGraw-Hill Book Company 1958, p. 84.

40. L. Salvo, M. Suery, and F. Decomps: Fabrication of Particulates Reinforced Composites, J. Masounave and F. G. Hamel, eds., ASM international, Materials Park, Ohio 44703, p. 139.

41. W.J. Boettinger, D. Shechtman, R.J. Schaefer, and F.S. Biancaniello: **Metallurgical Transactions**, 15A, 1984, p. 55.

42. H. Jones: **Philosophical Magazine**, 61B, 1990, p. 487.

43. A.F. Norman, and P. Tsakiropoulos: **International Journal of Rapid Solidification**, 6, 1991, p. 185.

44. J.L. Murray: **International Metals Review**, 30, 1985, p. 211.

45. J.A. Juarez Islas, H. Jones, and W. Kurz: **Materials Science and Engineering**, 98, 1988, p. 201.

46. M. Ruhr, E.J. Lavernia, and J.C. Baram: **Metallurgical Transactions, 21A**, (1990), p. 1785.

47. P.A. Beck, J.C. Kremer, L.J. Demer, and M.L. Holzworth: **Transactions of Metallurgical Society of AIME**, 175, 1948, p. 372.

48. P.A. Beck: **Journal of Applied Physics**, 19, 1948, p. 507.

49. P.A. Beck, J. Towers, and W.O. Manley: **Transactions of Metallurgical Society of AIME**, 175, 1951, p. 634.

50. R.L. Fullman: Metal Interfaces, American Society for Metals, 1952, p. 179.

51. P. Cotterill, and P.R. Mould: Recrystallization and Grain Growth in Metals, Surrey University Press, 1976, p. 275.

52. M. Gupta, F.A. Mohamed, and E.J. Lavernia: **Metallurgical Transactions B**, in press, 1992.

53. R.J. Arsenault, and N. Shi: **Materials Science and Engineering**, 1981, p. 175.

54. K.K. Chawla, A.H. Esmaeili, A.K. Datye, and A.K. Vasudevan: presented at TMS/ASM Fall meeting at Cincinnati, OH, Oct. 21-24, 1991.

55. H.J. Rack, and R.W. Krenzer: **Metallurgical Transactions**, 8A, 1977, p. 335.

56. Y. Wu, and E.J. Lavernia: **Journal of Metals**, 43 (8), 1991, p. 16.

57. The CRC Materials Science and Engineering Hand Book, J. Shakelford, and W. Alexander, eds., CRC press, Boca Raton, Ann Arbor, London, 1992.

58. A.R.E. Singer, and S. Ozbek: **Powder Metallurgy**, 28, 1985, p. 72.

59. C.M. Friend: **Journal of Materials Science**, 22, 1987, p. 3005.

60. S. Dionne, and M.R. Krishnadev: Fabrication of Particulates Reinforced Composites, J. Masounave and F. G. Hamel, eds., ASM international, Materials Park, Ohio 44703, p. 261.

61. M. Gupta, C. Lane, and E.J. Lavernia: **Scripta Metallurgica et Materialia**, 26, 1992, p. 825.

62. Y. Flom, and R.J. Arsenault: **Materials Science and Engineering**, 77, 1986, p. 191.

63. A.S. Argon, and J. Im: **Metallurgical Transactions**, 6A, 1975, p. 839.

64. D.L. McDanels: **Metallurgical Transactions**, 16A, 1985, p. 1105.

A COMPARISON OF CREEP BEHAVIOR OF 6061 Al AND SiC-6061 Al

Kyung-Tae Park, Enrique J. Lavernia,
and Farghalli A. Mohamed
Department of Mechanical and Aerospace Engineering
Materials Section
University of California
Irvine, Carlifornia

ABSTRACT

The stress dependence of the creep rate of 6061 Al, produced by powder metallurgy, has been studied at 648 K. The experimental data, which extend over six orders of magnitude of strain rate, show that the stress exponent, n, is high and increases with decreasing the applied stress. This finding is examined in the light of data reported for SiC-6061 Al, Al alloys, and dispersion-strengthened alloys.

1. INTRODUCTION

Metal-matrix composites consisting of silicon carbide, whiskers or particulates (SiC_w or SiC_p), in an aluminum alloy have received considerable attention due to several attractive characteristics that include: (a) low raw materials cost, (b) low fabrication costs combined with the ability to form the composites into useful shapes using conventional metal working processes, and (c) isotropic properties with substantially improved strength and modulus compared to unreinforced Al alloys.

In recent years, the high-temperature creep behavior of discontinuous silicon carbide reinforced aluminum alloys (SiC_w-Al or SiC_p-Al) has been the subject of creep investigations that aimed at assessing the potential of these composites for use as materials for high temperature applications. As a result of these investigations, several sets of experimental data [1-7] are now available. The data have revealed the following findings: (a) the stress dependence of the steady-state creep rate is high and variable, and (b) the temperature dependence of the steady-state creep rate is much larger than that for self-diffusion in aluminum. These findings are significant since they indicate that the creep behavior of discontinuous silicon carbide reinforced aluminum alloys is similar to that of dispersion-strengthened (DS) alloys [8, 9].

Despite the significance of the above findings, additional work is still needed in order to resolve the primary issue concerning the origin of high-temperature strengthening in discontinuous SiC-Al composites. For example, at the present there is no information that can be used to provide a comparison between the creep behavior of these composites and that of Al matrices over a wide range of strain rate. Such information is essential to establish whether the addition of SiC whiskers or particulates to Al matrices leads to high temperature strengthening. Accordingly, a detailed investigation has been undertaken: (a) to study the creep behavior of several Al alloys that have been used as matrices in the development of discontinuous SiC-Al composites, and (b) to examine the creep behavior of these Al alloys in the light of recent data and analyses documented for both discontinuous SiC-Al composites, solid-solution alloys, and DS alloys. It is the purpose of this paper to report and discuss some preliminary data obtained on 6061 aluminum in this investigation.

2. EXPERIMENTAL

The materials used in the present investigation was 6061 aluminum (unreinforced alloy). The alloy, which was supplied by Army Materials Technology Laboratory (AMTL), was prepared by powder metallurgy techniques and was received in the extruded condition.

Double-shear specimens [10] of shape and dimensions described elsewhere [11] were machined from PM 6061 Al. Prior to testing, the specimens were solutionized at 723 K in argon for 4 hr, water quenched, and left at room temperature for a minimum of one week. The creep tests were conducted on a suitably designed creep-testing machine operating at constant load; due to the shear configuration of the specimens, constant load implies constant stress. The tests were conducted in air at 648 K using a three-zone furnace in which the temperature was monitored and maintained constant to within ± 2 K. Steady-state creep rates were measured in the range of 10^{-8} to 10^{-1} s^{-1}. The strain during creep was measured with a linear variable differential transformer (LVDT) and amplifier, and monitored directly on a strip chart recorder.

3. EXPERIMENTAL RESULTS AND DISCUSSION

3.1 Creep Curves

Fig. 1, where the shear strain, γ, is plotted against time, t, provides an example of creep curves obtained for PM 6061 Al tested at 648 K. Examination of this creep curve, along with others, shows the following features: (a) the material exhibits the usual type of creep curve; a decelerating primary creep stage where $d\dot\gamma/dt < 0$, a secondary stage where $d\dot\gamma/dt = 0$, and a tertiary stage where $d\dot\gamma/dt > 0$; and (b) the steady-state stage is of short duration ($\gamma < 10\%$).

The above characteristic regarding the duration of the steady-state stage in PM 6061 Al contrasts with the behavior of Al based solid-solution alloys [11, 12, 13], such as Al-Mg, Al-Cu, and Al-Zn that, under creep conditions, exhibit an extensive, well-defined steady-state stage with steady-state shear strains of the order of 20-80%. On the other hand, the creep curves of PM 6061 Al are similar in trend to those reported recently for PM 30 vol.% SiC_p-6061 Al [6]. This similarity may be illustrated, for example, by comparing Fig. 1 to Fig. 2 which depicts the creep curve of PM 30 vol.% SiC_p-6061 Al tested in double shear at 648 K.

It is worth mentioning that the total creep strains shown by the creep curves of PM 6061 Al and PM SiC_p-6061 Al are much larger than those normally obtained in tension; this may be

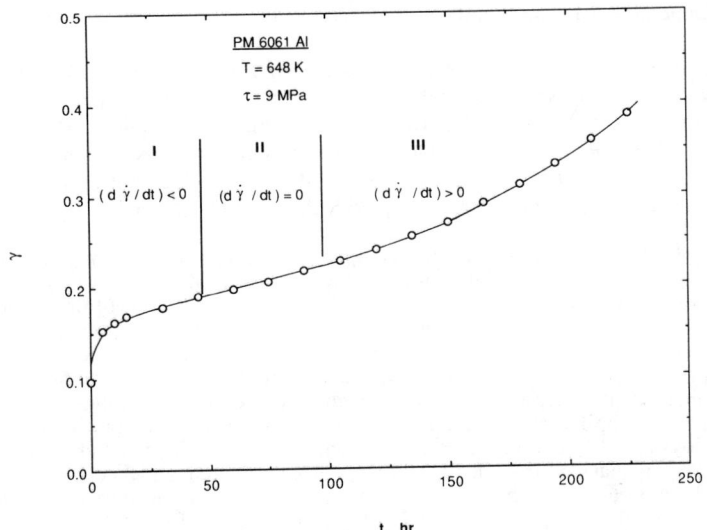

Fig. 1. Example of creep curve for PM 6061 Al.

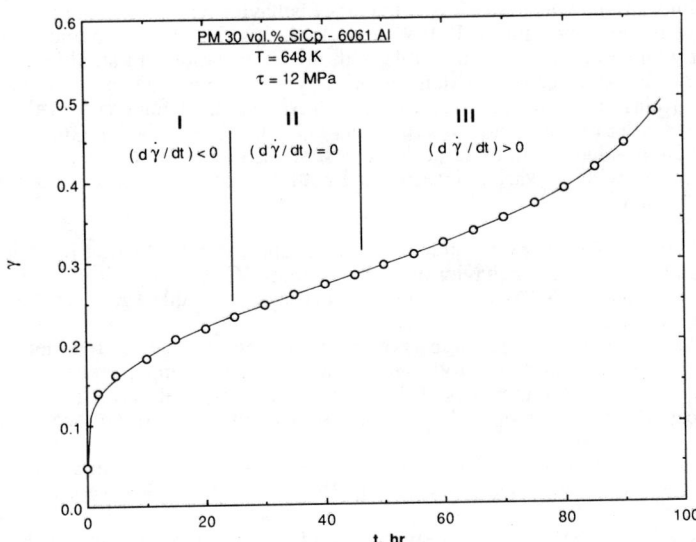

Fig. 2. Example of creep curve for 30 vol.% SiC$_p$-6061A [6].

attributed, in part, to the double shear configuration of the specimens tested in the present investigation and the investigation on the composite [6].

3.2 Stress Dependence of the Steady-State Creep Rate

The dependence of strain rate, $\dot{\gamma}$, on shear stress, τ, under steady-state conditions, was investigated by conducting a series of double-shear tests at a constant temperature on the creep machine and by plotting $\dot{\gamma}$ vs. τ on a logarithmic scale. Fig. 3 depicts this form of plot for PM 6061 Al at T = 648 K. Examination of the data of this figure reveals that the apparent stress exponent, n_a, $n_a = (\partial \ln \dot{\gamma}/\partial \ln \tau)_T$, for creep in PM 6061 Al does not remain constant, but increases with decreasing the applied stress from 25 MPa to 9.5 MPa.

3.2.1 Solid-Solution Alloys.
Earlier creep experiments [14, 15] showed the presence of two classes of creep behavior in solid-solution alloys: class I (alloy class) and class II (metal class). Class I alloys exhibit the characteristics of viscous glide control [16] that include a stress exponent of about 3 and brief primary creep. By contrast, class II alloys (metal class), like pure metals, exhibit the characteristics of climb control [17] that include a stress exponent close to 5 and extensive primary creep. In addition to this classification, it has been reported that under a favorable combination of materials parameters (such as the atom misfit ratio) and experimental variables (such as the applied stress), the creep behavior of some solid-solution alloys [15, 18], especially Al alloys [11, 13], exhibits a transition from that of class II (metal class) at low stresses to that of class I (alloy class) at intermediate stresses. Experimental evidence for such a transition was, for example, reported through a change in the stress exponent for creep in several Al alloys including Al-Mg [11, 13] and, Al-Cu alloys [13], from a value of 4.5 (characteristic of climb control in Al) to a value of about 3 (characteristic of viscous glide control) with increasing the applied stress.

The material used in the present investigation, PM 6061 Al, is an aluminum alloy in which several solute elements are present. These include Mg (1%), Cu (0.35%), Si (0.6%), and other minor impurities. At the testing temperature of 648 K, these solute elements should be in solid solution with Al. It is therefore expected that, depending on experimental conditions and materials parameter, PM 6061 Al either behaves as one of the two classes of solid-solution alloys (class I or class II) or exhibits a transition from class II to class I behavior with increasing the applied stress. Consideration of the present data on PM 6061 Al indicates that the value and variation of the stress exponent with stress are not compatible with this expectation. First, although the data on PM 6061 Al show the presence of an extensive primary stage, in agreement with one of the creep characteristics of class II alloys [14, 15, 19], the stress exponent for creep in the alloy, when measured over six order of magnitude of strain rate, is higher than 5. Second, the variation of the stress exponent from value of about 22 at the lowest stress to a value of about 10 at the highest stress is not in accord with the variation associated with the occurrence of a creep transition from class II to class I behavior.

3.2.2 6061 Al.
Creep results on 6061 Al were reported in two earlier investigations [2, 5] that focused mainly on the creep behavior of discontinuous SiC-Al composites. A summary of the results of investigation on 6061 Al are given in Table 1; also the Table 1 includes the results of an investigation on 2124 Al [4].

Nieh [2] conducted tensile creep tests on 6061 Al at 561 K. He has reported that the creep behavior of the material exhibits the following characteristics: primary creep is absent and the stress exponent is about 3. On the basis of these characteristics, Nieh has suggested that the rate controlling process during the creep of 6061 Al is viscous glide [14, 15], i.e., the alloy behaves as a class I alloy.

Morimoto et al. [5] studied the tensile creep of 6061 Al at 573 K over a stress range which was almost identical with that used by Nieh [2]. Replotting the data of Morimoto et al. [5] as creep rate vs. stress on a logarithmic scale yields a stress exponent of about 5. As mentioned previously, this value of the exponent is suggestive of climb-controlled creep behavior [14, 15, 19].

Consideration of the preceding discussion along with the data of Table 1 shows that the stress dependence of creep rate in 6061 Al as measured in the present investigation is not in agreement with that inferred from the data of Nieh [2] and Morimoto et al. [5]; in the present investigation, n, is not only higher than 5 but also variable. While the origin of the discrepancy in the value of the stress exponent between the present investigation and earlier investigations is not known, it is worth mentioning that in the present investigation, PM 6061 Al was creep tested at 648 K over six orders of magnitude of strain rate while in earlier investigations [2, 5] the creep

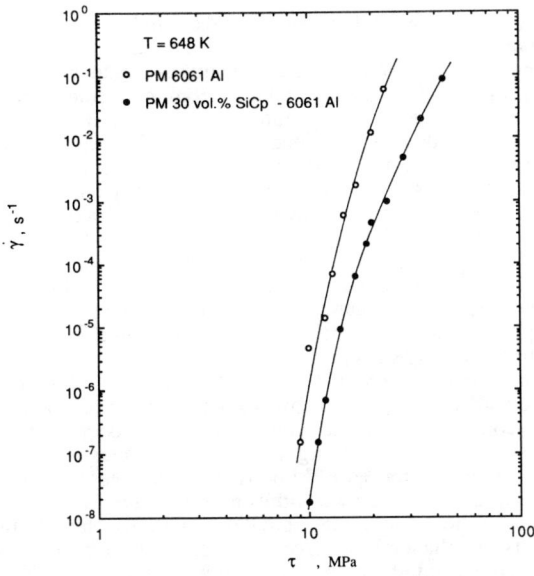

Fig. 3. Steady-state creep rate as a function of the applied stress (logarithmic scale) for PM 6061 Al and 30 vol.% SiC$_p$-6061 Al [6]; the experimental data on the alloy extend over six orders of magnitude of strain rate.

Table 1. Summary of Creep Investigations on Al Matrices

Investigators	Alloy	T, K	$\dot{\gamma}^*$, s^{-1}	n
Nieh (1984)	6061 Al	561	10^{-7} - 10^{-5}	3
Morimoto et al. (1988)	6061 Al	573	10^{-9} - 10^{-8}	5†
Nieh et al. (1988)	2124 Al	673	2×10^{-5} - 10^{-4}	5

*$\dot{\gamma} = 3/2\dot{\varepsilon}$

†Inferred from replotting the creep data.

behavior of the alloy was studied at 567 K or 573 K over two orders of magnitude of strain rate or less.

3.2.3 $\underline{SiC_p\text{-}6061\ Al}$. Several sets of creep data on SiC-6061 Al composites are available [2, 4, 5], and a summary of these results is given in Table 2. As indicated by Table 2 and Fig. 4, one important common characteristic of the creep studies conducted by Nieh [2], Nieh et al. [4], and Morimoto et al. [5] is that the experimental data from these studies describe the creep behavior of the composites over three orders of magnitude of strain rate. Because of this characteristic, it is difficult to establish whether the stress exponent is genuinely constant, as suggested by the above studies, or increases with decreasing the applied stress, as reported for dispersion-strengthened (DS) alloys. Very recently, this issue was addressed in a creep study by Park et al. [6] who creep tested a 30 vol.% SiC_p-6061 Al composite over seven orders of magnitude of strain rate (see Table 2). The steady-state creep data of Park et al. [6] at 648 K are plotted in Fig. 3 as shear strain rate, $\dot{\gamma}$, vs. shear stress τ on a logarithmic scale. As reported by Park et al. [6] the creep data on the composite can be divided into two regions: high-stress region and low-stress region. In the high-stress region, the stress exponent, n, increases very slowly with decreasing stress. In the low-stress region, the stress exponent not only is higher than that measured in the high-stress region but also increases rapidly with decreasing stress.

A comparison between the creep data of PM 6061 Al and those of 30% vol. SiC_p-6061 Al as plotted in Fig. 3 reveals two important observations. First, the stress exponent for creep in the alloy, like that in the composite, increases continuously with decreasing stress. Second, over the whole range of stresses used in the investigation, the creep rates of SiC_p-6061 Al are one to two orders of magnitude slower than those of PM 6061 Al, and, in general, the lower the creep rate, the greater the difference in creep rate between the composite and the alloy. The above observation demonstrates that, under the present experimental conditions, the addition of 30% vol. SiC_p to 6061 Al matrix results in enhancing the creep resistance of the matrix alloy, especially at lower stresses. This finding regarding the high temperature strengthening effect of SiC_p contrasts with the results reported [4] for 20 vol.% SiC_p-2124 Al and 2124 Al (see Table 1) which have indicated that there is a cross over in the logarithmic plot of $\dot{\gamma}$ vs. τ for the two materials and that at high strain rates ($\dot{\gamma} > 10^{-4} s^{-1}$ and T = 673 K), the composite is less creep resistant than the alloy. On the other hand, the increase in the creep resistance of SiC-6061 Al relative to that of 6061 Al is qualitatively consistent with the results of recent analytical treatments [5, 20, 21]. For example, the numerical results of Dragone and Nix [20], who studied the creep behavior of discontinuous metal matrix composites (MMCs) by a continuum mechanics treatment utilizing finite element techniques, have revealed the following finding: that large triaxial stresses develop in the matrix near the center of the reinforcement phase and that the presence of these stresses leads to a lower creep rate in the composite.

3.3 Explanations for the Variation in the Stress Exponent

As indicated by the present data, the stress exponent for creep in PM 6061 Al, when measured over a wide range of strain rate, is not constant but increases with decreasing the applied stress. There are two possible suggestions [22, 23] that may, in general, explain a continuous increases in the stress exponent with decreasing the applied stress. First, this type of variation in the stress exponent may be caused by the operation of two sequential deformation processes A and B having stress exponents of n_A and n_B, respectively, where $n_B > n_A$. In this case, as demonstrated elsewhere [22, 23], the slower process controls the creep behavior so that: (a) n approaches the limiting values of n_A and n_B at high and low strain rates, respectively, (b) a continuous variation in the stress exponent with stress occurs over an intermediate range of strain rates. However, an explanation for the variation of the stress exponent for creep in PM 6061 Al in terms of sequential processes cannot be adopted here, basically because the creep behavior of alloy is characterized by a stress exponent whose value (n = 20) at the lowest stress is too high to be accounted for by present creep theories or semiempirical approaches.

Second, there exists a threshold stress for creep, τ_0, in the material and as a result, the observed deformation is driven by an effective stress, τ_e, not by the applied stress, τ; $\tau_e = \tau - \tau_0$. In this case, the creep behavior is controlled by a rate equation of the form:

$$\dot{\gamma}_s = A \left(\frac{\tau - \tau_0}{G} \right)^n \exp\left(-\frac{Q_c}{RT}\right) \qquad (1)$$

where $\dot{\gamma}_s$ is the steady state creep rate, A is a constant that is sensitive to microstructure, G is the shear modulus, Q_c is the activation energy for creep, and R is the gas constant. The concept of the presence of a threshold stress during creep was proposed previously to explain the variation in the

Table 2.   Summary of Creep Investigations on Discontinuous SiC-Al Composites

Investigators	Reinf.	V_f,%	matrix	T, K	$\dot{\gamma}^{**}, s^{-1}$	n	Q_a, kJ/mol
Webster (1982)	SiC$_w$	25*	Al-3%Li	505-866	10^{-2}-10^0	--	---
Nieh (1984)	SiC$_w$	20*	6061Al	505-644	10^{-9}-10^{-6}	21	390
	SiC$_p$	30*	6061Al	561	10^{-9}-10^{-7}	--	---
Nardone & Strife (1987)	SiC$_w$	20	2124Al	450	10^{-9}-10^{-7}	9	431***
				561	10^{-9}-10^{-7}	21	227****
Nieh et al. (1988)	SiC$_p$	20	2124Al	623-723	10^{-7}-10^{-3}	9	400
Morimoto et al. (1988)	SiC$_w$	15	6061Al	573	10^{-9}-10^{-6}	--	---
Park et al. (1990)	SiC$_p$	30	6061Al	618-678	3×10^{-9}-9×10^{-2}	7-25	270-500

*   wt.%
** $\dot{\gamma} = 3\dot{\varepsilon}/2$
***   at 90 MPa
****   at 310 MPa

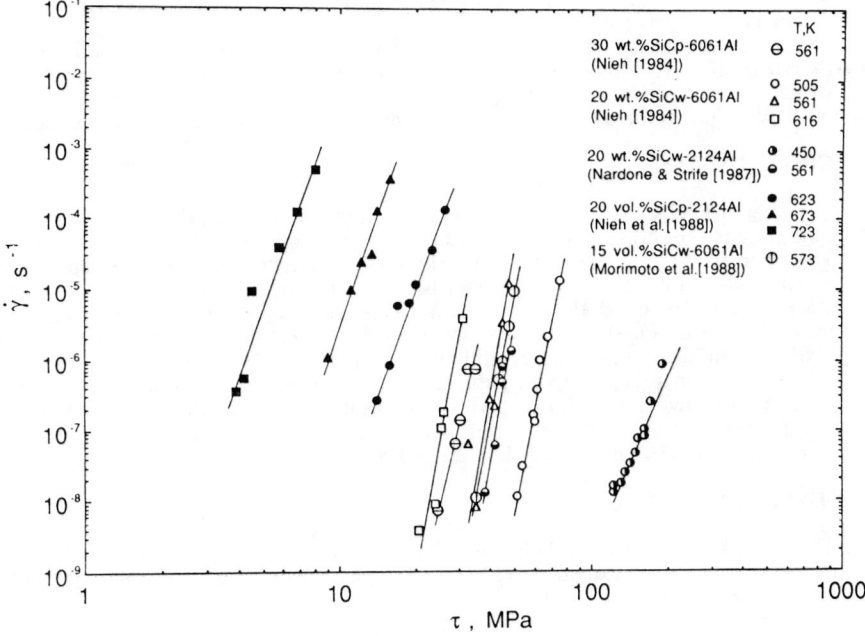

Fig. 4.   Data of creep investigations on discontinuous SiC-Al alloy composites plotted as shear strain rate, $\dot{\gamma}$, vs. shear stress, τ (logarithmic scale). The tensile data from refs. [2], [3] and [5] were transformed to shear data by using $\tau = \sigma/2$ and $\dot{\gamma} = 3\dot{\varepsilon}/2$.

stress exponent with the applied stress in two types of materials: superplastic alloys which have a fine-grained structure [24] and DS alloys which contain dispersoid particles [9]. For superplastic alloys, the origin of the threshold stress was attributed [25, 26] to strong impurity segregation at boundaries and their interaction with boundary dislocations. For DS alloys, three theoretical deformation models [27-29] that explain the origin and give the magnitude of threshold stresses were proposed. In these models, the threshold stress is equal to: (a) the stress required to cause dislocation bowing between particles [27] (the Orowan stress), τ_o, (b) the extra back stress, τ_b, required to create the additional dislocation line length as the dislocation segment climbs over a particle (local climb) [28], and (c) the stress required to detach the dislocation from the particle after climb is completed [29]. Recent considerations of the above three threshold stress models have shown that the detachment model is supported by both theoretical and experimental studies [30-33].

3.4 Origin of the Threshold Stress in PM 6061 Al

As mentioned in Section 2, 6061 Al used in the present creep investigation was prepared by powder metallurgy techniques. According to the general information for PM processing, aluminum powders, following atomization, are sieved to gather the powder with desirable size. The hot press followed by degassing in vacuum after sieving makes the powder bulk with approximately 95% of the theoretical density. Finally, the powder bulk is hot extruded to the desired geometry. Experimental evidence documented elsewhere [34, 35] indicates that during the atomization process, aluminum powder particles exhibit a stable surface oxide layer due to reactivity of aluminum with oxygen this oxide layer is an amorphous $Al_2O_3/Al_2O_3\text{-}3H_2O$ film that contains small amounts of MgO crystallites and is surrounded by physically absorbed H_2O/O_2 [34]. As a result of degassing, additional Al_2O_3 and crystalline MgO are formed. Hot pressing leads to densification of the degassed powders by eliminating porosity and by breaking surface oxide layer into Al_2O_3 fragments and MgO crystallites. Hot extrusion, following hot pressing, results in further breaking up of Al_2O_3 fragments and Al_2O_3/MgO clusters and in dispersing oxide particles over a wide area; in heavily extruded material, the oxide particles are, in general, expected to be uniformly distributed.

It is suggested that the fine incoherent oxide particles introduced during atomization in PM 6061 Al serve as effective barriers to dislocation motion and give rise to a threshold stress in PM 6061 Al, and that this threshold stress is the cause of the anomalous stress dependence of creep rate in the alloy. Work is currently in progress to examine in detail the validity of this suggestion.

4. CONCLUSIONS

1. The creep behavior of PM 6061 Al at 648 K exhibits two characteristics: (a) the steady-state stage is very brief, and (b) the stress exponent is high.
2. The creep characteristics of PM 6061 Al are not in agreement with those reported for Al based solid-solution alloys which either behave as one of the two classes of solid-solution alloys (class I or class II) or exhibit a creep transition from class II to class I with increasing stress. On the other hand, the anomalous stress dependence of creep rate in PM 6061 Al is similar to that noted recently in SiC_p-6061 Al and in DS alloys.
3. It is suggested that the fine oxide particles present in PM 6061 Al, as a result of processing the alloy by powder metallurgy, give rise to a threshold stress for creep, and that such a threshold stress leads to the anomalous stress dependence of creep rate in the alloy. Work is in progress to examine the validity of this suggestion.

ACKNOWLEGEMENTS

The authors wish to acknowledge the Army Research Office (grant No. DAAL03-92-G) for their support. Thanks are extended to Mr. Charles Lane for providing the materials and to Lisa Rehbaum for typing the manuscript.

REFERENCES

1. D. Webster: Metallurgical Transactions, A13, 1982, p.1511.
2. T.G. Nieh: Metallurgical Transactions, A15, 1984, p. 139.
3. V.C. Nardone and J.R. Strife: Metallurgical Transactions, A18, 1987, p.109.
4. T.G. Nieh, K. Xia and T.G. Langdon: Journal of Engineering Materials and Technology, 110, 1988, p. 77.
5. T. Morimoto, T. Yamaoko, H. Lilholt and M. Taya: Journal of Engineering Materials and Technology, 110, 1988, p. 70.
6. K.T. Park, E.J. Lavernia and F.A. Mohamed: Acta Metallurgica, 38, 1990, p. 2149.
7. F.A. Mohamed, K.T. Park and E.J. Lavernia: Materials Science and Engineering, A150, 1992, p. 21.
8. B.A. Wilcox and A.H. Clauer: Transactions AIME, 236, 1966, p. 570.
9. R.W. Lund and W.D. Nix: Acta Metallurgica, 27, 1976, p. 469.
10. B.Y. Chivouze, D.M. Schwartz and J.E. Dorn: Transaction AIME, 60, 1967, p. 51.
11. K.L. Murty, F.A. Mohamed and J.E. Dorn: Acta Metallurgica, 20, 1972, p. 1009.
12. A. Goel, T.J. Ginter and F.A. Mohamed: Metallurgical Transactions, A14, 1983, p. 2308.
13. P.K. Chaudhury and F.A. Mohamed: Metallurgical Transactions, A18, 1987, p. 2105.
14. P.M. Burke and O.D. Sherby: Progress in Materials Science, 13, 1968, p. 325.
15. F.A. Mohamed and T.G. Langdon: Acta Metallurgica, 22, 1974, p. 779.
16. J. Weertman: Journal of Applied Physics, 28, 1957, p. 1185.
17. J. Weertman: Journal of Applied Physics, 28, 1957, p. 362.
18. F.A. Mohamed: Materials Science and Engineering, 61, 1983, p. 149.
19. J.E. Bird, A.K. Mukherjee and J.E. Dorn: Quantitative Relation Between Properties and Microstructures, D.G. Brandon and A.Rosen (eds.), Israel University Press, Jersusalem, 1969, p. 255.
20. T.L. Dragone and W.D. Nix: Acta Metallurgica, 38, 1990, p. 1941.
21. S. Goto and M. Mclean: Acta Metallurgica, 165, 1991, p. 39.
22. F.A. Mohamed and T.J. Ginter: Journal of Materials Science, 16, 1981, p. 2890.
23. T.G. Langdon and F.A. Mohamed: Journal of Australian Institute of Metals, 22, 1977, p. 189.
24. R.H. Johnson: Metals Review, 15, 1979, p. 115.
25. F.A. Mohamed: Journal of Materials Science, 18, 1983, p. 582.
26. P.K. Chaudhury and F.A. Mohamed: Acta Metallurgica, 3, 6, 1988, p. 1099.
27. E. Orowan: Dislocations in Metals, M. Cohen (ed), AIME, New York, 1954, p. 531.
28. E. Arzt and M.F. Ashby: Scripta Metallurgica, 16, 1982, p. 1285.

29. E. Arzt and D.S. Wilkinson: Acta Metallurgica, 34, 1986, p. 1893.

30. D.J. Srolovitz, R.A. Petkovic-Luton and M.J. Luton: Acta Metallurgica, 31, 1983, p. 2151.

31. V.C. Nardone and J.K. Tien: Scripta Metallurgica, 17, 1983, p. 467.

32. J.H. Schroder and E. Arzt: Scripta Metallurgica, 19, 1985, p. 1129.

33. R.S. Herrik, J.R. Weertman, R. Petkovic-Luton and M.J. Luton: Scripta Metallurgica, 22, 1983, p. 1879.

34. Y-W. Kim, W.M. Griffith and F.H. Froes: Journal of Metals, 37, 1985, p. 27.

35. T.S. Srivatsan, E.J. Lavernia and F.A. Mohamed: International Journal of Powder Metallurgy, 16, 1990, p. 321.

EVALUATION OF ANALYTICAL AND NUMERICAL MODELS FOR THE ELASTIC-PLASTIC RESPONSE OF PARTICULATE COMPOSITES

David G. Taggart and Jialiang Qin
Department of Mechanical Engineering and Applied Mechanics
University of Rhode Island
Kingston, Rhode Island

Mark D. Adley
U.S. Army Corps of Engineers
Waterways Experiment Station
Vicksburg, Mississippi

ABSTRACT

The deformation and failure mechanisms of particle and whisker reinforced metal-matrix composites include inclusion/matrix debonding, ductile failure of the matrix and inclusion cracking. These mechanisms are influenced by processing induced microstructural features such as the inclusion aspect ratio, inclusion orientation distribution, spatial distribution of inclusions (clustering) and residual stresses. To understand the effects of these localized microstructural features on the composite response requires detailed numerical modeling. Numerous modeling techniques have been proposed for the elastic and elastic-plastic response of composites with discontinuous particle and whisker reinforcements. These models are primarily analytical and typically assume idealized microstructural features. Modeling of non-ideal, localized effects requires the development of appropriate numerical procedures. It is therefore of interest to review and compare existing idealized models in order to identify appropriate numerical models to investigate localized effects. In this paper, the composite elastic and elastic-plastic responses predicted by several models are evaluated and discussed. These results may be used to define appropriate numerical models to determine the effects of processing induced microstructural features on the macroscopic composite response.

1. INTRODUCTION

Metal-matrix composites with discontinuous reinforcements such as particles, whiskers and platelets offer the potential for successful application in a variety of structural components. A major advantage of these materials is their ability to be processed using conventional processing methods. They exhibit improved stiffness and strength as compared to corresponding unreinforced alloys. By varying the processing conditions and heat treatment parameters, one can control microstructural features such as inclusion orientation, inclusion aspect ratio, inclusion spatial distribution (clustering), matrix yield strength, matrix hardening characteristics, and residual stresses. The effect of these parameters on the overall composite behavior, however, is not well understood. Therefore, the development of micromechanics models to predict the effects of these microstructural parameters on the macroscopic composite response will aid in the development of improved material systems. In this paper, various models for both the elastic and elastic-plastic behavior of composites with discontinuous reinforcements are compared and evaluated.

For the elastic case, relevant models include variational bounds, the dilute approximation, the self-consistent approximation, the differential self-consistent method, generalized self-consistent method, composite assemblage models, the Mori-Tanaka method and periodic array models. The reliability of these models has received considerable attention in recent years and the range of applicability of each model is well characterized. For the elastic-plastic case, several analogous models have been proposed. These include dilute approximations, self-consistent approximations, the Mori-Tanaka approach, and periodic array (unit cell) models. In this paper, elastic and elastic-plastic predictions given by these models will be discussed and compared.

2. ELASTIC PROPERTY MODELS

Numerous models for the elastic properties of composites with discontinuous reinforcements have been developed in recent years. The limitations of many of these models are well known and comparisons of various model predictions have been discussed in several recent papers. While these comparisons are normally made for the case of spherical inclusions, some methods can also be extended to model other inclusion geometries. The major conclusions of the comparisons for the case of rigid spherical inclusions in an elastic matrix are summarized below. Similar comparisons are presented for the case of rigid spheroidal inclusions in an elastic matrix. These comparisons will serve as a basis for comparing the inelastic response predicted by analogous models.

Models for the elastic properties can be categorized into three types. For the case of elastic composites, absolute upper and lower bounds to the composite moduli are derived based on variational principles. The Voigt and Reuss bounds (also known as the Paul (1960) bounds) can be derived using the elasticity theorems of minimum potential energy and minimum complementary energy and are given by

$$E^{(+)} = E_i V_i + E_m V_m$$
$$E^{(-)} = \left[\frac{V_i}{E_i} + \frac{V_m}{E_m} \right]^{-1} \tag{1}$$

where $E^{(+)}$ and $E^{(-)}$ are the upper and lower bounds to the composite Young's modulus, E_i and E_m are the inclusion and matrix Young's moduli, and V_i and V_m are the inclusion and matrix volume fractions. These bounds are valid for any composite without regard to reinforcement geometry or orientation. For the case of statistically isotropic composites such as when the inclusions are spherical or are randomly oriented in space, refined bounds have be derived based on the Hashin-Shtrikman (H-S) variational principles (Hashin and Shtrikman, 1963). The H-S lower bounds for the composite bulk and shear modulus, $K^{(-)}$ and $G^{(-)}$ are given by

$$K^{(-)} = K_m + \frac{V_i}{\frac{1}{K_i - K_m} + \frac{3V_m}{3K_m + 4G_m}}$$
$$G^{(-)} = G_m + \frac{V_i}{\frac{1}{G_i - G_m} + \frac{6V_m(K_m + 2G_m)}{5G_m(3K_m + 4G_m)}} \tag{2}$$

where K_i and K_m are the inclusion and matrix bulk moduli ($K_i > K_m$) and G_i and G_m are the inclusion and matrix shear moduli ($G_i > G_m$). The H-S upper bounds are found by interchanging the subscripts i and m everywhere in Eq. 2. Bounds on other elastic properties, such as Young's modulus and Poisson's ratio can be found using the standard relations for isotropic materials.

The second category of micromechanics models are idealized geometry models. In these models, the inclusions are imagined to be arranged in a well-defined, idealized configuration. Examples of this type include periodic array models and assemblage models. In periodic array models, the inclusions are imagined to be arranged in regular array such as a cubic or hexagonal array. This assumption allows for the definition of a representative volume element (or unit cell) which can be analyzed analytically or numerically to determine the overall composite response. Models of this type have been formulated and solved by Nemat-Nasser and co-workers (Nemat-Nasser et al., 1982, Iwakuma and Nemat-Nasser, 1983). These models have been shown to be consistent with the rigorous Hashin/Shtrikman bounds.

For the case of composites with an elastic-plastic matrix material (such as metal-matrix composites), several researchers have used periodic array models to define unit cell geometries which are analyzed using finite element methods. It is of interest, therefore, to evaluate the predicted elastic behavior of such models. To avoid performing numerically intensive three-dimensional finite element analyses, a commonly used unit cell for these problems is an axisymmetric cylinder which contains either a spherical, cylindrical or ellipsoidal inclusion. As discussed by Christman et al. (1989a), this unit cell geometry closely approximates the hexagonal cell generated if the inclusions are imagined to be arranged in a hexagonal array. Hence, in the elastic property comparisons discussed below (for the case of uniaxial loading), the unit cell models considered consist of either a spherical or spheroidal inclusion embedded in a cylindrical unit cell. The unit cell aspect ratio (length/diameter) is chosen such that the lateral inclusion spacing ($2(r_c-r_i)$), where r_c and r_i are the unit cell and inclusion radii transverse to the loading direction, equals the vertical inclusion spacing (l_c-l_i), where l_c and l_i are the unit cell and inclusion lengths parallel to the loading direction. The inclusion and cell dimensions are chosen such that the inclusion volume fraction in the unit cell equals the overall composite inclusion volume fraction. Periodic boundary conditions are applied to the unit cell such that (for the case of uniaxial loading), the axial displacement at the top of the unit cell, u_z, is determined from the applied axial strain, ε_z^∞, using the relation $u_z(z=l_c/2)=(l_c/2)\varepsilon_z^\infty$. On the lateral boundary, $r=r_c$, the net radial and shear tractions are zero. As discussed by Christman et al. (1989b), the displacements on the lateral boundary can be either unconstrained or constrained so that the cell remains cylindrical. Although the cylindrical constraint condition is more realistic, results for both types of boundary conditions will be presented so that the effects of lateral constraint can be characterized.

Another idealized geometry model is the assemblage model introduced by Hashin (1962) for the case of spherical and cylindrical reinforcements. In this idealization, it is imagined that the composite is constructed of composite spheres (or cylinders) of diminishing diameter. Each composite sphere (or cylinder) consists of an inclusion embedded in a self-similar shell of matrix material of dimension to provide a local inclusion volume fraction equal to the macroscopic volume fraction. This idealization allows for the derivation of an exact composite bulk modulus which is equivalent to that given by the H-S lower bound (Eq. 2a) and bounds on the composite shear modulus. The upper bound for the composite spheres assemblage is given by

$$G^{(+)} = G_m + \left[\frac{G_m V_i}{\left(\dfrac{G_m}{G_i - G_m}\right) + A(1-V_i) - V_i \dfrac{(1-V_i^{2/3})^2}{BV_i^{7/3}+C}} \right] \quad (3)$$

where

$$A = \frac{2(4-5v_m)}{15(1-v_m)}$$

$$B = \frac{10(1-v_m)}{21} \frac{(7-10v_i)(7+5v_m)-(G_i/G_m)(7-10v_m)(7+5v_i)}{4(7-10v_i)(G_i/G_m)(7+5v_i)} \qquad (4)$$

$$C = \frac{10}{21}(7-10v_m)(1-v_m)$$

where v_i and v_m are the inclusion and matrix Poisson's ratios. Since the lower bound for the composite spheres assemblage geometry falls below the H-S lower bound (Eq. 2b) for any isotropic composite, the composite spheres lower bound can be taken to be the H-S lower bound.

The third type of elastic micromechanics models are various approximate models. Specific approximate models to be considered in this paper are the dilute approximation (Dewey, 1947), the self-consistent method (Hill, 1965 and Budiansky, 1965), the differential self-consistent method (Roscoe, 1973 and McLaughlin, 1977), the generalized self-consistent (three-phase) model (Christensen and Lo, 1979, 1986), and the Mori-Tanaka method (Mori and Tanaka, 1973, Tandon and Weng, 1984, and Benveniste, 1987). The dilute and self-consistent approximations can best be interpreted by considering the volume averaging result given by Willis (1981)

$$\sigma^\infty = [\, L_m + V_i(L_i-L_m) : A \,] : \varepsilon^\infty \qquad (5)$$

where σ^∞ and ε^∞ are the macroscopic stress and strain tensors, L_i and L_m are the inclusion and matrix elastic moduli tensors, and A is the strain concentration tensor which relates the inclusion average strain to the overall composite strain

$$\varepsilon_i = A : \varepsilon^\infty \qquad (6)$$

The term in the square brackets in Eq. 5 provides the overall composite moduli. Hence, if estimates of strain concentration tensor, A, can be obtained, then the composite moduli can be determined.

In the dilute approximation, interactions between neighboring inclusions are neglected. With this approximation, for the case of ellipsoidal inclusions, one can use Eshelby's (1957) solution to determine the strain concentration tensor. For the case of spherical inclusions, it is easy to show that the composite bulk (K^∞) and shear (G^∞) moduli are given by

$$K^\infty = K_m + V_i \frac{(K_i-K_m)}{1+[(K_i-K_m)/(K_m+\frac{4}{3}G_m)]} \qquad (7)$$
$$G^\infty = G_m + V_i \frac{G_m(G_i-G_m)}{\beta_m G_i + (1-\beta_m)G_m}$$

where

$$\beta_m = \frac{2}{15}\left(\frac{4-5v_m}{1-v_m}\right) \qquad (8)$$

Note the linear dependence of K^∞ and G^∞ on inclusion volume fraction.

The classical self-consistent model (Hill, 1965 and Budiansky, 1965) estimates the strain concentration tensor by assuming the inclusion to be embedded in an infinite material whose properties are those of the unknown composite effective properties. Again, Eshelby's solution is invoked, leading to composite moduli given by solving

$$\frac{V_i}{K^\infty - K_m} + \frac{V_m}{K^\infty - K_i} = \frac{\alpha^\infty}{K^\infty}$$
$$\frac{V_i}{G^\infty - G_m} + \frac{V_m}{G^\infty - G_m} = \frac{\beta^\infty}{G^\infty} \qquad (9)$$

where

$$\alpha^\infty = \frac{1}{3}\left(\frac{1+v^\infty}{1-v^\infty}\right), \quad \beta^\infty = \frac{2}{15}\left(\frac{4-5v^\infty}{1-v^\infty}\right) \qquad (10)$$

where v^∞ is the effective composite Poisson's ratio.

In the differential self-consistent model (Roscoe, 1973 and McLaughlin, 1977), the effective properties are estimated from the solution for a single inclusion of volume dV_i embedded in an infinite material whose properties are those of a composite with an inclusion volume fraction V_i. Using Eshelby's solution for the case of spherical inclusions leads to the system of differential equations

$$\frac{dK^\infty}{dV_i} = \frac{K_i - K^\infty}{1 - V_i} \frac{K^\infty + K^*}{K_i + K^*}$$
$$\frac{dG^\infty}{dV_i} = \frac{G_i - G^\infty}{1 - V_i} \frac{G^\infty + G^*}{G_i + G^*} \qquad (11)$$

where

$$K^* = \frac{4}{3} G^\infty$$
$$G^* = \frac{G^\infty}{6}\left(\frac{4K^\infty + 8G^\infty}{K^\infty + 2G^\infty}\right) \qquad (12)$$

Integration of Eqs. 11 with the initial conditions $K^\infty(V_i=0)=K_m$ and $G^\infty(V_i=0)=G_m$ yields the composite moduli $K^\infty(V_i)$ and $G^\infty(V_i)$.

The generalized self-consistent or three-phase model (Christensen and Lo, 1979, 1986) is similar to the classical self-consistent except that here the inclusion is imagined to be embedded in a shell of matrix material which in turn is embedded in an infinite material whose properties are the composite effective properties. The inclusion and matrix shell dimensions are chosen such that the "local" inclusion volume fraction equals the overall composite volume fraction. For the case of spherical inclusions, Christensen and Lo (1979) showed that the composite bulk modulus is the same as that given by the H-S lower bound and the composite shear modulus is given by solving the quadratic equation

$$A\left(\frac{G^\infty}{G_m}\right)^2 + 2B\left(\frac{G^\infty}{G_m}\right) + C = 0 \qquad (13)$$

where the constants A, B, and C are given by Christensen and Lo (1979). The shear modulus given by Eq. 13 lies between both the H-S bounds as well as the composite spheres assemblage model bounds.

The final approximate model to be considered is the Mori-Tanaka method. In the original formulation (Mori and Tanaka, 1973), the average internal stress in the matrix is determined by considering an inclusion eigenstrain. By combining the Mori-Tanaka concept of average stress with Eshelby's solution, one can estimate the overall composite properties.

For example, Tandon and Weng (1984) used this approach to determine the five independent elastic constants for a composite with aligned spheroidal inclusions. For the case of spherical inclusions, the composite moduli are given by Benveniste (1987) and Christensen (1990). For this case, the Mori-Tanaka predictions are identical to the H-S lower bound.

The composite Young's moduli predicted by these models for the case of rigid spherical inclusions are shown in Fig. 1. It can be seen that at low inclusion volume fractions, all models give consistent results. At increasing inclusion volume fractions, however, the models provide increasingly different predictions. As indicated by Budiansky (1965), the classical self-consistent model prediction for composites with rigid spherical inclusions becomes unbounded at an inclusion volume fraction of 0.5. The Mori-Tanaka and composite spheres assemblage (lower bound) coincide with the H-S lower bound. The generalized self consistent model predicts a composite stiffness between the H-S lower bound and the composite spheres assemblage upper bound and is therefore believed to provide reliable predictions. Note that the H-S and Paul upper bounds are infinite for the case of rigid inclusions. The dilute model underestimates the composite stiffness and falls below the Paul lower bound for $V_i < 50\%$ and the H-S lower bound for all non-zero inclusion volume fractions.

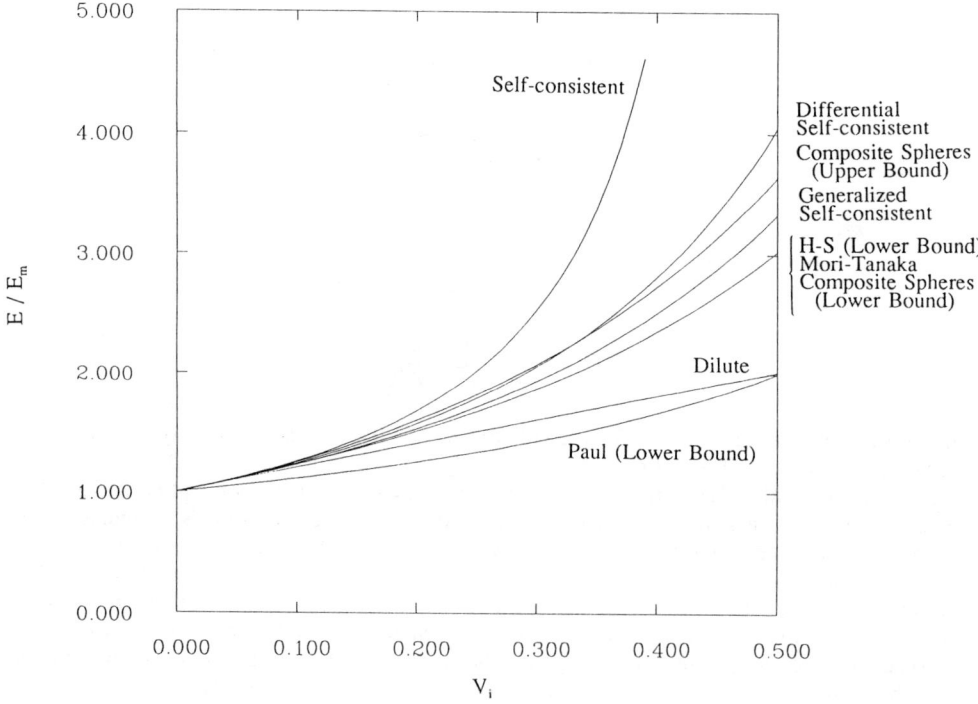

Figure 1. Comparison of predicted Young's moduli for composites with rigid spherical inclusions.

Since many commercial metal-matrix composites utilize whisker reinforcements, it is of interest to determine the relative predicted composite elastic response for the case of aligned rigid spheroidal inclusions. For this case, the models discussed above which can be adapted to spheroidal inclusion geometries are the dilute approximation, the Mori-Tanaka method, the generalized (three-phase) self consistent model, and a unit cell model. The geometries considered for the dilute, three-phase and unit cell models are shown in Fig. 2. For these models, several finite element calculations were performed to predict the composite

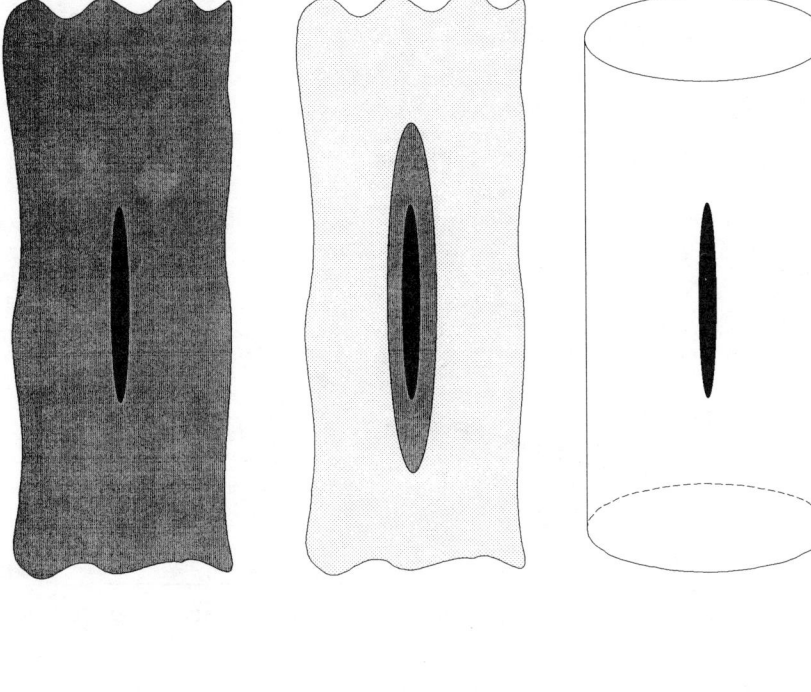

        Dilute Model   Three-Phase Model   Unit Cell Model

Figure 2.   Geometric models corresponding to the dilute approximation, the three-phase model and the unit cell models.

response for several inclusion volume fractions in the range $0 \leq V_i \leq 0.5$. As discussed above for the unit cell calculations, results for both cylindrical lateral constraint and no lateral constraint are presented. For all models, inclusion aspect ratios of 1, 2, and 5 were considered. These numerical results were compared to analytical results using the Mori-Tanaka method given by Tandon and Weng (1984) in Figs. 3-5. It can be seen that for low inclusion aspect ratios, the unit cell models predict the highest composite stiffness. At increasing inclusion aspect ratios, however, the three-phase and Mori-Tanaka models provide a stiffer response. At low inclusion volume fractions, all models are seen to asymptotically approach the dilute approximation.

3. ELASTIC-PLASTIC COMPOSITE RESPONSE

To study the elastic-plastic composite response, several of the models discussed above can be extended to include the effects of plastic deformation of the matrix. For the case of spherical inclusions in an elastic-plastic matrix, Mori-Tanaka's method has been applied by Tandon and Weng (1988) to provide the composite elastic-plastic response. Similarly for the case of spherical inclusions in an elastic-plastic matrix, Taggart and Bassani (1991) applied the three-phase model to determine composite response. To study the behavior of whisker reinforced composites, recent studies have assumed a spheroidal inclusion geometry where the inclusions are assumed to be aligned parallel to the loading direction. A dilute approximation for the elastic-plastic response for such a composite is given by Adley (1991). Similarly, the three-phase model was extended by Qin (1991) for the case of rigid spheroidal inclusions in an elastic-plastic matrix.

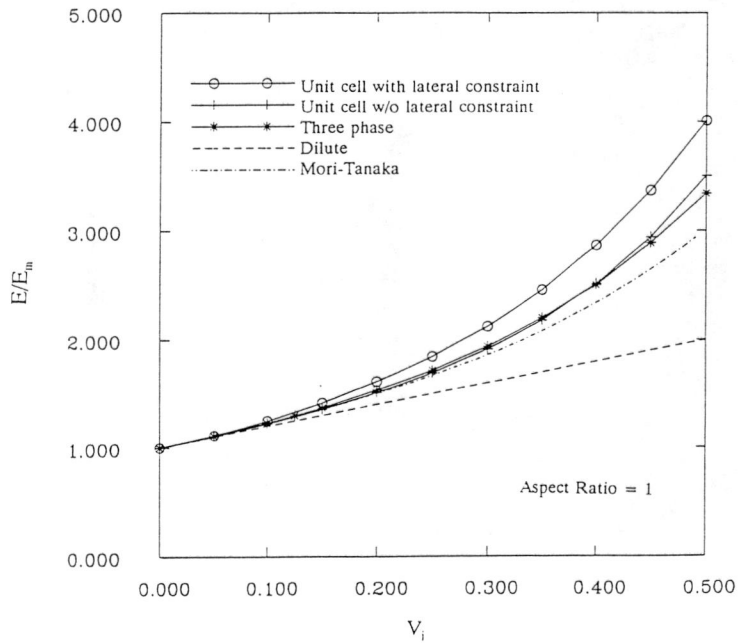

Figure 3. Comparison of the composite Young's moduli predicted by the dilute, three-phase, Mori-Tanaka and unit cell models for composites with rigid spherical inclusions.

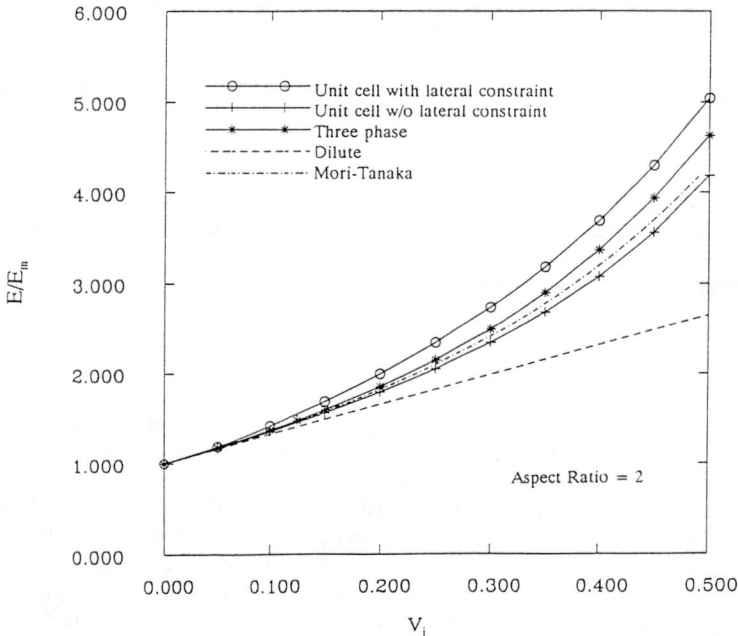

Figure 4. Comparison of the composite Young's moduli predicted by the dilute, three-phase, Mori-Tanaka and unit cell models for composites with rigid spheroidal inclusions (inclusion aspect ratio = 2).

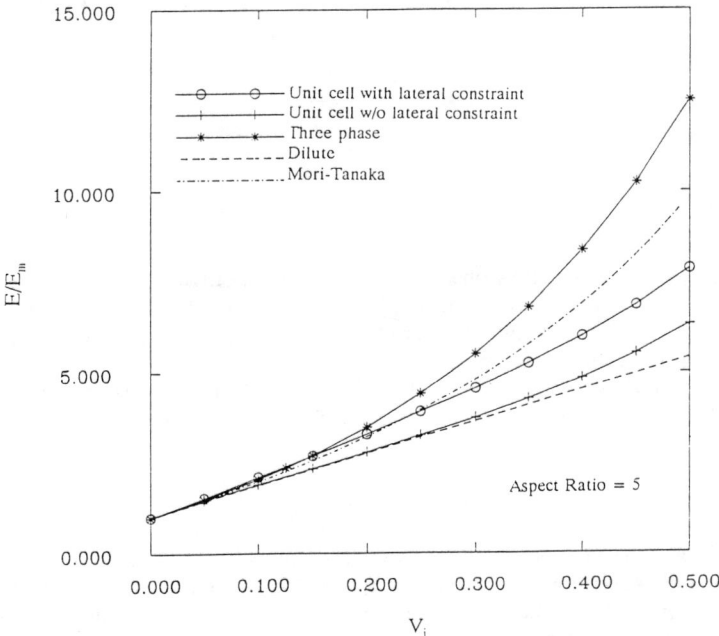

Figure 5. Comparison of Young's moduli predicted by the dilute, three-phase, Mori-Tanaka and unit cell models for composites with rigid spheroidal inclusions (inclusion aspect ratio = 5).

To compare the relative responses predicted by each of these models, composites with rigid spheroidal inclusions aligned in the loading direction are considered. The matrix is assumed to by elastic-plastic with the inelastic deformation described by a time independent J_2 flow theory. The uniaxial matrix response is assumed to be given by a power-law relation of the form

$$\sigma/\sigma_0 = \begin{cases} \varepsilon/\varepsilon_0, & \varepsilon \leq \varepsilon_0 \\ (\varepsilon/\varepsilon_0)^{1/n}, & \varepsilon > \varepsilon_0 \end{cases} \tag{14}$$

where σ_0 and ε_0 are the matrix yield stress and yield strain, respectively, $(E_m = \sigma_0/\varepsilon_0)$ and n is the strain-hardening exponent ($n \geq 1$). For the cases discussed below, a strain-hardening exponent, n=3, was selected to represent a moderate degree of strain hardening.

The Mori-Tanaka method prediction for the composite elastic-plastic response of composites with spherical inclusions is given by Tandon and Weng (1988). In this model, the secant composite bulk (K_s^∞) and shear (G_s^∞) moduli are given by

$$\frac{K_s^\infty}{K_m} = 1 + \frac{V_i(K_i - K_m)}{(1-V_i)\alpha_s(K_i - K_m) + K_m}$$

$$\frac{G_s^\infty}{G_m^s} = 1 + \frac{V_i(G_i - G_m^s)}{(1-V_i)\beta_s(G_i - G_m^s) + G_m^s} \tag{15}$$

where G_m^s is the matrix secant modulus which depends on the average matrix effective plastic strain and

$$\alpha_s = \frac{1}{3}\frac{(1+v_m^s)}{(1-v_m^s)}$$

$$\beta_s = \frac{2}{15}\frac{(4-5v_m^s)}{(1-v_m^s)}$$

(16)

where v_m^s is the matrix secant Poisson's ratio. Since this method is based on the average matrix stress, the effects of localized matrix plasticity which tend to "soften" the composite response, are neglected. As a result, the overall composite stress-strain response for this case will be stiffer than the other cases considered.

To extend the dilute and three-phase model to include the effects of matrix plasticity requires that the macroscopic stress-strain relation, Eq. 5, be rewritten in incremental form as

$$\dot{\sigma}^\infty = [\, L_m + v_i(L_i - L_m) : A\,] : \dot{\varepsilon}^\infty \tag{17}$$

where $\dot{\sigma}^\infty$ and $\dot{\varepsilon}^\infty$ are the macroscopic stress and strain increments, L_i is the inclusion elastic modulus tensor and L_m is now interpreted as the matrix tangent modulus tensor which is assumed to be a function of the matrix effective plastic strain. The "incremental" strain concentration tensor, A, relates the inclusion strain increment to the macroscopic strain increment and is determined numerically using an elastic-plastic axisymmetric finite element calculation. For the dilute model calculation, the boundary value problem consists of a single spheroidal inclusion embedded in a very large ("infinite") cylindrical region of elastic-plastic matrix. At each increment of macroscopic straining, the inclusion strain increment provides the incremental strain concentration which is used to compute the composite effective tangent moduli. A similar analysis is performed in the elastic-plastic three-phase calculation except that in this case, the inclusion is embedded in a matrix shell which is embedded in an infinite effective material. During the incremental elastic-plastic finite element calculation, the tangent moduli of the effective material are continuously updated based on the current incremental strain concentrations. Details of these numerical procedures are given by Taggart and Bassani (1991), Adley (1991) and Qin (1991).

The unit cell model for the elastic-plastic response of composites with spherical or spheroidal inclusions consists of a single inclusion embedded in an axisymmetric cylinder of elastic-plastic matrix material. As in the elastic unit cell models discussed above, the inclusion and cell dimensions are determined based on the overall inclusion volume fraction. Results are presented both for cylindrically constrained lateral boundary conditions and unconstrained lateral boundary conditions.

Comparisons of the predicted uniaxial elastic-plastic composite response for these models are shown in Figs. 6-8. For the case of spherical inclusions (Fig. 6), it can be seen that since the Mori-Tanaka method neglects the effects of local matrix plasticity, it predicts a higher yield stress. The three-phase and unit cell (with lateral constraint) predict comparable composite yield strengths but the three-phase model predicts a slightly higher degree of strain-hardening. It is interesting to note the strong effect of the unit cell lateral boundary conditions on the composite response. As discussed by Christman et al. (1989b), the cylindrically constrained lateral boundary condition induces a high level of hydrostatic stress which tends to reduce the amount of plastic deformation and leads to a much stiffer overall composite response. At increasing inclusion aspect ratios (Figs. 7 and 8), the unit cell model with lateral constraint is seen to be slightly stiffer than the three-phase model. The effect of the lateral constraint becomes increasingly important as the inclusion aspect ratio is increased. As shown in Figure 8 (inclusion aspect ratio = 5), the overall composite stress at a given

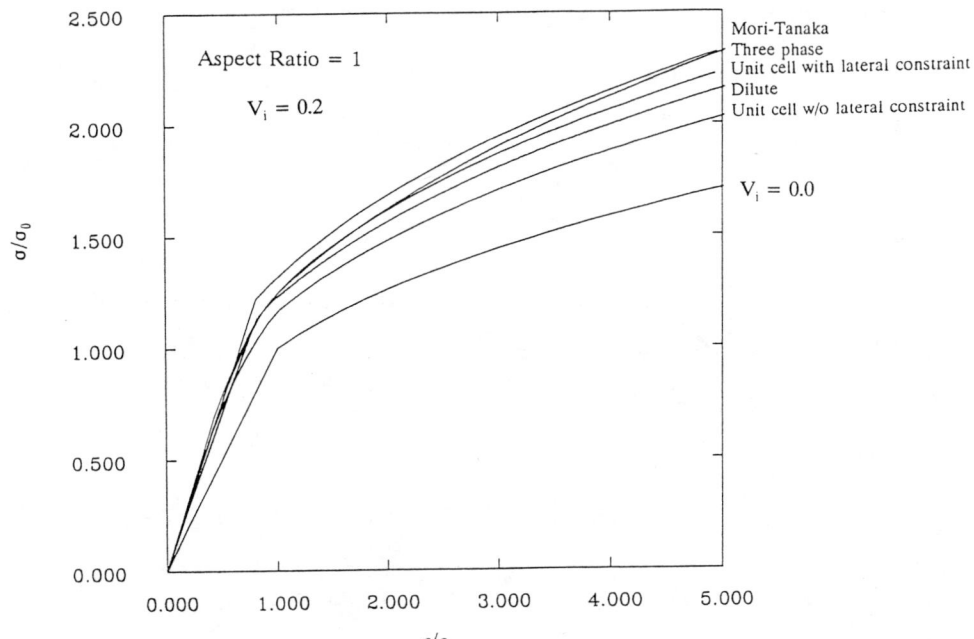

Figure 6. Comparison of the composite elastic-plastic response predicted by the dilute, three-phase, Mori-Tanaka and unit cell models for composites with rigid spherical inclusions.

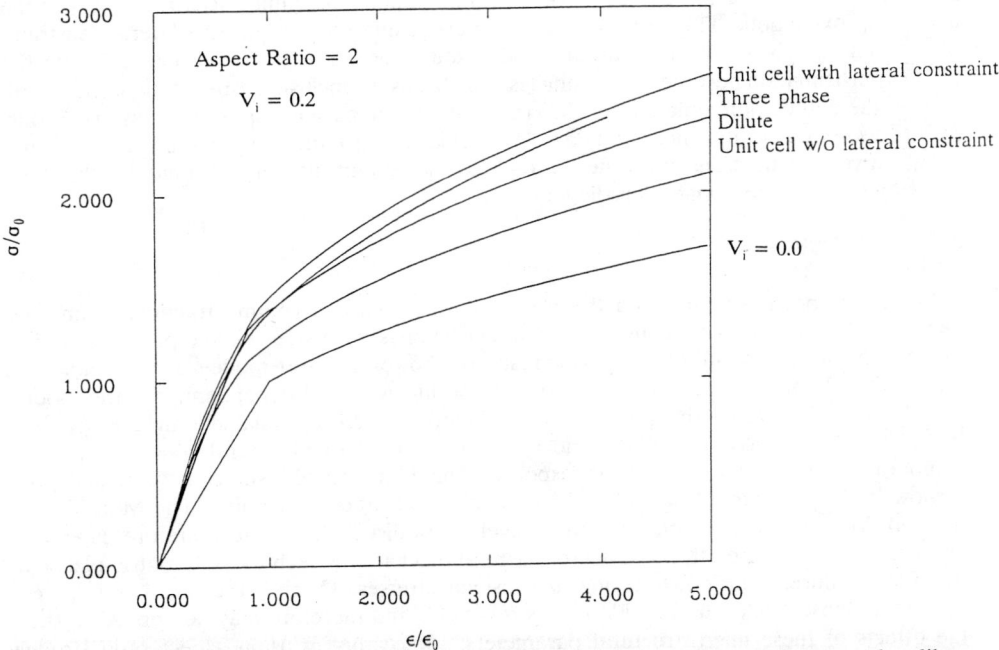

Figure 7. Comparison of the composite elastic-plastic response predicted by the dilute, three-phase and unit cell models for composites with rigid spheroidal inclusions (inclusion aspect ratio = 2).

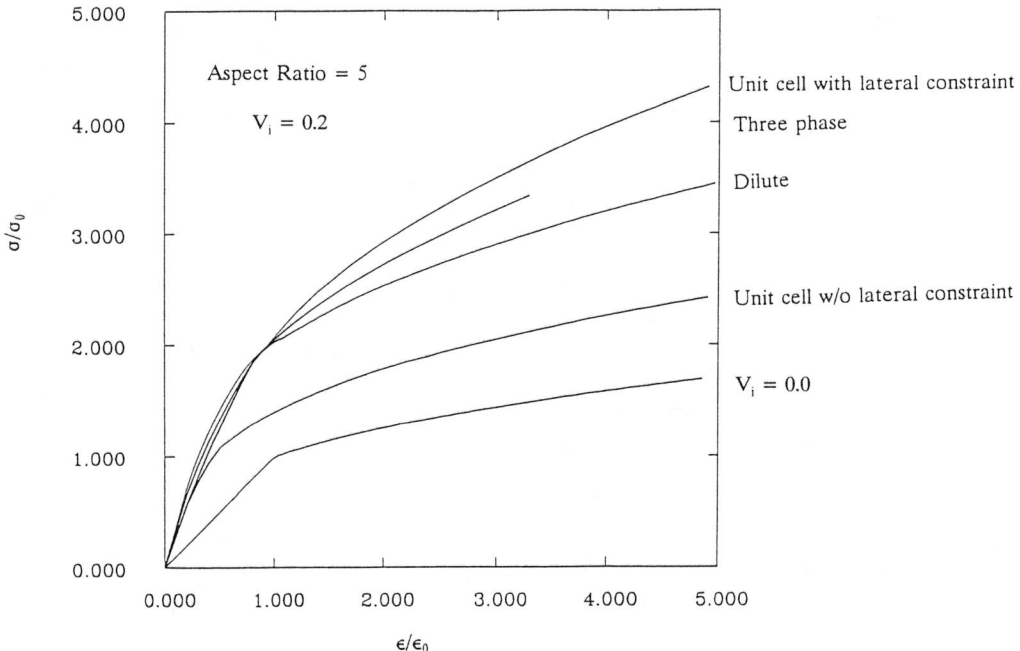

Figure 8. Comparison of the composite elastic-plastic response predicted by the dilute, three-phase and unit cell models for composites with rigid spheroidal inclusions (inclusion aspect ratio = 5).

applied strain level for the unit cell model with no lateral constraint falls far below even the dilute approximation. This result confirms the expectation that neglecting lateral constraint of the unit cell gives unrealistic predictions. Examination of Figs. 6-8 also reveals that the dilute approximation becomes increasing less reliable as the inclusion aspect ratio is increased. Hence, the dilute approximation for the composite elastic-plastic response is reasonable only for inclusion aspect ratios near unity and low inclusion concentrations ($<\sim10\%$). For higher inclusion volume fractions, both the three-phase model and the unit cell model with lateral constraint provide reasonable predictions.

4. CONCLUSIONS

It has been demonstrated that for moderate inclusion volume fractions, numerous micromechanics models for composites with elastic phases provide reliable predictions. For the case of the composite elastic-plastic response, however, fewer models are available and the predicted composite response is shown to be highly model dependent. Of the models examined in this study, the Mori-Tanaka method (currently available only for spherical inclusions), the three-phase model and the unit cell model with lateral constraint provide comparable composite elastic-plastic response. The dilute model is shown to be reliable only for low inclusion volume fractions and inclusion aspect ratios near unity. The Mori-Tanaka method does not provide microstructure level stress distributions and therefore cannot be applied to consider the effects of microstructural mechanisms such as local matrix plasticity, interfacial failure, inclusion clustering, and residual stresses. On the other hand, the unit cell and three-phase models do provide local stress fields and therefore may be applied to study the effects of these microstructural parameters. Since the inclusion aspect ratio strongly effects the local stress distributions, it is expected that the microstructural deformation and failure mechanisms will be very sensitive to inclusion aspect ratio. For example, Fig. 9 shows a three-phase model prediction for the local von Mises effective stress distribution for the

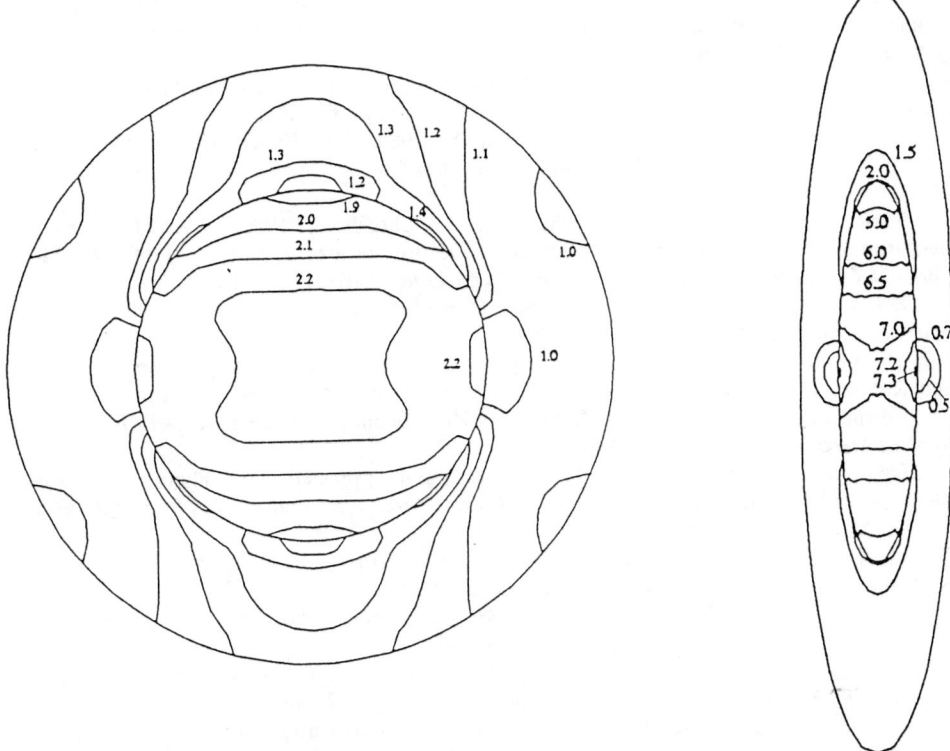

Figure 9. Comparison of the three-phase model microstructure level von Mises stress, σ_e/σ_0 distribution at a macroscopic strain level $\varepsilon^\infty = \varepsilon_0$ for spheroidal inclusion aspect ratios of 1 and 5.

cases of spherical and spheroidal (aspect ratio = 5) inclusions at a macroscopic strain level equal to the matrix yield strain. Clearly, the level and distribution of stresses is highly dependent on the inclusion aspect ratio. Hence, it is believed that future modeling studies of effects of processing induced microstructural features must include the effects of localized stress fields.

5. ACKNOWLEDGEMENTS

Support from an Air Force / Engineering Foundation Research Initiation Grant is gratefully acknowledged.

6. REFERENCES

Adley, M.D., 1991, "Computation of the Effective Properties of Materials with Microstructure," PhD. Dissertation, University of Rhode Island.

Benveniste, Y., 1987, "A New Approach to the Application of Mori-Tanaka's Theory in Composite Materials," *Mechanics of Materials*, Vol. 6, pp. 147-157.

Budiansky, B. 1965, "On the Elastic Moduli of Some Heterogeneous Materials," *Journal of the Mechanics and Physics of Solids*, Vol. 13, pp. 223-227.

Christensen, R. M. and Lo, K. H., 1979, "Solutions for Effective Shear Properties in Three Phase Sphere and Cylinder Models," *Journal of the Mechanics and Physics of Solids*, Vol. 27, pp. 315-330.

Christensen, R. M. and Lo, K. H., 1986, "Erratum - Solutions for Effective Shear Properties in Three Phase Sphere and Cylinder Models," *Journal of the Mechanics and Physics of Solids*, Vol. 34, p. 639.

Christensen, R. M., 1990, "A Critical Evaluation for a Class of Micromechanics Models," *Journal of the Mechanics and Physics of Solids*, Vol. 38, p. 379-404.

Christman, T., Needleman, A. and Suresh, S., 1989a, "An Experimental and Numerical Study of Deformation in Metal-Ceramic Composites," *Acta Metallurgica*, Vol. 37, No. 11, pp. 3029-3050.

Christman, T., Needleman, A., Nutt, A., and Suresh, S., 1989b, "On Microstructural Evolution and Micromechanical Modelling of Deformation of a Whisker-reinforced Metal-Matrix Composite," *Materials Science and Engineering*, A107, pp. 49-61.

Dewey, J. M., 1947, "The Elastic Constants of Materials Loaded with Non-Rigid Fillers," *Journal of Applied Physics*, Vol. 18, p. 578.

Eshelby, J. D., 1957, "The Determination of the Elastic Field of an Ellipsoidal Inclusion, and Related Problems," *Proceedings of the Royal Society*, Vol. A, No. 241, pp. 376-396.

Hashin, Z, 1962,"The Elastic Modulus of Heterogeneous Materials," *ASME Journal of Applied Mechanics*, Vol. 29, pp. 143-150.

Hashin, Z. and Shtrikman, S., 1963, "A Variational Approach to the Theory of the Elastic Behavior of Multiphase Materials," *Journal of the Mechanics and Physics of Solids*, Vol. 11, pp. 127-140.

Hill, R., 1965, "A Self Consistent Mechanics of Composite Materials," *Journal of the Mechanics and Physics of Solids*, Vol. 13, pp. 213-222.

Iwakuma, T. and Nemat-Nasser, S., 1983, "Composites with Periodic Microstructure," *Composites & Structures*, Vol. 16, No. 1-4, pp. 13-19.

McLaughlin, R., 1977, "A Study of the Differential Scheme for Composite Materials," *International Journal of Engineering Science*, Vol. 15, pp. 237-244.

Mori, T. and Tanaka, K., 1973, "Average Stress in Matrix and Average Elastic Energy of Materials with Misfitting Inclusions," *Acta Metallurgica*, Vol. 21, pp. 571-574.

Nemat-Nasser, S. Iwakuma, T. and Hejazi, M., 1982, "On Composites with Periodic Microstructure," *Mechanics of Materials*, Vol. 1, pp. 239-267.

Paul, B., 1960, "Prediction of Elastic Constants of Multiphase Materials," *Trans. of the Metallurgical Society of AIME*, Vol 218, pp. 36-41.

Qin, J., 1991, "Elastic-Plastic Micromechanics Modeling of Metal-Matrix Composites and Porous Materials," M.S. Thesis, University of Rhode Island.

Roscoe, R., 1973, "Isotropic Composites with Elastic or Viscoplastic Phases: General Bounds for the Moduli and Solutions for Special Geometries," *Rheologica Acta*, Vol. 12, pp. 404-411.

Taggart, D. G. and Bassani, J. L., 1991, "Elastic-Plastic Behavior of Particle Reinforced Composites - Influence of Residual Stress," *Mechanics of Materials*, Vol. 12, pp 63-80.

Tandon, G. P. and Weng, G. J., 1984, "The Effect of Aspect Ratio of Inclusion on the Elastic Properties of Unidirectionally Aligned Composites," *Polymer Composites*, Vol. 5, No. 4, pp. 327-333.

Tandon, G. P. and Weng, G. J., 1988, "A Theory of Particle-Reinforced Plasticity," *Journal of Applied Mechanics*, Vol. 55, pp. 126-135.

Willis, J. R., 1981, "Variational and Related Methods for the Overall Properties of Composites," *Advances in Applied Mechanics*, Vol. 21, pp. 1-78.

ELEVATED TEMPERATURE BEHAVIOR OF FINE-GRAINED Mg-9Li-B$_4$C COMPOSITES

J. Wolfenstine
Department of Mechanical and
Aerospace Engineering
Materials Section
University of California, Irvine
Irvine, California

G. Gonzalez-Doncel
Centro Nacional de
Investigaciones Metalurgicas
Madrid, Spain

O. D. Sherby
Department of Materials Science and Engineering
Stanford University
Stanford, California

ABSTRACT

A fine-grained Mg-9Li-5wt.%B$_4$C particulate composite prepared using a foil metallurgy technique was superplastic in the temperature range 150 to 300°C with a stress exponent equal to about 2 and an elongation-to-failure as high as 390%. At high temperatures in the n≈2 region the activation energy for superplastic flow for the Mg-9Li-5wt.%B$_4$C particulate composite is about equal to 108 kJ mol^{-1} and decreases to about 56 kJ mol^{-1} at low temperatures. It is postulated that the change in activation energy is a result of a change from grain boundary sliding controlled by lattice diffusion to grain boundary sliding controlled by grain boundary diffusion.

1. INTRODUCTION

It was recently shown that a fine-grained two phase Mg-9wt.%Li (Mg-9Li) material prepared using a foil metallurgy processing technique with true grain sizes, d, between 6 to 35 μm exhibited superplastic behavior in the temperature range 150 to 250°C with an elongation-to-failure as high as 450% [1]. The Mg-9Li alloy in the superplastic region exhibited a stress exponent of about two, a grain size exponent of about two and an activation energy for superplastic flow of 104 kJ mol^{-1} which was related to the activation energy for lattice diffusion in the majority beta phase. It was postulated that the dominant deformation mechanism in the fine-grained Mg-9Li alloy in the superplastic region is grain boundary sliding accommodated by slip controlled by lattice diffusion in the majority beta phase.

It is the intent of this paper to show that a Mg-9Li-5wt.%B$_4$C particulate composite prepared in a similar manner to the Mg-9Li alloy with finer grain sizes (3.6 to 8.5 μm) and tested in a similar temperature range as that for the Mg-9Li alloy was also superplastic but, as the temperature of testing was decreased the dominant deformation mechanism in the superplastic region switched from grain boundary sliding controlled by lattice diffusion to grain boundary sliding controlled by grain boundary diffusion as result of the finer grain sizes in the composite than in the Mg-9Li alloy.

2. EXPERIMENTAL PROCEDURE

The method used to prepare the Mg-9Li-5wt.%B$_4$C particulate composite has been previously described [2]. The Mg-9Li matrix alloy was prepared by vacuum casting. The Mg-9Li alloy consists of two phases, α (hexagonal-close-packed structure) and β (body-centered-cubic structure). The α phase making up about 30 volume percent of the material is elongated and dispersed within the β matrix. Scanning electron microscopy revealed that the B$_4$C particles were typically less than 20 μm in size. Prior to use the B$_4$C particles were soaked in a 5%HCl, 2%HF and 93% ethanol solution to remove any surface contamination. The fine-grained Mg-9Li-5wt.%B$_4$C particulate composite was prepared from the as-cast Mg-9Li alloy and B$_4$C particles using a foil metallurgy technique. The as-cast material was cut into plates, which were given a repeated sequence of cold-rolling and annealing (recovery only, no recrystallization) until foils of about 0.20 mm thick were obtained. The total reduction of the individual foils was 200 to 1. The B$_4$C particles were suspended in a ethanol solution which was then painted on one side of the foils. The foils were then dried and stacked together and press-bonded at 200°C (0.55 T$_m$, where T$_m$ is the absolute melting temperature, 861 K). The amount of reduction during press-bonding was about 5:1. The amount of B$_4$C was determined by the change in weight of the foils before and after application of the suspension containing the B$_4$C particles.

Two types of mechanical tests were used to investigate the superplastic behavior of the Mg-9Li-5wt.%B$_4$C particulate composite. Strain-rate-change (SRC) tests were conducted to determine the value of the stress exponent, n. Strain-rate-change tests were preformed in compression. Samples were deformed perpendicular to the pressing direction. The compression specimens had a height to width ratio of approximately 2:1 with dimensions of a typical sample being 2 x 2 x 4 mm. Boron nitride was used as a lubricating compound for the compression tests. The samples for SRC testing were given a prestrain of 15% at a strain rate of about 3 x 10^{-4} s^{-1}. The purpose of prestraining is to allow dynamic grain growth to occur in this step, so that no additional grain growth occurs during the strain-rate-change test and thus, a condition of constant structure (stable grain size is maintained) [1,3]. This is particularly important for a fine-grained material when tested at high temperature where dynamic grain growth is possible. Each step in the strain-rate-change tests was carried out in such a way that a steady-state flow stress was observed. Approximately 3% strain was used per step. The SRC tests were preformed in the strain rate range between 3 x10-5 to 5 x 10-2 s-1 in a temperature range between 155 to 300°C (0.50 to 0.67 Tm) in air. Constant strain rate tests in tension were conducted to determine the ductility of the Mg-9Li-5wt.%B$_4$C particulate composite in the superplastic region. Microstructural observations were conducted using light optical microscopy prior to and after deformation. The samples were etched using a 10% hydrochloric acid and 90% ethanol mixture.

3. RESULTS AND DISCUSSION

The microstructure of the Mg-9Li-5wt.%B_4C particulate composite after press-bonding is shown in Figure 1. From Figure 1 several important points are revealed. First, good bonding between the foils was achieved. Second, a highly dense composite is obtained. The composite had a relative density of greater than 99%. Third, the B_4C particles are randomly distributed within the matrix. The particles are typically larger in size than the foil spacing and are primarily located at foil interfaces. Fourth, the Mg-9Li matrix has a fine-equiaxed grain structure after press-bonding. The Mg-9Li matrix shown in Figure 1 exhibits a true grain size of 3.6 µm.

The results of the compression strain-rate-change tests at the various testing temperatures are shown in Figure 2. The data are plotted as logarithm of steady-state strain rate, $\dot{\varepsilon}$, versus logarithm of flow stress, σ. The data in Figure 2 were analyzed using the following relation:

$$\dot{\varepsilon} = K \left(\frac{b}{d}\right)^p \left(\frac{\sigma}{E}\right)^n \exp\left(-\frac{Q_c}{RT}\right) \tag{1}$$

where b is the Burgers vector, p is the grain size exponent, E is the dynamic Young's modulus, Q_c is the activation energy for plastic flow, T is the absolute temperature, R is the gas constant and K is a material constant. The slope of the curves yields the stress exponent according to equation 1. From Figure 2 it is observed that two different stress exponent regions are exhibited as function of the flow stress at all temperatures of testing. At high values of the flow a stress exponent of between 5 to 7 is exhibited. In this region the rate-controlling deformation mechanism is likely a diffusion-controlled dislocation process [4,5]. As the flow stress is decreased a stress exponent of about 2 is observed at all temperatures. Such a value of the stress exponent is generally associated with a deformation process controlled by grain boundary sliding and in this region superplasticity is expected [6,7]. In addition, the 175, 200 and 222°C tests reveal an increase in the stress exponent from 2 at the lowest strain rates, a trend often observed in superplastic materials [7]. Similar behavior was exhibited by the Mg-9Li alloy over the same testing temperature range.

Microstructural investigation of the composite after strain-rate-change testing revealed that the initial grain size of d=3.6 µm remained the same for the three lowest temperatures of testing (155, 175 and 200°C). At the three highest temperatures of testing (222, 250 and 300°C), however, grain growth occurred during prestraining (dynamic grain growth). Grain sizes were measured in the deformed samples in a region of the gage section which was representative of the grain size during strain-rate-change testing. The grain size was 4.6 µm at 222°C, 5.5 µm at 250°C and 8.5 µm at 300°C.

Constant true strain rate tests were performed in a strain rate range where n≈2 and hence, superplastic behavior is expected. The tensile true stress-true strain curve for the Mg-9Li-5wt.%B_4C particulate composite tested at 200°C at a constant true strain rate of 2×10^{-3} s^{-1} is shown in Figure 3. From Figure 3 two important points are noted. First, an elongation-to-failure of 390% is achieved. This value is within the range of 200-1000% that is typical for superplastic metallic alloys [6]. Second, extensive work hardening is exhibited up to a true strain of about 1.1 to 1.2. The extensive work hardening in superplastic metallic alloys is generally attributed to dynamic grain growth that occurs during deformation [3,8].

The activation energy for superplastic flow for the Mg-9Li-5wt.%B_4C particulate composite cannot simply be determined from a plot of logarithm $\dot{\varepsilon}$ versus 1/T at constant stress as a result of the variation in grain size (3.6 to 8.5 µm) with testing temperature. However, it is possible to determine the activation energy for superplastic flow for the Mg-9Li-5wt.%B_4C particulate composite if the grain size exponent in equation 1 is known. It was previously shown that the steady-state strain rate for the Mg-9Li alloy with true grain sizes between 6 to 35 µm in the n≈2 region is inversely proportional to the square of the grain size (p=2) when tested over a similar temperature range as that used in this study. Thus, the activation energy for superplastic flow for the Mg-9Li-5wt.%B_4C particulate composite can be obtained from a plot of logarithm of grain size-compensated strain rate, $\dot{\varepsilon}d^2$, versus 1/T. Figure 4 is such a plot for the composite in the n=2 region at σ=10 MPa. From Figure 4 it is observed that two different activation energies for superplastic flow are exhibited as a function of temperature. At high temperatures, T>222°C, an activation energy for superplastic flow equal to 108 kJ mol^{-1} is exhibited. A modulus correction

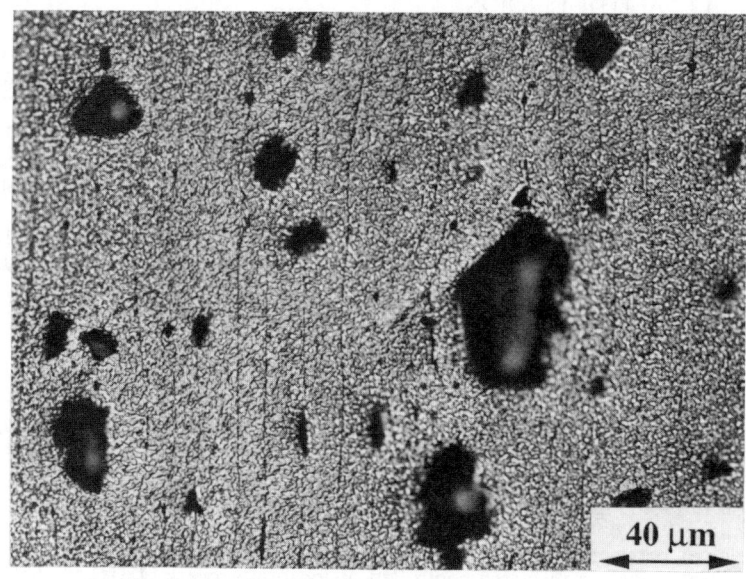

Figure 1. Microstructure of the fine-grained Mg-9Li-5wt.%B_4C particulate composite after press-bonding.

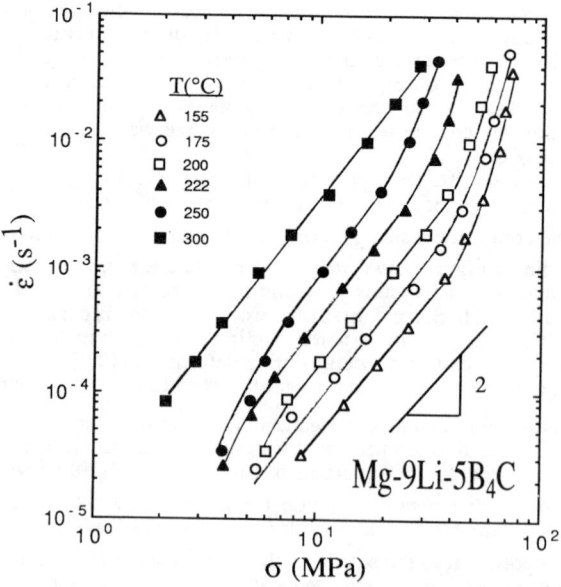

Figure 2. Strain rate vs. flow stress for the fine-grained Mg-9Li-5wt.%B_4C particulate composite at various temperatures.

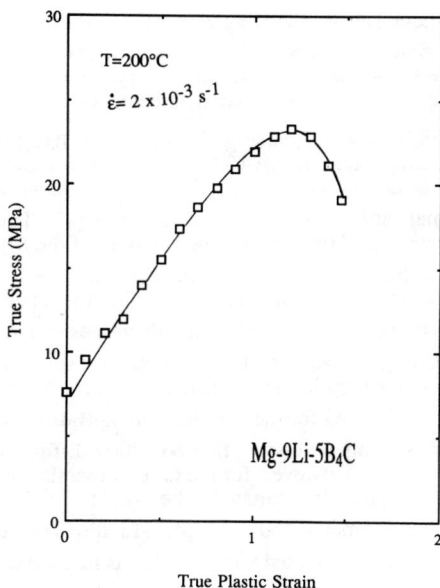

Figure 3. True stress-true strain curve at a constant strain rate for the fine-grained Mg-9Li-5wt.%B$_4$C particulate composite.

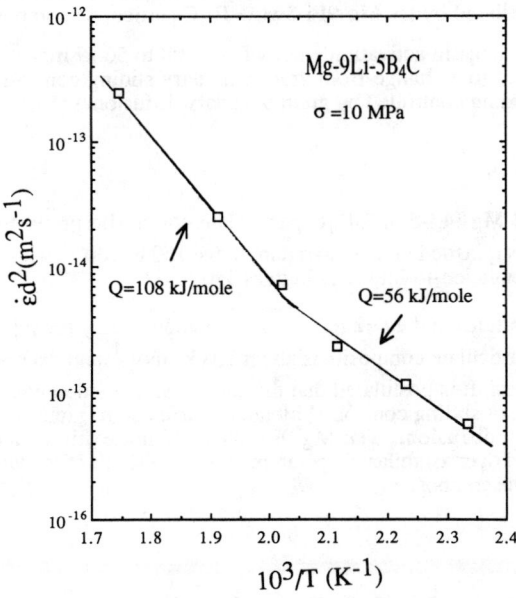

Figure 4. Normalized strain rate vs. inverse temperature for the fine-grained Mg-9Li-5wt.%B$_4$C particulate composite.

factor was not used to calculate Q_c because this correction is small at low homologous temperatures [9]. This value is in excellent agreement with the activation energy for superplastic flow exhibited by the fine-grained Mg-9Li material of 104 kJ mol^{-1} tested over a similar temperature range. The value for the activation energy for superplastic flow for the Mg-9Li-5wt.% B$_4$C particulate composite is also in good agreement with the activation energy for plastic flow of 98 kJ mol^{-1} exhibited by the coarse-grained (d≈70 to 100 μm) as-cast Mg-9Li in the n=5 to 7 region tested in the temperature range 150 to 250°C. The activation energy for plastic flow for the fine-grained and the as-cast Mg-9Li materials was correlated with the activation energy for lattice diffusion in the majority beta phase of 103 kJ mol^{-1} [1]. Thus, it is highly likely that at testing temperatures above 222°C the superplastic behavior of the Mg-9Li-5wt.%B$_4$C particulate composite is controlled by lattice diffusion in the beta phase.

At lower temperatures, T<222°C, the activation energy for superplastic flow for the Mg-9Li-5wt.% B$_4$C particulate composite in the n=2 region decreases to about 56 kJ mol^{-1}. It has been observed that at as the testing temperature and or the grain size is decreased that grain boundary sliding can switch from control by lattice diffusion to control by grain boundary diffusion [6]. The value of 56 kJ mol^{-1} obtained for the Mg-9Li-5wt.% B$_4$C particulate composite studied in this investigation cannot be compared with grain boundary diffusion data for the Mg-9Li alloy, since no such data is available. However, for the case of metals the activation energy for grain boundary diffusion, Q_{gb}, is typically estimated to be equal to 0.6 Q_L (where Q_L is the activation energy for lattice diffusion) for metals [10]. For Mg-9Li alloys this corresponds to an estimated Q_{gb} of about 61 kJ mol^{-1}. The estimated value of Q_{gb} is in good agreement with the measured activation energy of 56 kJ mol^{-1} below 222°C. In addition, phenomenological relations have been developed for determining the activation energy for grain boundary diffusion in metallic alloys. For example, Hwang and Balluffi [11] predict that Q_{gb} is proportional to the absolute melting temperature, with a constant of proportionality equal to 9.35R for 0.42<T/T$_m$<1.0. For Mg-9Li alloys the predicted activation energy for grain boundary diffusion from Hwang and Balluffi is 67 kJ mol^{-1}. The value predicted by Hwang and Balluffi is in reasonable agreement with the value exhibited by the Mg-9Li-5wt.% B$_4$C particulate composite at T<222°C. Thus, it appears that the change in activation energy from 108 to 56 kJ mol^{-1} as the testing temperature decreases is related to a change from grain boundary sliding controlled by lattice diffusion to grain boundary sliding controlled by grain boundary diffusion.

4. CONCLUSIONS

1. A fine-grained Mg-9Li-5wt.%B$_4$C particulate composite prepared using a foil metallurgy technique was superplastic in the temperature range 150 to 300°C with a stress exponent equal to about 2 and an elongation-to-failure as high as 390%.

2. At high temperatures in the n≈2 region the activation energy for superplastic flow for the Mg-9Li-5wt.%B$_4$C particulate composite is about 108 kJ mol^{-1} and decreases to about 56 kJ mol^{-1} at low temperatures. It is postulated that the change in activation energy is a result of a change from grain boundary sliding controlled by lattice diffusion to grain boundary sliding controlled by grain boundary diffusion. The Mg-9Li alloy did not exhibit such a change in activation energy when tested over a similar temperature range as a result of the larger grain sizes in the Mg-9Li alloy than in the composite.

REFERENCES

1. Metenier, P., Gonzalez-Doncel, G., Ruano, O. A., Wolfenstine, J., and Sherby, O. D., 1990, "Superplastic Behavior of a Fine-Grained Two-Phase Mg-9wt.%Li Alloy," *Materials Science and Engineering*, Vol. 125 A, pp. 195-202.

2. Gonzalez-Doncel, G., Wolfenstine, J., Metenier, P., Ruano O. A., and Sherby, O. D., 1990, "The Use of Foil Metallurgy Processing to Achieve Ultrafine Grained Mg-9Li Laminates and

Mg-9li-5B$_4$C Particulate Composites," *Journal of Materials Science*, Vol. 25, pp. 4535-4540.

3. Walser, B., and Sherby, O. D., 1979, "Mechanical Behavior of Superplastic Ultrahigh Carbon Steels at Elevated Temperature," *Metallurgical Transactions*, Vol. 10A, pp. 1461-1471.

4. Mukherjee, A. K., Bird J. E., and Dorn, J. E., 1969, "Experimental Correlations for High-Temperature Creep," *Transactions ASM*, Vol. 62, pp. 155-179.

5. Weertman, J., 1957, "Dislocation Climb Theory of Steady-State Creep," *Journal of Applied Physics*, Vol. 28, pp. 362-364.

6. Sherby, O. D., and Wadsworth, J., 1982, "Development and Characterization of Fine-Grain Superplastic Materials," *Deformation Processing and Structure*, G. Krauss, ed., ASM, Metals Park, OH, pp. 355-389.

7. Edington, J. W., Melton, K. N., and Cutler, C. P., 1976, 'Superplasticity" *Progress Materials Science*, Vol. 21, pp. 63-170.

8. Nieh, T. G., and Wadsworth, J., 1990, "Superplastic Behaviour of a Fine-Grained, Yittria-Stabilized Tetragonal Zirconia Polycrystal (Y-TZP)," *Acta Metallurgica et Materialia*, Vol. 28 [6], pp. 1121-1131.

9. Barrett, C. R., Ardell A. J., and Sherby, O. D., 1964, "Influence of Modulus on the Temperature Dependence of the Activation Energy for Creep at High Temperatures," *Transactions AIME*, Vol. 230, pp. 200-204.

10. Oikawa, H., and Langdon, T. G., 1985, "The Creep Characteristics of Pure Metals and Metallic Solid Solution Alloys," *Creep Behavior of Crystalline Solids*, B. Wilshire and R. W. Evans, ed., Pineridge, United Kingdom, pp. 33-82.

11. Hwang, J. C. M., and Balluffi, R. W., 1978, "On a Possible Temperature Dependence of the Activation Energy for Grain Boundary Diffusion in Metals," *Scripta Metallurgica*, Vol. 12, pp. 709-714.

EFFECTS OF NON-HOMOGENEOUS COMPACTION ON DENSIFICATION OF CERAMIC MATRIX COMPOSITES

L. R. Dharani and Wei Hong
Department of Mechanical and Aerospace Engineering
and Engineering Mechanics
University of Missouri, Rolla
Rolla, Missouri

ABSTRACT

A viscoelastic finite element method is applied to analyze the effect of non-homogeneous compaction on densification of ceramic matrix composites. The heterogeneous matrix of the composite is represented by a two region matrix geometry. The free sintering potential can be expressed in terms of a shrinkage coefficient. The shrinkage coefficient is derived from the relationship of relative density versus sintering time, measured from the pure matrix phase sample. The shrinkage coefficient is a prerequisite for conducting the finite element analysis. Since the information needed to deduce the shrinkage coefficient is not available for all portions of the heterogeneous matrix, two different relationships are assumed, based on the concepts of neck formation and neck growth between contacting particles, and pore coarsening. The results of the finite element analysis show that a non-uniform compaction of the composites generally has a detrimental effect on the densification process.

1. INTRODUCTION

Pressureless sintering densification is a conventional method of processing ceramics. Sintering is a heat treatment process in which ceramic powder compacts are fired in a certain range of temperatures below their melting points. Although its densification rate is relatively low, pressureless sintering has some advantages, such as efficiency for complex shape formation and low cost in fabrication, compared to pressure assisted sintering (hot isostatic pressing, hot pressing and sinter forging) methods. The presence of inert, rigid inclusions (particles or whiskers) usually reduces the sinterability of the matrix phase of ceramic matrix composites by exerting constraints on the matrix (De Jonghe, 1986, Bordia, 1988a and Tuan, 1989). The drastically reduced sinterability has been attributed to many reasons, such as inclusion-induced backstress (Raj, 1984 and Hsueh 1986), heterogeneity (De Jonghe, 1989), network formation (Lange, 1987) and sintering damages (Lange, 1989a). However it has been found that the inclusion induced backstress is too small for the densification rate to reduce so much (Scherer, 1987, De Jonghe, 1988 and Bordia, 1988b,c,d). Sintering damages and network formation are not usually observed in sintering experiments. While the fine grained pure matrix powder can be compacted homogeneously if the particle size distribution is controlled in a narrow band, it is often observed that the density distribution in the matrices is quite heterogeneous when the dispersed second inert, rigid inclusion phase exists in the matrix (Fan, 1992).

A viscoelastic finite element method and a computer code have been developed and utilized to analyze isothermal sintering problems (Hong, 1992). The advantages of finite element analysis over analytical solutions are in dealing with the changeable material parameters, non-homogeneous density distribution, nonlinearity, and the ability to simulate the sintering process over the entire sintering time, if needed. In this paper the modeling covers only the first fifty minutes of sintering time, the initial stage and some portion of intermediate stage, for computational efficiency. For these sintering times, the difference in the final relative densities between the constrained matrix and the unconstrained matrix is about 25%. For the ZrO_2/ZnO inclusion/matrix system considered, the inclusion particle size is about $14\mu m$ and the inclusion/matrix particle ratio is about 47 (Fan, 1992). Although the finite element method is useful in dealing with many features of the sintering problem, our initial concern was about the inclusion induced backstresses. After the finite element analysis showed that the backstresses alone cannot account for the large reduction in densification observed in experiments (Hong, 1992), the efforts are being made to incorporate other factors. While the pure matrix samples are always compacted very homogeneously, it is often observed that the density distributions in the matrix is quite heterogeneous when the dispersed inert, rigid inclusions are present in the matrix. Each of the inclusions is surrounded by a layer of low density matrix region as shown in Fig.1. The reason may be attributed to the mismatch in shape and size between matrix particles and inclusions. In this paper a finite element viscoelastic model is developed to investigate the role of initial compaction of a ceramic matrix composite consisting of viscoelastic matrix and uniformly dispersed rigid inclusions. This is an extension of authors' earlier work (Hong, 1992) in which the details of finite element formulation, numerical solution scheme and the general description of sintering process are described. In the following section a brief description of the features specific to the problem of non-homogeneous compaction is presented.

2. VISCOELASTIC FINITE ELEMENT FORMULATION

The following formulation can be also applied to hot isostatic pressing (HIP) as long as the external pressures are accounted for appropriately. From continuum mechanics viewpoint, the ceramic powder compacts undergoing sintering can be assumed to be viscoelastic. Hence the

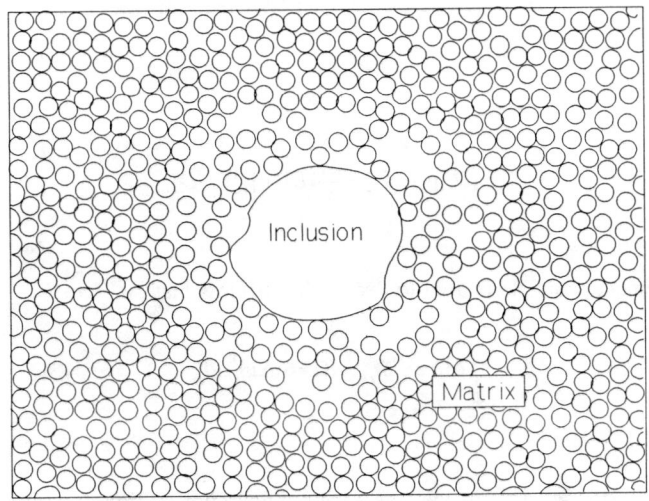

Fig.1. The inclusion is surrounded by a layer of low density region.

total strain rate $\{\dot{\varepsilon}\}$ is a combination of elastic $\{\dot{\varepsilon}^e\}$, viscous $\{\dot{\varepsilon}^v\}$, and shrinkage $\{\dot{\varepsilon}^s\}$ strain rates

$$\{\dot{\varepsilon}\}=\{\dot{\varepsilon}^e\}+\{\dot{\varepsilon}^v\}+\{\dot{\varepsilon}^s\} \tag{1}$$

where

$$\{\dot{\varepsilon}^e\}=[D]^{-1}\{\dot{\sigma}\} \tag{2}$$

$$\{\dot{\varepsilon}^v\}=[\eta]^{-1}\{\sigma\} \tag{3}$$

$$\{\dot{\varepsilon}^s\}=\{\alpha,\alpha,\alpha,0,0,0\}^T \tag{4}$$

where $[D]$ and $[\eta]$ are the elastic and viscosity coefficient matrices respectively and α is the linear shrinkage coefficient.

For an updated Lagrange formulation, the differentials are written in incremental form so that the stress increments can be written as

$$\{\Delta\sigma\}=[D^*]\{\Delta\varepsilon\}-[D^*][\eta]^{-1}\{\sigma\}\Delta t-[D^*]\{\dot{\varepsilon}^s\}\Delta t \qquad (5)$$

where

$$[D^*]=([\,I\,]+\frac{\Delta t}{2}[D][\eta]^{-1})^{-1}[D] \qquad (6)$$

Virtual displacement principle is used to derive the stiffness matrix:

$$\int_{\Omega_t}\delta\{\Delta\varepsilon\}_t^T\{S\}_{t+\Delta t}d\Omega=\delta\{\Delta u\}_t^T\{R\}_{t+\Delta t} \qquad (7)$$

Since the Cauchy stresses $\{\sigma\}_t$ in the reference configuration at time t are same as the 2nd Piola-Kirchholff stresses $\{S\}_t$, then

$$\{S\}_{t+\Delta t}=\{\sigma\}_t+\{\Delta S\}_t \qquad (8)$$

The increments of 2nd Piola-Kirchholff stresses are

$$\{\Delta S\}_t=[D^*]_t\{\Delta\varepsilon\}_{t+\Delta t}-[D^*]_t[\eta]_t^{-1}\{\sigma\}_t\Delta t-[D^*]_t\{\dot{\varepsilon}^s\}_t\Delta t \qquad (9)$$

The resulting 2nd Piola-Kirchholff stresses are converted to Cauchy stresses for use in next time step.

$$[\sigma]_{t+\Delta t}=\frac{\rho_{t+\Delta t}}{\rho_t}[\chi][S]_{t+\Delta t}[\chi]^T \qquad (10)$$

where $[\chi]$ is the deformation tensor.

3. SHRINKAGE COEFFICIENT

In the formulation described earlier, the sintering potential is expressed in terms of the shrinkage coefficient. If the relative density versus sintering time data obtained from isothermal sintering experiments on neat matrix of ZnO powder compact can be expressed as $\rho = f(t)$, this data will be used in determining the shrinkage coefficient, as defined in equation (4), in the matrix phase of the composite as follows:

Time dependent shrinkage coefficient:

$$\alpha = -\frac{f'(t)}{3f(t)} \tag{11}$$

Density dependent shrinkage coefficient:

$$\alpha = -\frac{f'(f^{-1}(\rho_c))}{3\rho_c} \tag{12}$$

where ρ_c is the local relative density in the matrix of the composite.

Density-grain size dependent shrinkage coefficient:

$$\alpha = -\frac{f'(f^{-1}(\rho_c))}{3\rho_c}\left(\frac{1+kf^{-1}(\rho_c)}{1+kt}\right)^{\frac{a}{3}} \tag{13}$$

where k is a constant associated with grain growth and a is a coefficient between 3 and 4 (Hsueh, 1986).

4. MATERIAL PARAMETERS

The material parameters, Young's modulus, viscosity and Poisson's ratio are assumed to depend on the current density as

$$E = E_f \exp[-b_1(1-\rho_c)] \tag{14}$$

$$\eta = \eta_f \exp[-b_2(1-\rho_c)] \tag{15}$$

$$\nu = \frac{1}{2}\sqrt{\frac{\rho_c}{3-2\rho_c}} \tag{16}$$

The parameters used in this paper are: $E_f=125.0 GPa$, $\eta_f=5.833 GPa$, $b_1=1.6$ and $b_2=2.5$. For such a range of Young's modulus considered, the contribution from the elastic strains may be insignificant.

5. NON-HOMOGENEOUS MATRIX

For simplicity in the finite element modeling of the heterogeneous matrix, it is assumed that the matrix is composed of two different regions, the lower density region and the higher density region. The geometry of the lower density region is taken as an annular ring with its width equal to the radius of the inclusion as shown in Fig.2.

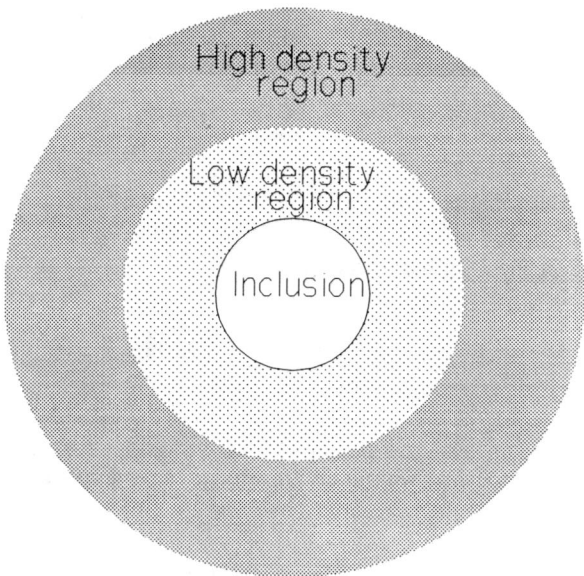

Fig.2. The geometry of the low density region is taken as a ring with its width equal to the radius of the inclusion.

To deduce the sintering potentials of the two regions with different initial density, two homogeneous pure matrix samples, with corresponding densities are needed. However, due to some practical difficulties such information has not been available. This is because that the green samples with much lower density are susceptible to disintegrate while handling although such low density regions can remain intact within a heterogeneous matrix. For this reason, it is necessary to make certain assumptions as to the relationship between the relative density and sintering time for the regions whose initial densities are different from that of the known pure matrix sample. We have proposed the following two models. Fig.3 shows the schematic of the first model in which the solid portion of the curve is from sintering experiment, while the dotted portion of the curve is assumed based on the theory of neck formation and growth between matrix particles.

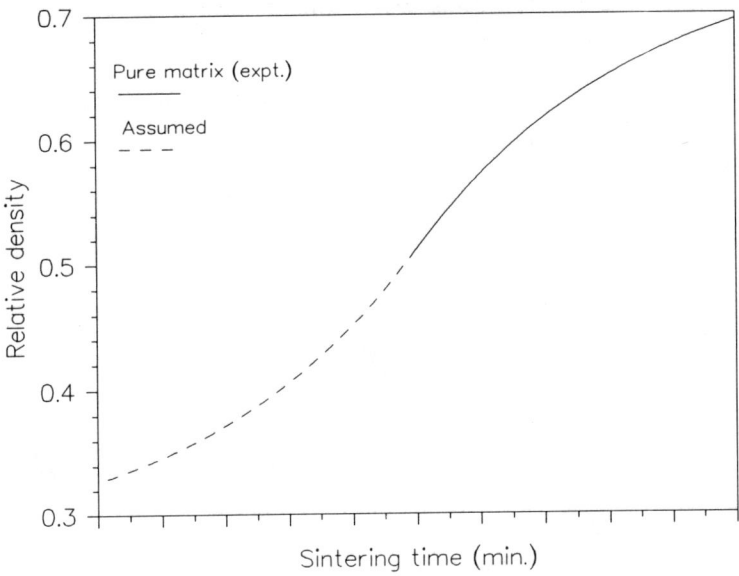

Fig.3. Relation between the relative density and sintering time for neat matrix.

As shown in Fig.4, the neck formation and growth between two spheres results in the centers of these spheres getting closer. Therefore, the average number of the contact points and the size of the contact area of each particle become the dominant factors for the sintering potential. The fewer the average number of the contacting points, the smaller the contact area and hence the weaker the sintering potential. Also, it has been pointed out that high porous materials may actually be less dense after they have been sintered (Reed, 1988). The unusual anti-symmetric shape of the curve reflects these concepts.

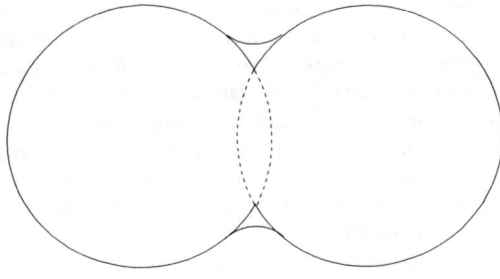

Fig.4. The neck formation and growth between two spherical ceramic particles.

The other assumed relationship for the relative density versus sintering time is shown in Fig.5 in which the solid curve is from sintering experiment, and the rest are assumed.

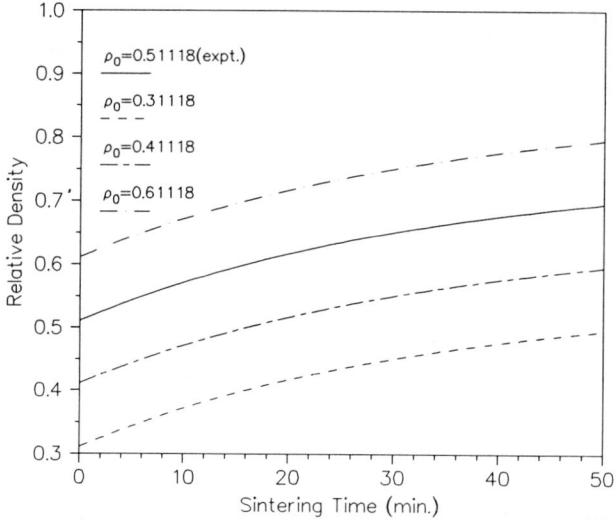

Fig.5. A parallel relation in which the curves of relative density versus sintering time for samples with different initial density are parallel in free sintering.

The curves with different initial densities are assumed to be approximately parallel. Although there have been no data for samples whose initial relative densities are below 0.4, there has been some evidence to support this assumption (Rahaman, 1991). Consideration of pore coarsening is also used in the finite element analysis with the later assumed relationship between relative density and sintering time. Pore coarsening is a phenomenon in which the big pores eat up small pores and get bigger. If a region is constrained in some way, the pore coarsening will be aggravated (Lange, 1989b).

It is postulated that the high porous region may slow down or even stop shrinking under tensile or positive bulk stresses so that the inclusion induced backstresses would not only cause volume creep but also aggravate the pore coarsening. Fig.6(a) shows a cell in the matrix, in which only small pores exist at interstices of matrix particles, and the cell shrinks well since there is no pore coarsening. On the other hand, Fig.6(b) shows a cell in the matrix, in which bigger pores are coarsening, and the cell may slow down or even stop shrinking. In this case, the portions of the cell containing only interstitial pores are still shrinking while the big pores become bigger by acting as a sink for the interstitial pores. To incorporate these concepts to the finite element analysis, it is assumed that

$$\text{if} \quad \sigma_{\text{eff}} \geq \sigma_{cr}(\rho), \text{ then } \alpha = 0.$$

where σ_{eff} is the effective stress, $\sigma_{cr}(\rho)$ is a critical stress and ρ is the local relative density.

6. RESULTS AND DISCUSSION

The results of finite element analysis adopting the two assumed relationships of relative density versus sintering time, the single curve relation (Fig.3) and the parallel relation (Fig.5), using the density dependent shrinkage coefficient, are shown in Fig.7. The critical stress is chosen as $\sigma_{cr}(\rho)=18.10\rho(MPa)$, the initial relative density for the higher density region $\rho_0^h=0.55$. The initial relative density for the lower density region ρ_0^l is then calculated to make the overall initial relative density for the matrix sample $\rho_0=0.5112$.

The difference in the two models are mainly in the nature of the assumed relationships for relative density versus sintering time. However both are based on the same microstructural mechanisms and experimental data, and predict significant reductions in densification rates for the non-homogeneous matrix. The pore coarsening is significant in low initial density regions even in the initial stage of the sintering. However, the model lacks verification and precise details of the mechanism bridging the gap between microstructural behavior and macroscopic parameters. A systematic experimental investigation into this issue is needed.

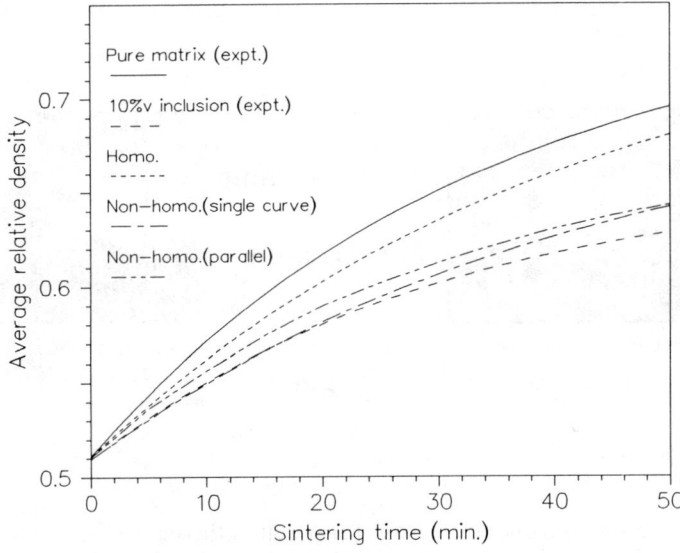

Fig.7. Predicted relative density versus sintering time for the heterogeneous matrix.

7. CONCLUSION

Non-homogeneity in the green compacts may be one of the reasons for the reduction in sinterability of a reinforced matrix. The heterogeneous matrix has less potential to shrink than a homogeneous pure matrix whose overall average density is same as that of the heterogeneous one. The inclusion induced viscoelastic backstress alone cannot directly reduce the densification rates very much, however it may indirectly reduce the densification rates by aggravating the pore coarsening.

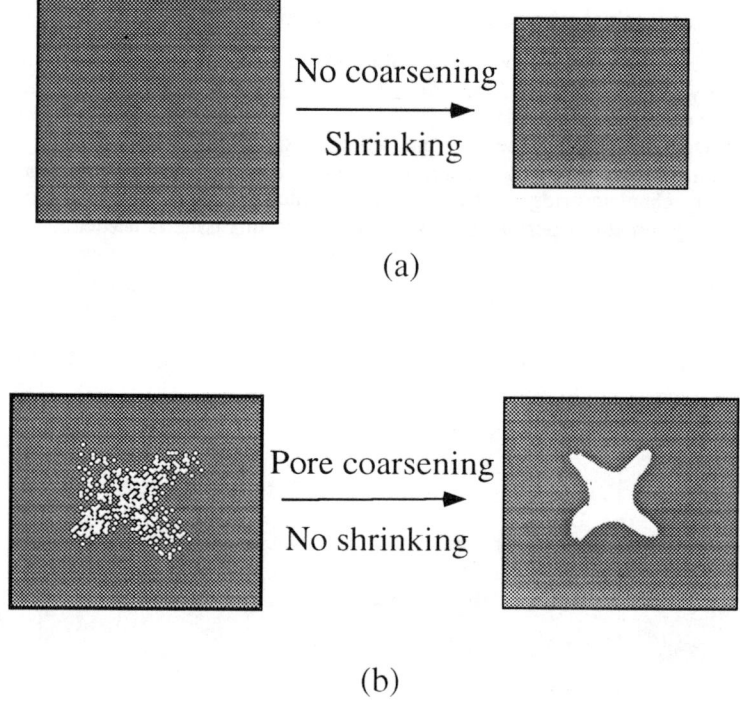

Fig.6. Two extreme cases: (a) A cell in the matrix, containing only interstitial pore, shrinks well in sintering. (b) A cell in the matrix, containing big proes, may slow down or even stop shrinking if it is somehow constrained.

ACKNOWLEDGEMENTS

The authors gratefully acknowledge the support of U.S. Air Force Office of Scientific Research (Grant AFOSR G-90-0267). The authors also express their gratitude to Dr. M. N. Rahaman, University of Missouri-Rolla, for his invaluable discussions and for providing the experimental data on sintering.

REFERENCES

Bordia, R. K. and Raj, R., 1988a, "Sintering of TiO_2 - Al_2O_3 Composites: A model Experimental Investigation", *Journal of American Ceramic Society*, Vol. 71 (4), pp. 302-310.

Bordia, R. K. and Scherer G. W., 1988b, "On constrained Sintering-I. Constitutive Model For A Sintering Body", *Acta Metallurgy*, Vol. 36 (10), pp. 2393-2397.

Bordia, R. K. and Scherer G. W., 1988c, "On Constrained Sintering-II. Comparison of Constitutive Models", *Acta Metallurgy*, Vol. 36 (10), pp. 2399-2409.

Bordia, R. K. and Scherer G. W., 1988d, "On Constrained Sintering --III. Rigid Inclusions", *Acta Metallurgy*, Vol. 36 (10), pp. 2411-2416.

De Jonghe, L. C., Rahaman, M. N., and Hsueh, C. H., 1986, "Transient stresses in Bimodal Compacts During sintering", *Acta Metallurgy*, Vol. 34 (7), pp. 1467-1471.

De Jonghe, L. C., Rahaman M. N., 1988, "Sintering Stress of Homogeneous and Heterogeneous Powder Compacts", *Acta Metallurgy*, Vol. 36 (1), pp. 223-229 (1988).

De Jonghe, L. C. and Rahaman, M. N., 1989, "Densification of Particulate Ceramic Composites: The Role of Heterogeneities", pp. 353-361 in *Material Research Society Symposium Proceedings, Vol 155: Processing of Advanced Ceramics*. Edited by Aksay, I. A., The Materials Research Society, Pittsburgh, PA.

Fan, C. L. and Rahaman, M.N., 1992, "Factors controlling for the sintering of Ceramic Particulate Composites", I: Conventional Processed Composites", Communicated to *Journal American Ceramic Society*.

Hong, W. and Dharani, R. L., 1992, "Modeling constrained Sintering Problem by A Finite Element Method", Communicated to *Mechanics of Materials*.

Hsueh, C. H., Evans, A. G., Cannon, R. M., and Brook, R. J., 1986, "Viscoelastic Stresses and Sintering Damage in Heterogeneous Powder Compacts", *Acta Metallurgy*, Vol. 34 (5), pp. 927-936.

Lange, F. F., 1987, "Constrained Network Model for Predicting Densification Behavior of Composite Powders", *Journal of Materials Research*, Vol. 2 (1), pp. 59-65.

Lange, F. F., 1989a, "Densification of Powder Rings Constrained by Dense Cylindrical Cores", *Acta Metallurgy*, Vol. 37 (2), pp. 697-704.

Lange, F. F., Lam, D. C. C. and Sudre O., 1989b, "Powder Processing and Densification of Ceramic Composites", Materials Research Society Symposium Proceedings. Vol. 155, pp. 309-318.

Raj R. and Bordia R. K., 1984, "Sintering of Bimodal Powder Compacts", Acta Metallurgy, Vol. 32 (7), pp. 1003-1019.

Reed, J. S., 1988, *Introduction to the Principles of Ceramic Processing*, p. 440, Wiley Interscience.

Scherer, G. W., "Sintering with Rigid Inclusions", *Journal of American Ceramic Society*, Vol. 70 (10), pp. 719-725.

Rahaman, M. N. and De Jonghe, L. C. 1991, "Sintering of Ceramic Particulate Composites: "Effect of Matrix Density", *Journal American Ceramic Society*, Vol. 74 (2), pp. 433-536.

Tuan, W. H., Gilbart, E., and Brook, R. J., 1989, "Sintering of Heterogeneous Ceramic Compacts, Part I: Al_2O_3 - Al_2O_3", *Journal of Material Science*, Vol. 24, pp. 1062-1068.

HIGH TEMPERATURE RUPTURE MECHANISMS IN A PARTICULATE REINFORCED INTERMETALLIC MATRIX COMPOSITE

H. K. Kim, X. Liang, Enrique J. Lavernia, and J. C. Earthman
Department of Mechanical and Aerospace Engineering
Materials Science and Engineering
University of California, Irvine
Irvine, California

ABSTRACT

The mechanisms that facilitate high temperature rupture in a spray deposited Ni_3Al matrix composite were studied. In this work, creep specimens were tested to failure under constant stress conditions ranging from 180 to 350 MPa. Metallographic examinations of tested specimen cross-sections reveal that cavitation damage is favored on the grain boundaries of the Ni_3Al matrix as opposed to large Ni_3Al/TiB_2 interfaces. The results also suggest that the cavities nucleate at fine carbides that result from the decomposition of the SiC_p. These observations account for relatively short rupture times observed for the $Ni_3Al/SiC/TiB_2$ composite compared to those for monolithic Ni_3Al.

1. INTRODUCTION

Intermetallic compounds have recently been the subject of considerable interest for high temperature applications. In particular, Ni_3Al is one of the more promising compounds as it exhibits increasing strength with temperature up to about 1030K, a relatively low density and good oxidation resistance [1-5]. In addition, the tensile ductility of this compound is significantly improved with trace additions of boron [2]. Intermetallic-matrix composites (IMCs) are potentially useful at elevated temperatures since they combine matrix properties of oxidation resistance and high temperature stability with reinforcement properties of high specific strength and modulus [6-8]. However, these materials can not be used successfully until a thorough understanding of their relevant high temperature properties is achieved.

It has been shown that creep rupture of Ni_3Al is facilitated by the nucleation, growth and coalescence of grain boundary cavities [9]. It follows that cavitation is also a controlling mechanism in the rupture of particulate reinforced Ni_3Al matrix composites at elevated temperatures. However, it is not immediately clear whether the most damaging cavitation occurs on the matrix grain boundaries or on the reinforcement/matrix interfaces. One of the factors adding to the difficulty of predicting where the dominant damage develops in intermetallic composites at high temperatures is the formation of interfacial reactions between matrix and reinforcement during processing and service. It has been shown that the development of interface reaction zone with some thickness is desirable for establishing good interfacial bonding, while overgrowth of the interfacial layer is detrimental to the mechanical properties of composites [10].

The objective of the present study was to develop a better understanding of the nature of creep rupture mechanisms in a Ni_3Al matrix composite produced by spray atomization and co-deposition. The selection of SiC particulates was prompted by the results of Chou and Nieh [11] which showed that strong solid state reactions developed between SiC and Ni_3Al, and multi-reaction layers were always generated in the interdiffusion zone. Similarly, titanium diboride was chosen because is also known to react with a Ni_3Al [12]. It was also selected on the basis of the work by other investigators that suggests that this type of particulate provides attractive combinations of properties in IMCs [6]. One of the aims of the present work was to determine if particle/matrix reactions could be avoided by rapid solidification processing. Another goal, which is addressed in the present paper, was to determine the effect interfacial reaction zones have on the high temperature mechanical properties of an intermetallic matrix composite. This goal was achieved by comparing the high temperature properties of the monolithic Ni_3Al and a $Ni_3Al/SiC/TiB_2$ IMC. Emphasis is placed on the roles of the reinforcement/matrix reaction products in controlling the creep life of specimens.

2. EXPERIMENTAL PROCEDURE

2.1. *Materials*

The nominal composition of the Ni_3Al used in the present study (designated IC-396) is 8.0 Al, 7.7 Cr, 0.85 Zr, 3.0 Mo, 0.005 B, bal Ni (in wt. pct.). The reinforcement SiC particulates (HCP α–phase) had an average size of 3 μm, whereas the TiB_2 particulates (C32 structure) had approximate average size of 15 μm prior to processing. During spray atomization and co-deposition, the Ni_3Al alloy was disintegrated into a dispersion of droplets using high energy N_2. Simultaneously, a mixture of TiB_2 and SiC particulates with the same volume fraction ratio was co-injected into the atomized spray at a previously determined spatial location. The primary experimental variables used in the processing procedure are given in [13].

The resulting Ni_3Al IMC was then hot extruded 9:1 at 1473K to eliminate micrometer sized pores that normally develop during spray deposition. The volume fraction of ceramic particulates was quantitatively characterized using two techniques: (a) image analysis and (b) acid dissolution. The results from both techniques indicate a total ceramic particulate volume fraction of 12.7% in the composite. It is not possible to differentiate between the SiC and TiB_2 particulates strictly on the basis of appearance. However, it is worth noting that the large particulates are most likely TiB_2, considering the average size of both SiC and TiB_2. Also, it is reasonable to assume that the composite contains approximately equal volume fractions of TiB_2 and SiC [13]. For comparison, the same alloy was spray deposited without the reinforcing particulates and extruded under the same conditions used for the composite.

2.2. High Temperature Testing

Tensile creep specimens were ground to size with a gage diameter of 3.2 mm and a gage length of 15.9 mm. Constant tensile stress conditions ranging from 180 to 350 MPa were imposed using an automated high temperature testing system. The specimens were tested at 1033K in a vacuum environment of 10^{-3} Pa. Longitudinal sections of tested specimens were examined using scanning electron microscopy (SEM) to determine the extent and location of creep damage within the composite microstructure. In making this determination, an ImageSet image analysis system (Dapple Systems Inc.) was used to measure the size of over 300 grain boundary cavities at different locations on each longitudinal section.

3. EXPERIMENTAL RESULTS AND DISCUSSION

3.1. Microstructural Observations

Figure 1 shows the microstructure of the hot-extruded $Ni_3Al/SiC/TiB_2$ composite revealing a relatively homogeneous distribution of both types of the reinforcing particles in the Ni_3Al matrix. The matrix grains recrystallized during hot extrusion, as indicated by their equiaxed morphology. It has been shown in an earlier work that a large number of carbides exist on the grain boundaries in the composite [13]. The observation that TiB_2 particulates are generally present at the grain boundaries of the matrix suggests that these particulates hindered grain growth in the matrix during hot extrusion.

The chemical reactions at the Ni_3Al/TiB_2 interface are evident from Figure 2, which shows TiB_2 particulates surrounded by well-defined reaction zone layers. The reaction zone appears to have increased in thickness by consuming both the matrix and the TiB_2 particulates, leading to the formation of two distinct interfacial reaction layers (labeled 1 and 2). Interfacial reaction layer 1 formed inside the TiB_2 particulate (i.e., at the TiB_2/reaction zone interface), while interfacial reaction layer 2 formed at the reaction zone/Ni_3Al matrix interface. Energy Dispersive X-ray Spectroscopy (EDS) analysis has shown that interfacial reaction layer 1 primarily consists of Ti, Ni, and Cr [13]. Furthermore, selected area diffraction using transmission electron microscopy (TEM) has indicated that this layer is Ni_3Ti which has a hexagonal crystal structure (DO_{24}--type). Interfacial reaction layer 2 is characterized by a very high Ni content and a relatively low Al content [13]. Finally, it is noted that microvoids were observed within the interface region that were 0.4 - 0.9 µm long and 0.1 - 0.4 µm wide.

Although X-ray diffraction studies conducted on the composite revealed the presence of SiC [13], these particulates were not evident from scanning electron microscope (SEM) observations combined with and EDS analysis. It appears that the SiC particulates were substantially reduced in size or eliminated during both the spray deposition and extrusion processes. In related studies, Chou and Nieh [11] noted that SiC reacts extensively with pure Ni_3Al during annealing at 1273K leading to the formation of multiple reaction layers in the interdiffusion zone at the interface. It appears that the temperatures during deposition (1100-1600K) provide sufficient thermal energy for the alloying elements Cr, Mo, Zr to segregate to the Ni_3Al/SiC interface. These alloying elements are then able to replace the Si in the SiC lattice leading to the formation of other carbides. Concurrently, diffusion of Si into the Ni_3Al matrix may also lead to the formation of compounds with Ni and Al, as reported by Chou and Nieh [11]. In addition, the high surface to volume ratio associated with the fine (3 µm) SiC particulates used in the present experiment may also have contributed towards their high reactivity.

3.2. Creep Behavior

Typical creep curves for specimens tested under 300 MPa are illustrated in Figure 3. Data for spray deposited and extruded monolithic Ni_3Al specimens are included for comparison. Steady state strain rate is plotted against applied stress in Figure 4 for both composite and monolithic Ni_3Al specimens. Least square fit analyses of the data in this figure give Norton law constants of

$$\dot{\varepsilon} = 3.4 \times 10^{-16} \sigma^{3.44} \text{ for monolithic } Ni_3Al$$
$$= 1.2 \times 10^{-15} \sigma^{3.50} \text{ for } Ni_3Al/SiC/TiB_2$$

where strain rate, $\dot{\varepsilon}$, and applied stress, σ, are in units of s^{-1} and MPa, respectively. The experimental data reveal that the stress exponent for power law creep for both materials is about 3.5, suggesting that the creep deformation for both materials is predominantly controlled by dislocation glide processes. However, the composite specimens crept about 6 times faster than the monolithic Ni$_3$Al specimens. It is not immediately clear why the composite creeps faster than unreinforced Ni$_3$Al under a given stress. Composite hardening is known to be due to a difference in plastic strain of the components compensated by elastic strains that give rise to back stresses. It is often assumed that there is no difficulty in transferring load between matrix and reinforcement in order to prevent the development of local incompatibility stresses. The shear-lag theory [14] can be used to predict composite strength, even though the reinforcing phase possesses a small aspect ratio as in the case of a particulate reinforced composite [15]. However, weak interfaces can have a substantial effect on this load transfer and, as a result, on the creep behavior of composites. In the limit of zero interface strength, where no load can be transferred to the reinforcement, a composite is weaker than the matrix alone [16]. As mentioned before, the processing of the composite apparently results in the decomposition of the SiC particulates and the formation of other carbide phases, such as Cr_3C_2--type carbides, and the formation of the Ni-rich interfacial phase at the Ni$_3$Al/TiB$_2$ interface. This could give rise to a weakening of the interfaces present in the composite.

Flinn [17] has proposed that the steady-state creep of Ni$_3$(Al, Fe) polycrystals for the temperature range used in the present investigation is controlled the viscous drag on dislocation pairs associated with the antiphase boundary (APB) connecting the members of the pair. Since the rate-controlling process is the diffusion of atoms to and from the APBs, impurity elements affecting lattice diffusion could play a role in the creep behavior of Ni$_3$Al. In this case, impurity elements in a composite may increase lattice diffusion and therefore increase the creep rate. It is likely that the present composite contains significant amounts of Si, resulting from the decomposition of SiC, which could then increase lattice diffusion. It is also possible that the precipitation of another soft phase containing Si or C might attribute to the weaker creep strength of the composite compared to that for the monolithic material.

The grain size of the composite is about 10 μm, whereas the grain width of the monolithic Ni$_3$Al is about 25 μm with grain aspect ratio of 5. The finer and more equiaxed grain size present in the composite relative to that of the monolithic material may also account for the faster creep rate of the composite, if grain boundary sliding makes a significant contribution to the overall straining of the specimens. The shear stresses on grain boundaries would be generally lower in the monolithic material due to the more columnar grain morphology. Thus grain boundary sliding would likely have a negligible role in contributing to the creep rate and redistributing stresses from inclined sliding inclined boundaries to cavitating boundaries that are transverse to the applied tensile stress.

3.3. *High Temperature Rupture*

Figure 5 shows a double logarithmic plot of creep rupture lifetime for both the monolithic and particulate reinforced Ni$_3$Al materials as a function of applied stress. The monolithic material exhibits a longer lifetime than the composite for the same applied stress in this plot. This finding suggests that the mechanism that causes rupture is assisted by the presence of the reinforcing particulates. We note for the monolithic material that the rupture time and the steady state strain rate have a similar exponential dependence on stress. In this case it is reasonable that the creep and fracture processes are closely linked. The data for the composite, on the other hand, indicate that the exponential dependence of rupture time on stress is somewhat different from the exponential dependence of strain rate on stress for this material.

3.4. *Microstructural Observations*

Cavity nucleation is generally favored at locations where diffusion is rapid and atomic bonding is relatively weak. Accordingly, reinforcement/matrix interfaces as well as the matrix grain boundaries represent possible preferred cavity nucleation sites in the present IMC. Thus, in order to develop a better understanding of the rupture mechanisms in this material it is first necessary to determine at which of these locations cavities nucleate preferentially. Longitudinal sections of ruptured specimens have been examined using SEM to determine the extent and location of creep damage within the gage sections. Observations were also made on tested material from the grip sections of specimens which experienced a relatively low stress (< 25% of the stress in gage section). This was done to avoid difficulties in identifying the preferred cavity nucleation sites in the heavily damaged gage sections.

A typical SEM micrograph of tested grip section material from a composite specimen tested under 300 MPa is shown in Figure 6. As indicated in this micrograph, larger cavities are observed at the Ni$_3$Al grain boundaries as opposed to TiB$_2$/Ni$_3$Al interfaces of comparable size and orientation. As mentioned previously, a large number of carbides exist on the grain boundaries in the composite [13]. Apparently, the grain boundary cavities nucleated at fine carbides that resulted from the decomposition of SiC particulates. It is possible that the nucleation of grain boundary cavities at the carbides occurs easily due to a relatively weak interfacial cohesive strength. On the other hand, the rare presence of large cavities at the Ni_3Al/TiB_2 reaction zones indicates that either cavity nucleation is difficult or cavity growth is slow at these interfaces. This finding is consistent with the microstructural observations of coherent interfacial bonding along the Ni_3Al/TiB_2 interface reported in an earlier work [13].

A representative micrograph of gage section material is shown in Figure 7 for a specimen tested to rupture under 180 MPa. Again in this figure we note that more cavities appear to lie on the matrix grain boundaries than on the Ni_3Al/TiB_2 interfaces. It is also apparent that the grain boundary cavities are larger than those on the interfaces. This observation is confirmed in a plot of relative cavity frequency versus cavity size for both grain boundaries and matrix/TiB_2 interfaces shown in Figure 8. A mean cavity size of 1.3 μm was measured for the matrix grain boundaries while a value of 0.7 μm was measured for cavities at the Ni_3Al/TiB_2 interfaces. However, it can be seen in Figure 8 that the overall frequency of cavities at matrix/TiB_2 interfaces was somewhat higher than that for the matrix grain boundaries. This is reasonable considering that the frequency decreases with increasing cavity size due to the increased likelihood of cavity coalescence. Thus, although significant cavity nucleation does occur on the matrix/reinforcement interfaces, it appears the cavity growth at the matrix grain boundaries is significantly faster. Thus, the microstructural measurements indicate that cavitation at matrix grain boundaries is responsible for the relatively short rupture times observed for the composite.

The overall results indicate that it is the Ni_3Al grain boundaries, and not the Ni_3Al/TiB_2 interfaces, that are most susceptible to cavitation damage in the present composite under creep conditions. One possible explanation for this finding is that the carbide precipitates resulting from the dissolution of SiC enhance the cavity nucleation process at the grain boundaries. Cavity nucleation at iron carbides has been well documented for several steel alloys [18, 19]. Grain boundary carbides in the present IMC could also serve as preferred nucleation sites. Although large interfacial reaction zones form at the Ni_3Al/TiB_2 interfaces, they do not appear to cavitate as readily as the matrix grain boundaries. Thus, it is expected that longer lifetimes could be achieved if TiB_2 is co-injected as the sole reinforcement phase for the present Ni_3Al alloy, even though there is a reaction between the particulates and the matrix.

4. CONCLUSIONS

i) The stress exponents for power law creep for an unreinforced and particulate reinforced Ni_3Al alloy are about 3.5, suggesting that the creep mechanisms for both materials are primarily controlled by dislocation glide processes. However, the creep rate for a composite specimens under a given stress is greater than that for an unreinforced Ni_3Al specimen. It appears that differences in the grain morphology are responsible for this difference in creep behavior. Also, soft phases resulting from the decomposition of SiC in the composite could contribute to this disparity.

ii) Monolithic Ni_3Al also exhibits longer creep lifetimes than that for the present Ni_3Al matrix composite. A difference in the magnitude of the stress exponents for steady-state strain rate and rupture time for the composite indicates that mechanisms other than dislocation creep have a significant effect on the rupture time.

iii) Microstructural observations indicate that the cavitation damage is more extensive on Ni_3Al grain boundaries compared to the cavitation on Ni_3Al/TiB_2 interfaces. Since grain boundary cavities appear to nucleate at carbides resulting from the dissolution of SiC, it is apparent that longer lifetimes could be achieved if TiB_2 is co-injected as the sole reinforcement phase for the present Ni_3Al alloy.

Acknowledgments

The authors would like to thank the Air Force Office of Scientific Research (grant no. AFOSR-90-0366) for their support of this work. Funding from the National Science Foundation under grant no. MSS 8957449 for the acquisition of the image analysis system used in the present investigation is also gratefully acknowledged.

References

1. N.S. Stoloff, *Inter. Mat. Rev.*, 34, 153 (1989).
2. K. Aoki and O. Izumi, *J. Jpn. Inst. Met.*, 43, 1190 (1979).
3. V.K. Sikka, in *Casting of Near Net Shape Products*, ed. Y. Sahai, J.E. Battles, R.S. Carbonara and C.E. Mobley, AIME, Warrendale, PA, 315 (1988).
4. C.T. Liu and V.K. Sikka, *J. Met.*, 5, 19 (1986).
5. K. Aoki, Mater. Trans. *JIM*, 31, 443 (1990).
6. J.M. Yang, W.H. Kao and C.T. Liu, *Mater. Sci. Eng.*, 107A, 81 (1989).
7. S. Nourbakhsh, F.L. Liang and H. Margolin, *Adv. Mat. & Manuf. Pro.*, 34, 57 (1988).
8. G.L. Povirk, J.A. Horton, C.G. McKamey and T.N. Tiegs, *J. Met. Sci.*, 23, 3945 (1988).
9. J.H. Schneibel and L. Martinez, *Acta Metall.*, 37, 2237 (1989).
10. T.W. Chou, A. Kelly and A. Okura, *Composite*, 16, 187 (1985).
11. T.C. Chou and T.G. Nieh, *J. Mater. Res.*, 5, 1985 (1990).
12. P. Angelilni, P.F. Becher, J. Bentley, C.B. Finch and P.S. Sklad, in J.H. Crawford, Y. Chen and W.A. Sibley (eds.), *Dect Properties and Processing of High-Technology Nonmetallic Materials*, Materills Research Society, Pittsburgh, PA, 299 (1984)
13. X. Liang, H.K. Kim, J.C. Earthman and E.J. Lavernia, *J. Mater. Sci. & Eng.*, in press.
14. A. Kelly and K.N. Street, *Proc. R. Soc. Lond.*, 328A, 267 (1972).
15. K.T. Park, PH.D. Disertation, University of California, Irvine, (1992).
16. S. Goto and M. McLean, *Acta Metall.*, 39, 165 (1991).
17. P.A. Flinn, *Trans. Met. Soc. AIME*, 218, 145 (1960).
18. E.P. George, P.L. Li and D.P. Pope, *Proc. of the Conference on Creep and Fracture of Engineering Materials and Structures*, Swansea, Inst. of Metals, London (1987).
19. G. Eggeler, J.C. Earthman, N. Nilsvang and B. Ilschner, *Acta Metallurgica*, 37, 49 (1989).

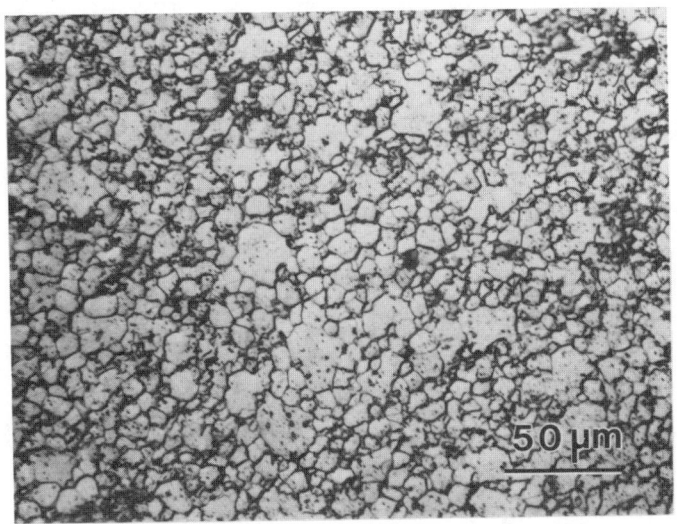

Figure 1. Microstructure of spray deposited Ni3Al/SiC/TiB2 following hot extrusion at 1473K.

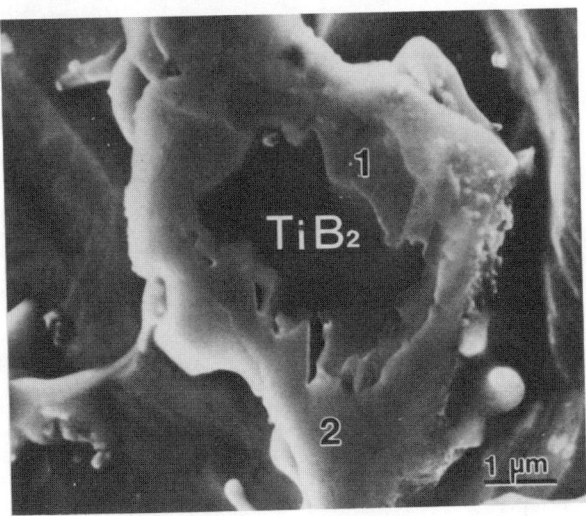

Figure 2. SEM micrograph showing the interfacial reaction zones that form around the TiB2 particulates.

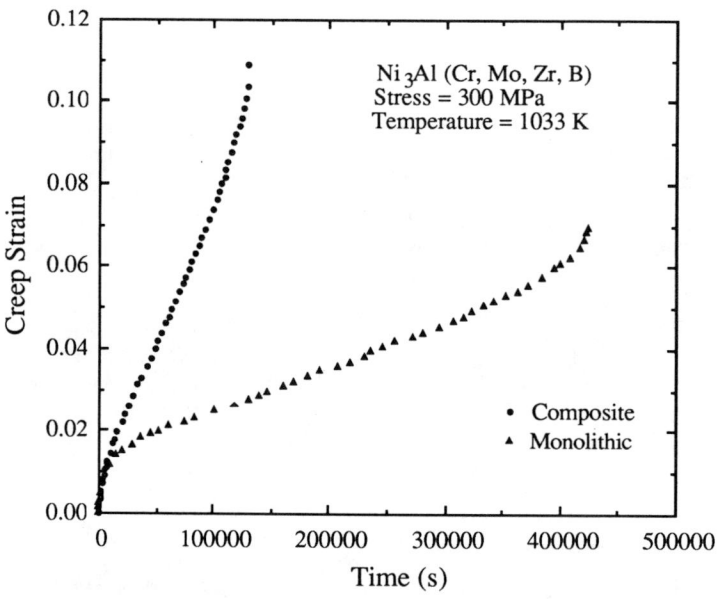

Figure 3. Typical creep curves for the present monolithic and composite materials.

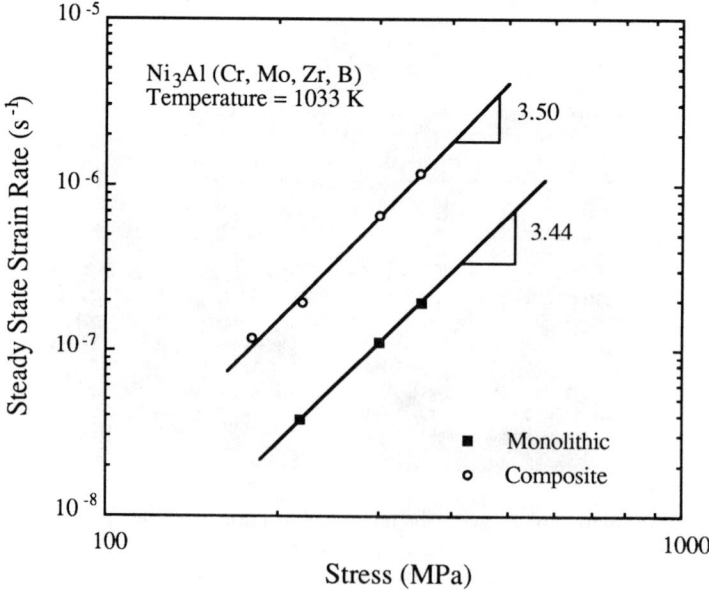

Figure 4. Strain rate versus applied stress for the monolithic and composite materials.

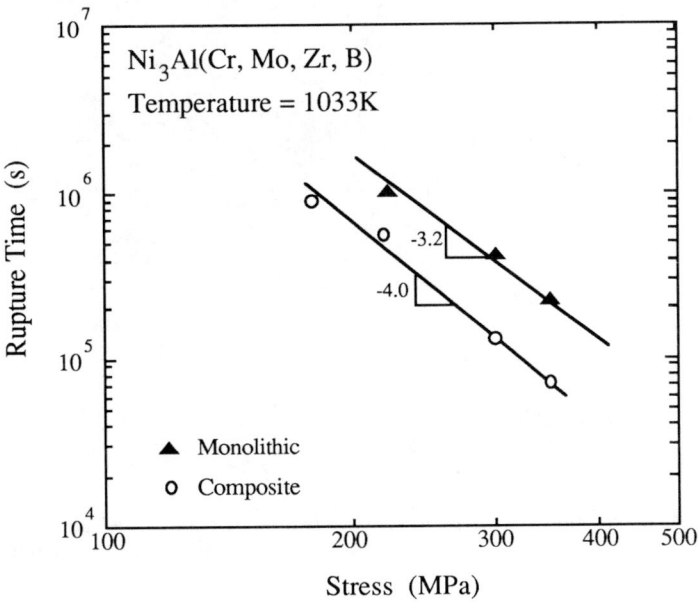

Figure 5. Rupture time versus applied stress for the monolithic and composite materials at 1033K.

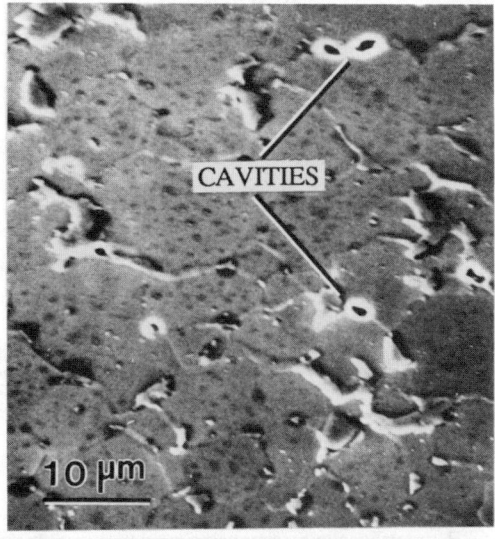

Figure 6. SEM micrograph of a region in the grip section of a composite specimen tested under 300 MPa. The tensile stress axis is in the vertical direction for this micrograph.

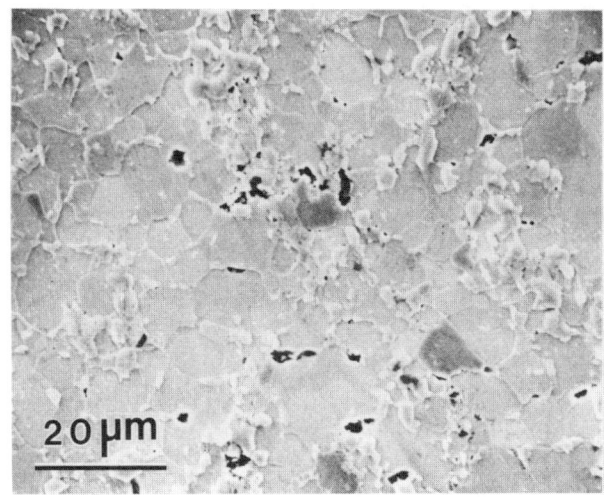

Figure 7. SEM micrograph of a region in the gage section of a composite specimen tested under 180 MPa. The stress axis is along the vertical edge of this micrograph.

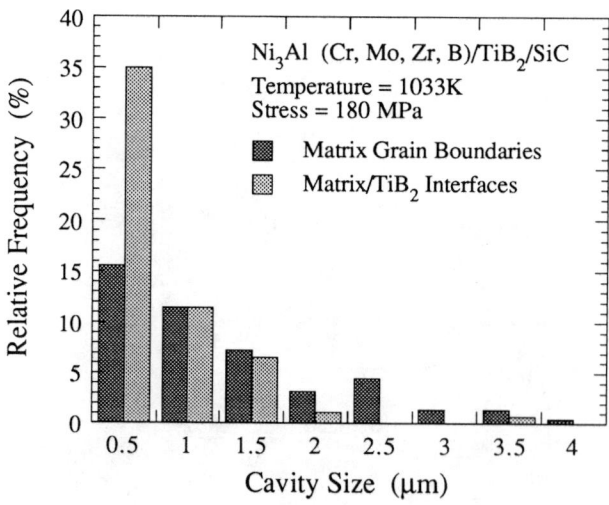

Figure 8. Relative frequency versus cavity size for both matrix grain boundaries and Ni_3Al/TiB_2 interfaces in a composite specimen tested under 180 MPa.

ROLE OF BACK STRESSES IN THE CREEP OF MOLYDISILICIDE COMPOSITES

K. Sadananda and C. R. Feng
Materials Science and Technology Division
Physical Metallurgy Branch
Washington, D.C.

ABSTRACT

Back stresses or internal stresses that arise during deformation of composites play a dominant role in resisting the deformation of the composite. These stresses can vary with applied stress and temperature thereby affecting the experimental evaluation of thermal and athermal contributions to the flow stress. In some cases, these stresses are considered as threshold stresses below which composite does not deform significantly or deforms anelastically, and hence their evaluation is important from design considerations. In addition, since these stresses are a function of applied stress and temperature, caution should be exercised in interpreting the experimental evaluation of strain rate-stress exponents and activation energies.

The magnitude of back stresses were determined for molybdenum disilicide composites and their role on the creep deformation was examined. It was shown that for the systems under consideration, the back stresses were small and did not contribute significantly to creep deformation. The results imply that anelastic strains are also expected to be small for molydisilicide whisker reinforced composites.

I. INTRODUCTION

There is an increasing realization that strength and toughness at high temperatures can not be obtained using simple monolithic systems and that a composite approach is the only viable means to obtain the desired properties. For high temperature applications, materials should not only possess strength and reasonable toughness, but also microstructural stability, oxidation resistance and creep resistance. At temperatures below 1000^0C, nickel base superalloys are generally adequate for most of the existing needs, and currently these are the best materials available for gas turbine applications. On the other hand, for temperatures above 1000^0C, ceramic based systems are the major contenders. But in recent years, molybdenum disilicide based systems are being explored as alternatives to ceramics because of their metallic nature, oxidation resistance and significant ductility at these high temperatures. Attempts [1-3] are being made to optimize the strength of molybdenum disilicide using particulate and whisker composites.

Since creep strength is an important requirement for the long term high temperature structural applications, an extensive study has been made recently [4-5] on the creep deformation behavior of molybdenum disilicides as function of stress and temperature (1100 - 1400^0C) for several systems including monolithic $MoSi_2$, a solid solution of 50%$MoSi_2$ and 50%WSi_2, and their composites containing 20% volume fraction of SiC whiskers. The presence of reinforcements are found to decrease the creep rates significantly. Since SiC at these temperatures can not deform plastically, the reinforcements enhance the strength by providing obstacles to dislocation glide in the matrix. Normally this builds up internal stresses or back stresses, and an understanding of these back stresses are important in evaluating a) the existence of threshold stress for creep, and b) modification of creep exponents and activation energies for creep.

2. MATERIALS AND METHODS

Four materials were selected for the study; monolithic $MoSi_2$, 20% volume fraction of SiC whisker reinforced $MoSi_2$, monolithic 50:50 $MoSi_2$ and WSi_2 alloy, and 20% volume fraction of SiC reinforced $MoSi_2$ - WSi_2 alloy. These facilitated an evaluation of the effect of alloying, reinforcement and their combined effect on the creep behavior of $MoSi_2$.

Extensive details of materials and methods are presented in the recent papers by the authors [4-5]. In summary, the materials tested were obtained from Los Alamos National Laboratory in the form of disks 30 mm in diameter with either 6.5 or 13 mm thickness made by hot pressing in the temperature range of 1600 - 1900^0C. Figure 1 shows the microstructure of both monolithic and SiC reinforced composites. The grain size of monolithic $MoSi_2$ increased from 10 to 35 μm with an increase in hot press temperature. The average grain size in the composite is around 18 μm although there were occasional large grains, 35 μm. While the aspect ratio of the whiskers was larger before processing (50), it reduced to an average of 10 with many of the particles being equiaxed.

Specimens of 12.7X4.8X4.8 mm were wire EDM cut from the original disks. These were tested under compression under constant load in the temperature range of 1100 - 1450^0C, and the displacement of the loading train was measured outside the furnace using a LVDT.

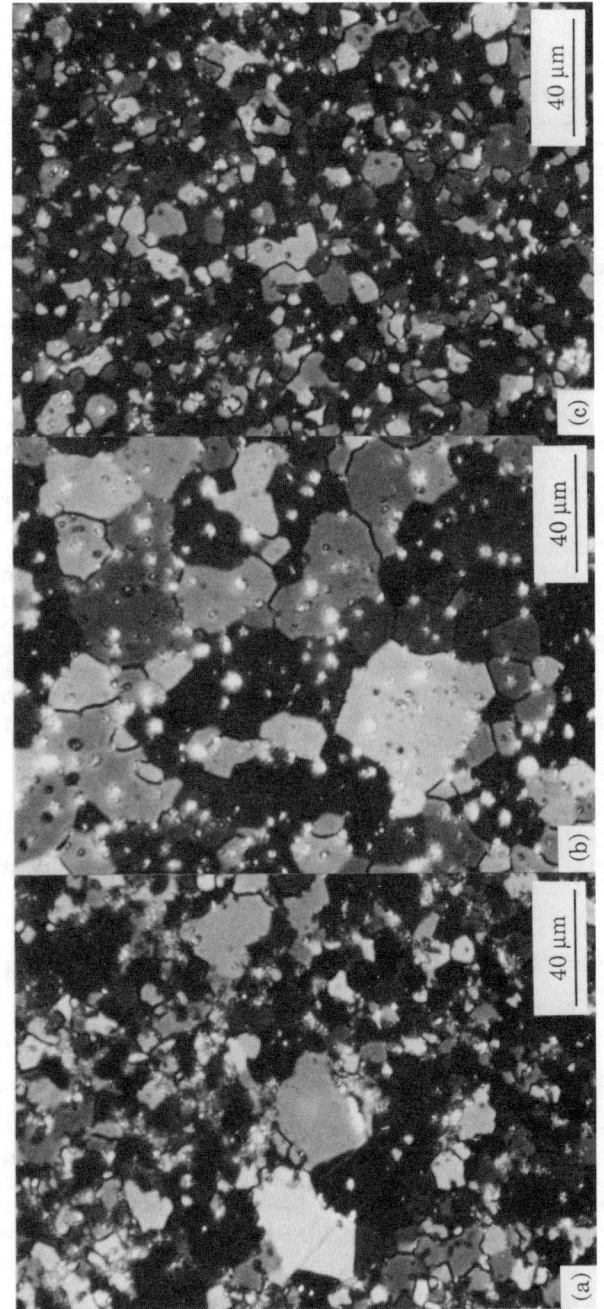

Figure 1. Microstructure of monolithic and composite $MoSi_2$. a) $MoSi_2$ + 20% Vol. Frac. SiC hot pressed at 1820°C with 18μm grain size, b) $MoSi_2$ hot pressed at 1820°C with 35 μm grain size and c) $MoSi_2$ hot pressed at 1620°C with 10 μm grain size.

3. RESULTS AND DISCUSSION

Figure 2 shows minimum or steady state creep rate for monolithic $MoSi_2$ (10 μm) as a function of applied stress at three temperatures. The stress exponents are about 2 and decrease approximately to 1 at higher temperatures and stresses. The results for $MoSi_2$ with SiC reinforcement are shown in Fig. 3. Except for the 1100^0C test, the strain rate - stress curves are parallel with the exponent equal to 3.3. At 1100^0C the curve is steeper with the stress exponent increasing to 5.2. On the average, the stress exponent varies from 3 - 5, which are characteristic values for power-law creep in metallic systems.

Figure 4. shows the creep rates in $MoSi_2$ - WSi_2 alloy as a function of stress and temperature. The stress exponent is constant for temperatures below 1400^0C and is of the order of 2.4. At 1400^0C, it increases to 3.6. Thus stress exponents are higher in the alloy compared to the $MoSi_2$, understandably due to increased resistance to dislocation motion.

Finally, Fig. 5 illustrates the preliminary results for the $MoSi_2$ - WSi_2 with 20% SiC reinforcement at 1100^0C. In comparison to monolithic alloy (Fig. 2), the stress exponent is slightly higher in the composite. Thus in both $MoSi_2$ and its alloy, the presence of reinforcement increases the stress exponents.

Using the data in Figs. 2 to 4, apparent activation energies for creep deformation in $MoSi_2$, its composites and for the alloy can be determined and are shown in Fig. 6. The activation energies for $MoSi_2$, $MoSi_2$ with 20%SiC and $MoSi_2$ + WSi_2 alloy are 433, 596 and 536 kJ/mole, respectively. The values are nearly independent of stress in the range investigated. There is an increase in the apparent activation energy with alloying or the addition of reinforcement, where the latter had a larger effect. Alloying with higher melting silicide such as WSi_2 is expected to increase the activation energy. Reinforcement need not increase the activation energy if the reinforcement deforms only elastically and the mechanism remains the same. Even under such cases, large activation energies [6-8] have been reported in systems such as dispersion strengthened alloys, and have been attributed to the development of temperature dependent back stresses which obstruct the creep process [9]. Their source and their role in thermal activation process therefore need to be evaluated to understand the creep deformation of composites.

3.1 Role of Back Stresses

Before discussing the relevance of the stress exponents and the activation energies to identify the micromechanisms of creep, it becomes important to subtract the back stresses which may be stress and temperature dependent. It has been shown that in many dispersion and precipitation strengthened alloys the experimentally determined stress exponents as well as the activation energies are significantly high [9]. Back stresses or internal stresses are thought to be due to formation of dislocation pileups or due to formation of dislocation networks or cell walls. In such cases, the effective stress, σ_e, contributing to creep is less than the applied stress, σ_a, and is given by

$$\sigma_e = \sigma_a - \sigma_0 \qquad (1)$$

where σ_0 is the back stress. In many of the particulate reinforced composites, the back stress is found to be significant and contributes to a large increase in the stress exponent, n, and the

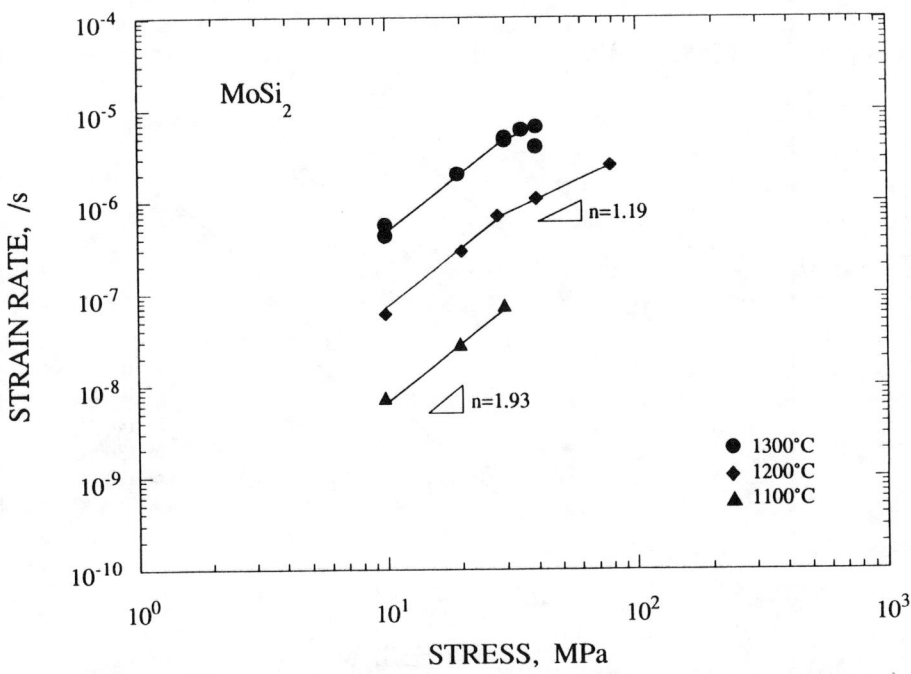

Figure 2. Steady state creep rates as a function of applied stress and temperature for MoSi$_2$

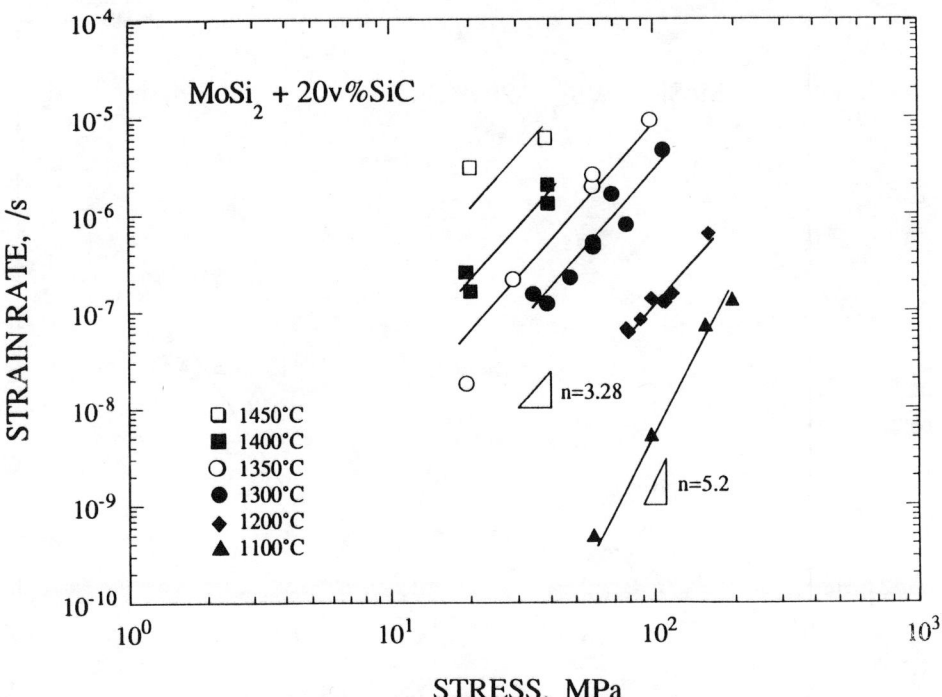

Figure 3. Steady state creep rates a function of applied stress and temperature for MoSi$_2$ with SiC

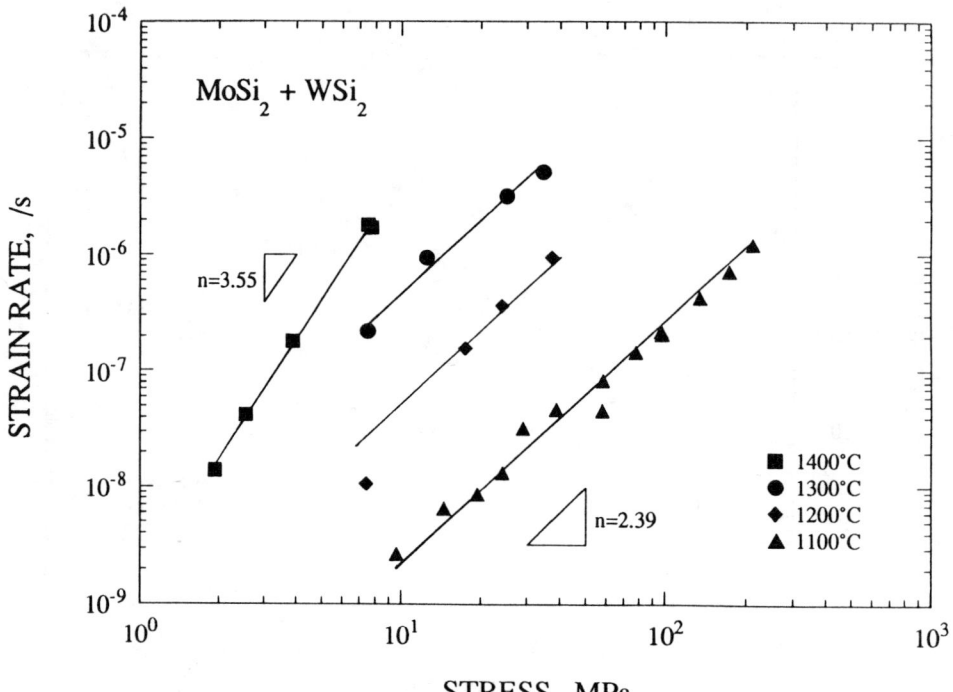

Figure 4 Steady state creep rates as a function of applied stress and temperature for MoSi$_2$ - WSi$_2$ alloy.

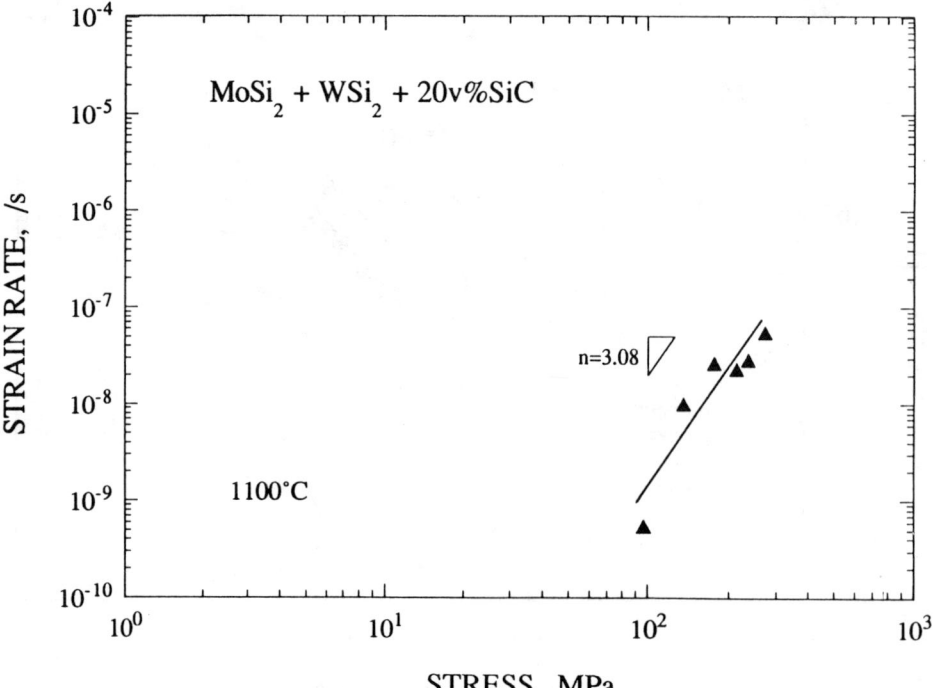

Figure 5. Steady state creep rates as a function of applied for MoSi$_2$ - WSi$_2$ alloy with SiC.

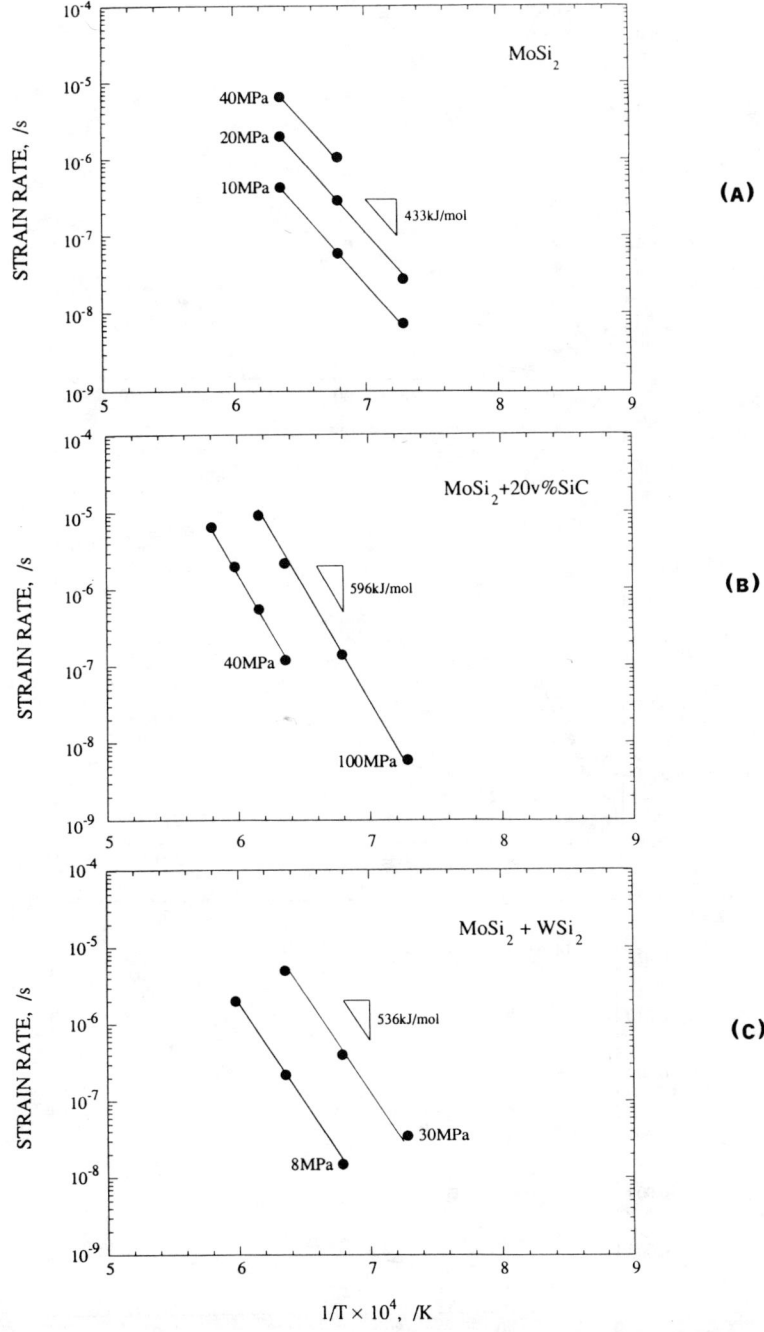

Figure 6. Arrhenius plots for the determination of apparent activation energy for creep for a) $MoSi_2$, b) $MoSi_2$ with 20 Vol% SiC, and c) $MoSi_2$ + WSi_2 alloy.

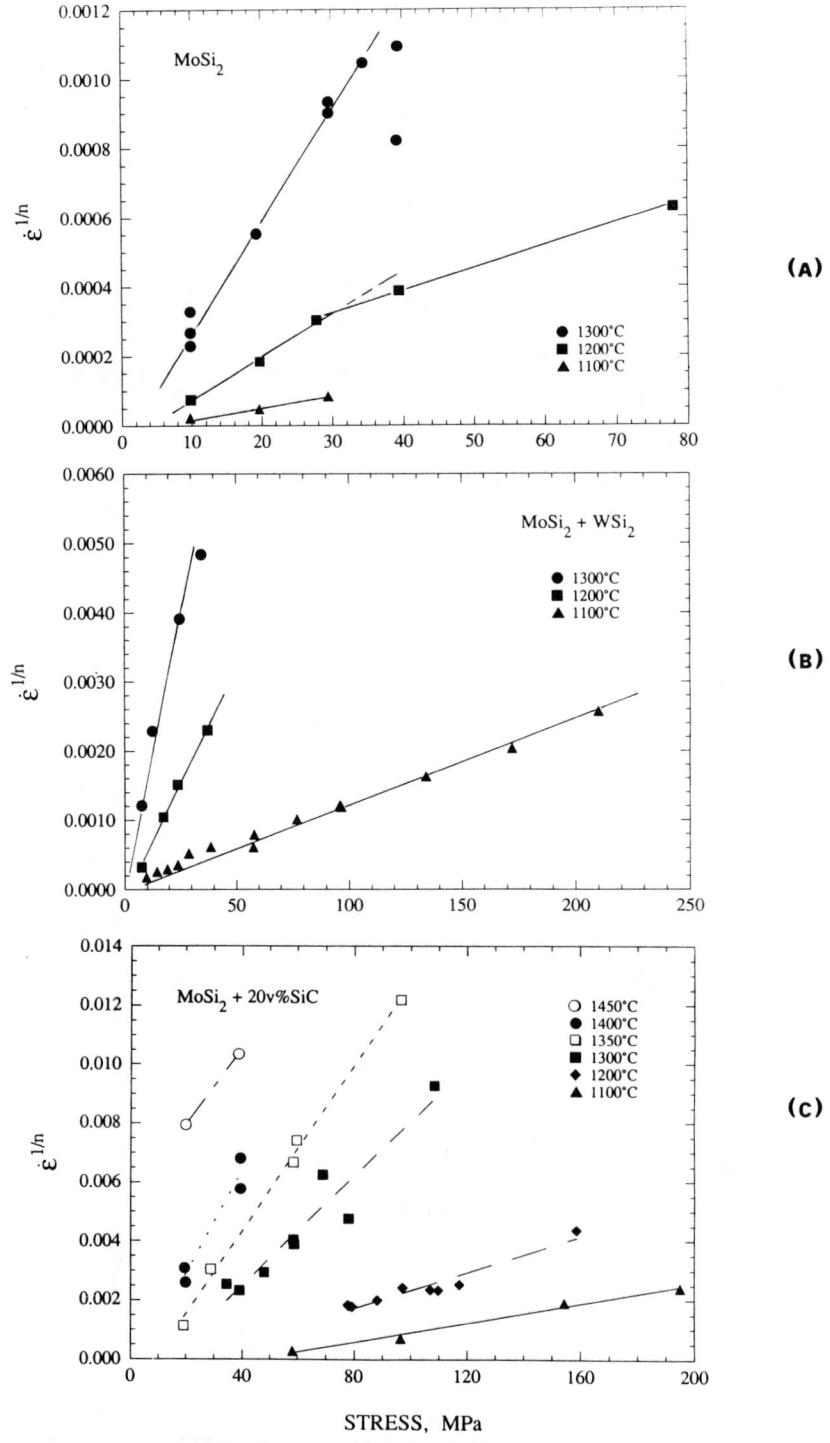

Figure 7. Determination of back stresses for a) $MoSi_2$ b) $MoSi_2$ and WSi_2 alloy and c) $MoSi_2$ with SiC reinforcement.

activation energy for creep. The back stresses can be determined in several ways [10]; we use here a simplified approach by plotting $\dot{\epsilon}^{1/n}$ versus stress on a linear scale and determine the stress at which the strain rate goes to zero. These plots are shown in Fig. 7 a, b, and c respectively for $MoSi_2$, $MoSi_2$ + WSi_2 alloy, and SiC whisker reinforced $MoSi_2$. The back stresses as a function of temperature are determined, and these are tabulated in Table I. In Fig. 7a, the 1200°C

TABLE I MAGNITUDE OF THE BACK STRESSES FOR THE THREE MATERIALS AT DIFFERENT TEMPERATURES

TEMPERATURE °C	σ_0 (MPa)		
	$MoSi_2$	$MoSi_2$ + 20v%SiC	MoSi2 + WSi_2
1100	6.0	45.0	4.0
1200	4.0	22.0	2.0
1300	2.0	12.0	1.0
1350		9.0	
1400		4.0	0.1

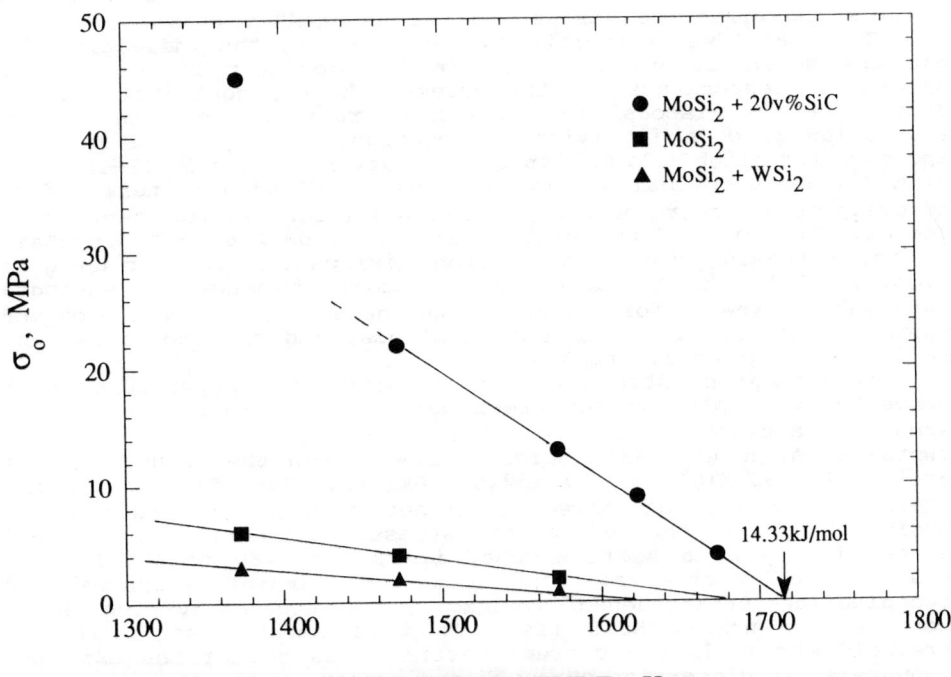

Figure 8. Variation of back stress in $MoSi_2$, SiC reinforced $MoSi_2$, and $MoSi_2$ - WSi_2 alloy

data for $MoSi_2$ fall on two lines; one for low stresses and the other for high stresses. The change in the slope and the absence of any threshold stress for the high stress regime is consistent with a change in the micromechanism of deformation. Figure 8 shows back stress as a function of temperature for the three materials. The back stresses are higher for the composite than those for the monolithic systems and increase more rapidly with the decrease in the temperature. The back stresses are sometimes referred to as threshold stresses since in some cases (oxide dispersion strengthened alloys) these are the minimum stresses required for creep. Interestingly, extrapolation of the data to high temperatures shows that the back stresses approach zero at 1625, 1670 and 1718K for the three materials. The thermal energy available at these temperatures is given by kT where k is the Boltzmann's constant. Thus, in terms of thermal energies, the above temperatures correspond to 13.6, 13.9 and 14.33 kJ/mole, respectively. These results imply that at these temperatures, the thermal energy is sufficient to overcome the back stresses. Reexamination of creep data in Fig. 4 for the $MoSi_2$ - WSi_2 alloy in view of the absence of back stresses at 1400^0C, confirms that the creep behavior at 1400^0C is indeed different from that at lower temperatures. Absence of any back stress or threshold stress at 1400^0C contributes to different stress exponent. Similarly the data at 1450^0C for the $MoSi_2$ composite, Fig. 3, although limited, may correspond to a different stress exponent than that shown in the figure.

Consideration of the effective stress by subtracting the back stresses can change the stress exponents as shown in Fig.9 a, b and c for the three materials. Note that for the monolithic $MoSi_2$, the change in the slope observed at 1200^0C in Fig. 2 is now smoothed out in terms the effective stress and the exponent varies from 1.2 to 1.5. Thus, at lower temperatures and stresses, the indications are that the Newtonian viscous flow is the dominant process. With increase in temperature and the stress, the exponent increases. In addition to the viscous flow, the creep rates are governed by the deformation process involving dislocation glide plus climb. The exponents for glimb (glide plus climb) vary from 3 to 5 [11-12]. For the $MoSi_2$ - SiC composite, Fig. 9b, the 1100^0C data in terms of the effective stresses are more in line with the data at high temperature (compare Fig. 2), and the slope increases from 2.6 to 3.5 probably due to increasing contribution from dislocation climb plus glide process. At 1450^0C, the data are too limited to make any meaningful statement. Likewise for the alloy, the data, Fig. 9c, fit much more smoothly in terms of the effective stress, and the exponents again are in the range of 1.9 to 3.5.

Since the back stress is more sensitive to temperature in the composite, we replotted the creep data for the $MoSi_2$ composite in terms of the effective stress. This is shown in Fig. 10. All the constants, A, n and Q are slightly lower than the values obtained earlier (A=1.93 X10^4, n=2.63 and Q=460kJ/mol, Ref. 5) when the back stresses were not considered. Current results show that by the consideration of the effective stress, the activation energy decreased only by a small amount, i.e., from 460 to 420 kJ/mole. This can be contrasted with the behavior of dispersion or precipitation strengthened systems where the activation energy decreases by nearly half [10]. In that sense, back stress or threshold stress in the current materials is not of the same type encountered in dispersion strengthened system, where the size of the dispersoids are much smaller than the dislocation mean free path. In the present case, size of the reinforcement and the spacing between the reinforcement are comparable to the mean dislocation length; therefore, conventional Orowan loop mechanism does not operate.

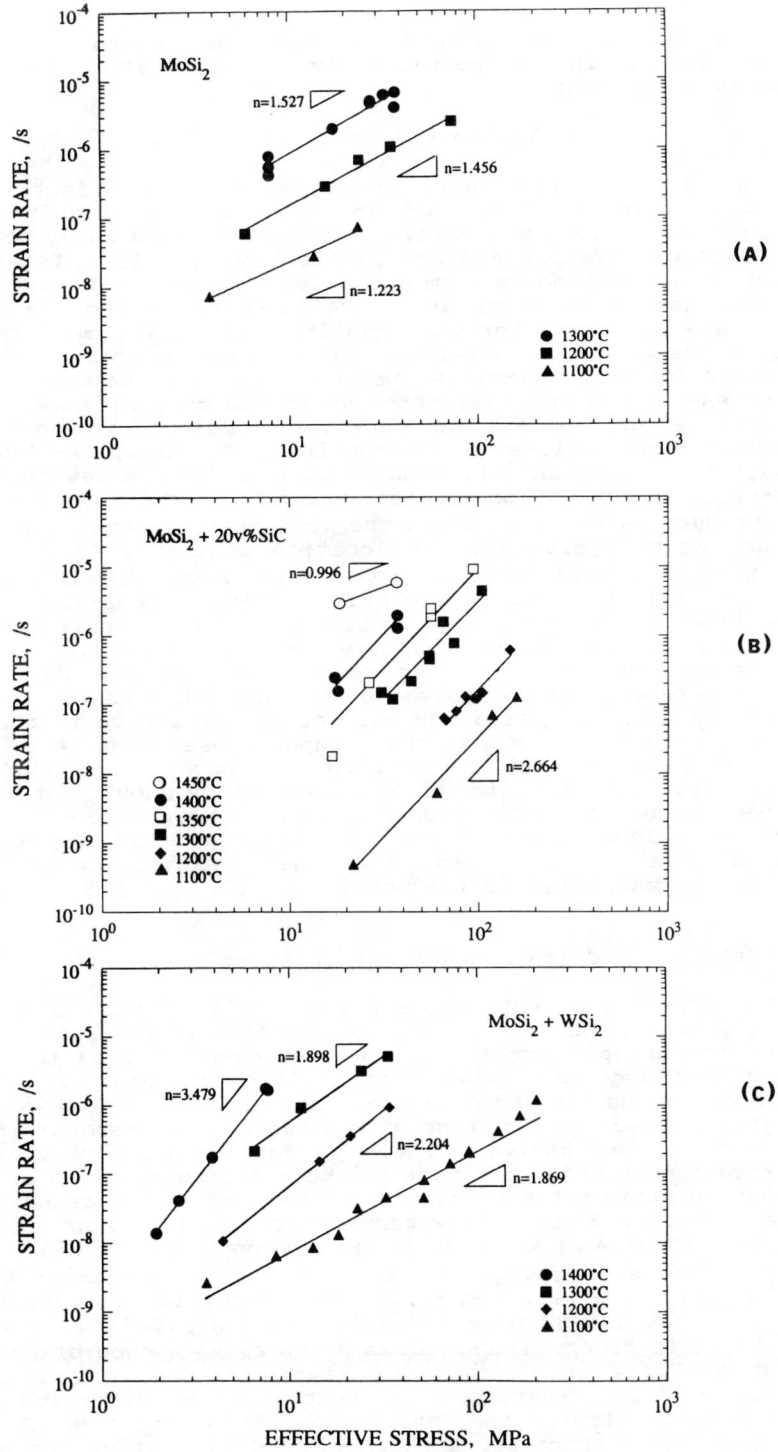

Figure 9. Strain rates in terms of effective stress for
a) $MoSi_2$, b) $MoSi_2$ + SiC and c) $MoSi_2$ - WSi_2 alloy.

Finally, for completeness, we have determined the activation volumes involved in the thermal activation processes. The activation volume is defined as

$$V^* = kT \{d\ln \epsilon/d\sigma\}_T, \qquad (2)$$

where σ can be either applied stress or the effective stress. Activation volumes normalized in terms of the magnitude of the Burgers vector ([100] dislocation) were determined using both applied and effective stresses and are plotted in Fig. 11. Interestingly, the activation volumes are significantly larger in terms of applied stresses than in terms of effective stresses. Also the activation volumes are different for the monolithic and the composite. But in terms of the effective stresses, the activation volumes fall on the same curve for both materials, implying that (a) the magnitude of the back stress or the internal stresses in the composite are mostly due to reinforcements and (b) these activation volumes are mostly related to mobile dislocations that contribute to creep in the matrix. Normally for a dislocation climb as well as for the Newtonian viscous processes, the activation volumes should be of the order of 1 b^3, where b is the Burgers Vector. Since the activation volumes are 100 to 700 b^3, they must involve long dislocation segments. This may be an indication that more than one process is involved. This could mean that in addition to dislocation climb, thermally activated dislocation glide involving double kink formation either in the lattice or in the grain boundary may be involved. There were significant reductions in the activation volumes when effective stresses were considered instead of the applied stresses. This means that the calculated activation volumes are sensitive to the presence of any secondary process that superimpose over the thermally activated process that is dominant. Because of such extreme sensitivity, identification of micromechanisms should not be based on the magnitude of activation volume. Nevertheless, identical activation volumes for both the monolithic and the composite on the basis of effective stresses imply that deformation processes are essentially restricted to the matrix.

3.2 Dislocation Climb and Activation Energies

The stress exponents approaching 3 with increasing temperature generally implies that the dislocation climb is the rate controlling process for creep. For the cases involving 'n' less than 3, it is assumed that they corresponds to transition regions; transition from Newtonian viscous flow to power-law creep involving dislocation climb with glide; a weighted average of the two process determining the net creep rate. The activation energy for dislocation climb should therefore corresponds to either the bulk diffusion of Mo or Si or the dislocation pipe diffusion observed sometimes at low temperature and generally referred to as "low temperature creep". Activation energy data for diffusion of Mo in $MoSi_2$ are not available in the literature. The self diffusion energy for Mo in Mo is of the order of 420-430 kJ/mole [13]. The reported activation energy for diffusion of Si in $MoSi_2$ is about 250 kJ/mole [14,15]. When a [100] dislocation that contributes to creep [5] undergoes climb along a cube plane, it moves through one layer of Mo and two layers of Si, in sequence. The dislocation climb, therefore, is controlled by diffusion couple in series and the slowest one, namely the diffusion of Mo controls the kinetics. The current crude estimation of the activation energy for Mo in $MoSi_2$ (350 - 540 kJ/mole) is in the range of experimentally determined activation energies for creep, Figs. 6 and 10. We, therefore, conclude that the kinetics of Mo, diffusion control the

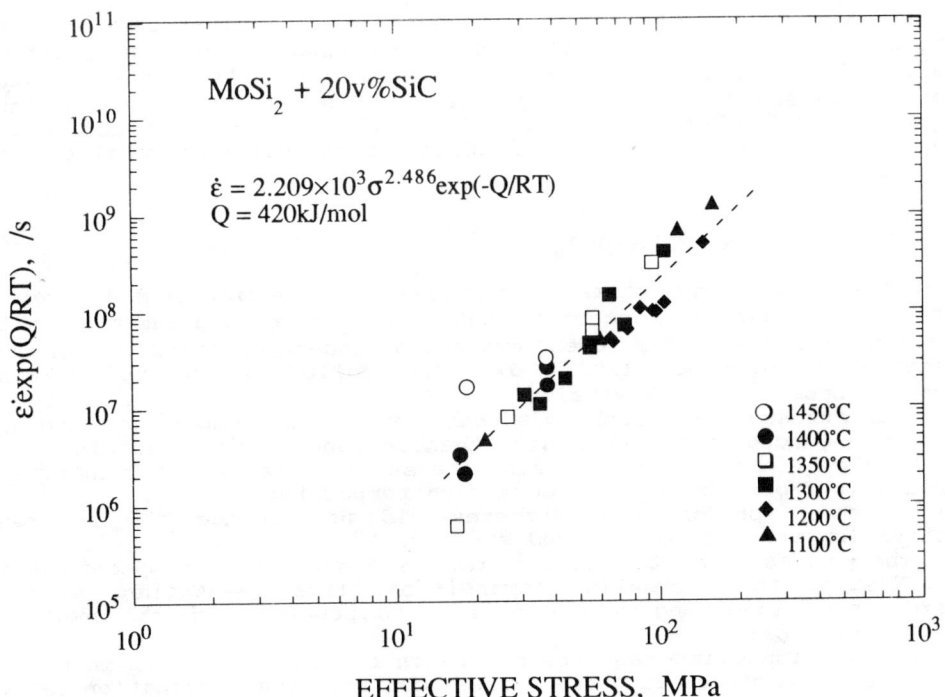

Figure 10. Constitutive equation for MoSi$_2$ with SiC in terms of effective stress.

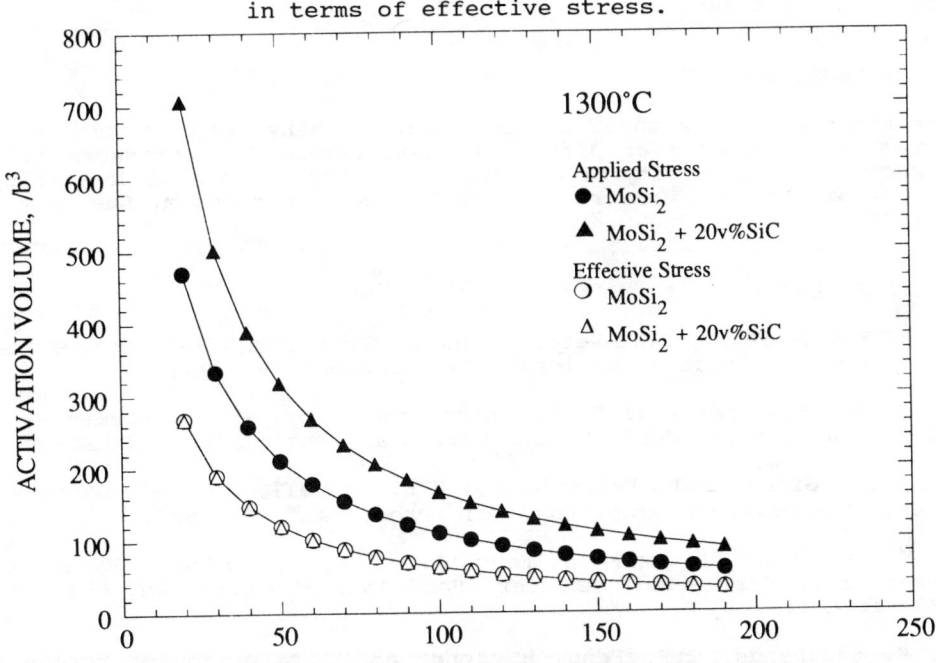

Figure 11. Activation volume as a function of both the applied stress and the effective stress

kinetics of creep in MoSi$_2$. The increase in the activation energy for the alloy, MoSi$_2$ - WSi$_2$ is to be expected since the melting point of WSi$_2$ is higher than that of MoSi$_2$. Likewise a small increase in the activation energy for the composite could arise from contributions from internal stresses. But in all these cases the activation energies are within the range expected for diffusion of metal ions in the silicides.

4. SUMMARY AND CONCLUSIONS

The role of back stress in the creep deformation of MoSi$_2$, MoSi$_2$ with SiC whisker reinforcement, MoSi$_2$ + WSi$_2$ alloy and the alloy with SiC reinforcement has been evaluated under compression in the temperature range of 1100 - 1450^{0}C. Following conclusions were reached based on the analysis:
a) Back stresses or threshold stresses vary with temperature and are more pronounced in the composite than in monolithic materials.
b) On the effective stress basis, the strain rate exponent increases gradually from 1 to 3 with increasing temperature.
c) Substraction of back stresses did not change the apparent activation energies determined earlier.
d) The increase in the strain rate exponent with temperature is attributed to increasing transition from Newtonian viscous deformation involving self diffusion to power-law creep involving dislocation climb.
e) Dislocation climb requires diffusion of both Mo and Si in series and diffusion of Mo being the slowest controls the deformation rates.
f) Estimation of activation energy for the diffusion of Mo in MoSi$_2$ agrees reasonably with the experimentally determined activation energies for creep.

5. ACKNOWLEDGEMENTS

The authors express their appreciation for many helpful comments to Dr. A. K. Vasudevan of Office of Naval Research, Professors A. K. Mukherjee of University of California, R. Raj of Cornell University, W. Nix of Stanford University, and Drs. N. Louat and M. Duesbery of Fairfax Materials Group.

6. REFERENCES

1. A. K. Vasudevan, J.J. Petrovic and K. Sadananda, "High Temperature Structural Silicide composites", RISO conference, 1991

2. A. K. Vasudevan and J. J. Petrovic, "A Comparative Overview of Molybdenum Disilicides", Mater. Sci. & and Engrg. 1992, In Press.

3. W. S. Gibbs, J.J. Petrovic and R.E. Honnell, "SiC Whisker-MoSi$_2$ Matrix Composite", Ceram. Eng. Sci. Proc., **8**, 645 (1987).

4. K. Sadananda, H. Jones, J. Feng, J.J. Petrovic and A.K. Vasudevan, Ceram. Eng. Sci. Proc., Am. Cer. Soc. Inc., Westerville, Ohio, **12**, 1671-1678 (1991)

5. K. Sadananda, C.R. Feng, H. Jones and J.J. Petrovic, "Creep of Molydisilicide Composites", Mat. Sci. Engrg., 1992, In Press.

6. B. A. Wilcox, and A.H. Clauer, Met. Sci. J., **3** pp 26-33 (1969).

7. W. Oliver and W. D. Nix, Acta Met., **27** pp. 1335-1347 (1982).

8. J. D. Whittenberger, Met. Trans. A., **15A** pp. 1753-1762 (1984).

9. S. Purushothaman and J. K. Tien, Acta Met., **26** pp. 519-528 (1978).

10. Sai V. Raj, "Creep amd Fracture of Dispersion Strengthened Materials", NASA CR 185299, NASA Lewis Research Center, Cleveland, (1991).

11. J. Weertman, Trans. AIME **233,** 2069 (1965)

12. J. Weertman, in Proc. Second INt. Conf. Creep and Fracture in Engineering Materials and Structures, Eds. B. Wilshire and R.R.J. Owen,, Pineridge Press, Swansea, p. 1, (1984).

13. J. Askill, TRACER DIFFUSION DATA FOR METALS, John Wiley and Sons, New York, NY 1966.

14. R.W. Bartlett, P. R. Gage and P. A. Larssen, Trans. AIME, **230** 1528 (1964).

15. P. Kofstad, HIGH TEMPERATURE OXIDATION OF METALS, John Wiley and Sons Inc., New York, NY (1966).

ON THE NOTCH-SENSITIVITY AND TOUGHNESS OF A CERAMIC COMPOSITE

Keith T. Kedward
Department of Mechanical Engineering
University of California
Santa Barbara, California

Peter W. R. Beaumont
Department of Engineering
University of Cambridge
Cambridge, United Kingdom

ABSTRACT

A Nicalon TM/SiC fabric (8-harness satin) reinforced alumina matrix (CMC) was loaded in tension to fracture. A coating on the surface of the fibre reduced the strength of the fibre-matrix bond, increased the fibre pull-out length and toughness of the CMC by more than 3 times, and raised the unnotched strength by more than 2 times.

1. INTRODUCTION

The development of ceramic-matrix composites (CMC) for general engineering usage must address the inherent notch-sensitive characteristics for which most unreinforced ceramic materials are notorious. For most polymeric matrix composites (PMC) the subject of notch sensitivity is quite well understood and characterized (1). With the more rigid matrix of CMC's the general design approaches must not be too strongly influenced by PMC experience and the possibility of closer attention to the microstructural characteristics, and particularly the fibre/matrix interface, is required. One example of a potential CMC-specific design criterion is the importance placed on limiting the tensile stress or strain to a level below which matrix cracking will be precluded, particularly when the ambient service environment degrades the reinforcing fibres.

2. MATERIALS

For a preliminary experimental evaluation specimens of a Nicalon TM/SiC fabric (8-harness satin) reinforced alumina matrix CMC was obtained. The CMC was produced by an aluminium oxide-based slurry process wherein the ceramic fibre fabric is drawn through a matrix bath to form preimpregnated layers. These layers are stacked and heated under pressure to establish a well-bonded laminate before firing to the finished CMC condition. The final CMC comprises typically 45% fibre volume fraction with 15-20% void fraction (porosity). In some samples, the Nicalon fibres were coated with boron nitride (BN). Specimens were cut from four plates of 1.5 mm nominal thickness using a conventional water-jet system.

Specimen configurations of unnotched and notched type illustrated in fig. 1 included a standard "dogbone" and two double edge notched (DEN) geometries. For comparative purposes, a tensile specimen containing a central hole was tested.

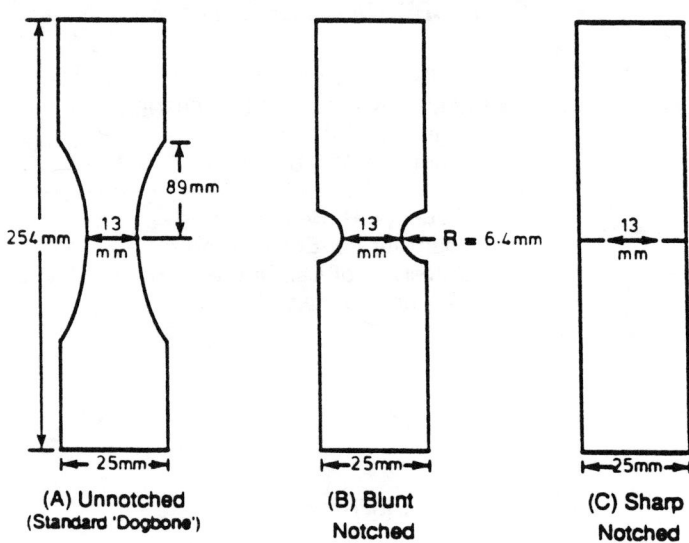

Fig. 1 Specimen Configurations

In the experiments tensile loading was introduced using adhesively bonded aluminium end tabs and an MTS servohydraulic universal testing machine or an Instron servomechanical machine. Displacement control was adopted in both cases using an unclamped gauge length of 150 mm and an actuator velocity of 0.05 mm/min. Strain/displacement measurement was recorded by centrally mounted 10 mm clip gauges. Three replicates were used for both the notched and unnotched specimen configurations.

3. RESULTS

Results of the CMC testing are presented in Table 1 and Figure 2 and, despite the limited data, clearly indicate the superior performance of the coated fibre system relative to the uncoated system. The coating essentially provides a reduced fibre/matrix bond strength shown by longer fibre pull-out lengths (fig. 3). A similar ratio of unnotched to notched strength for the semicircular notch configuration of both strengths are much higher for the coated fibre condition. Furthermore, the sharp slit notched DEN specimen results indicate no significant effect of notch acuity for the coated fibre condition (see Table 1). The data point on the fig. 2 representing a ratio of flaw size-to-specimen width of 0.2 was for the configuration of a centrally located open hole. Of interest here is the location of the data points relative to the notch-insensitive and ideally notch-sensitive curves. Observe that the uncoated CMC specimens with the slit DEN configuration exhibit classical notch-sensitivity.

	σ_o, MPa Unnotched Configuration (A)	σ_n / σ_o Blunt Notched R = 6.4mm Configuration (B)	σ_n / σ_o Sharp Notched R ≡ 0.15mm Configuration (C)
CMC / Uncoated Filaments	61	0.35	0.24
CMC / Coated Filaments	143	0.34	0.33

Table 1 Results of exploratory investigation of the notch sensitivity of bi-directionally reinforced CMC.

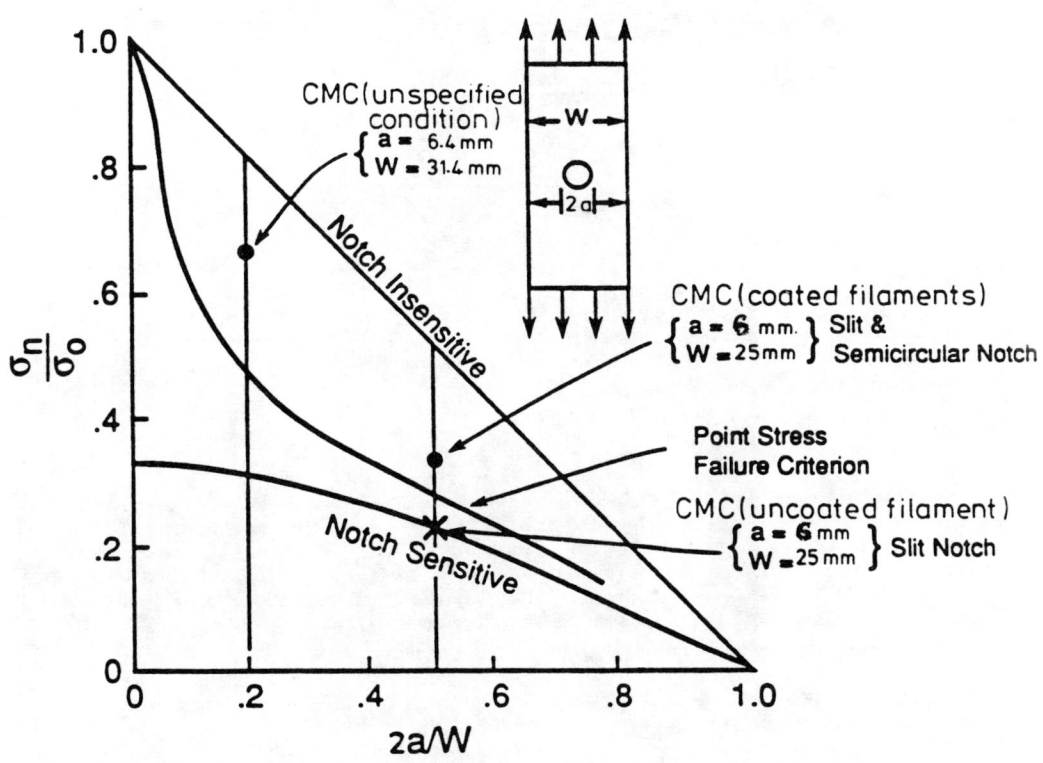

Figure 2   Results for preliminary notch sensitivity assessment of CMC.

All specimens tested displayed substantial nonlinearity in stress/strain (for the unnotched condition) and stress/displacement (for the notched condition) but post-failure examination by scanning electron microscopy revealed surface roughness with extensive fibre pullout for the coated specimens (fig. 3). Failure surfaces for the uncoated samples were considerabley smoother (fig. 3).

4. DAMAGE MECHANICS AND FRACTURE TOUGHNESS

A further point of interest concerns the notch tensile strength which, for the DEN configuration with a slit (sharp) notch of length a, is given by the equation (assuming isotropy):

$$\sigma_n = \frac{K_C}{Y\sqrt{a}} \qquad (1)$$

where the finite specimen width correction factor $Y_{DEN} = 1.98 + 0.36 \left(\frac{2a}{W}\right) - 2.12 \left(\frac{2a}{W}\right)^2 + 3.42 \left(\frac{2a}{W}\right)^3$ and K_C = Fracture Toughness.

For the center cracked plate (CCP), $Y_{CCP} = 1.77 \left[1 - 0.1 \left(\frac{2a}{W}\right) + \left(\frac{2a}{W}\right)^2 \right]$

By substituting the nominal dimensions of the DEN specimens, i.e. $\frac{2a}{W} = 0.5$, we obtain;

$$Y_{DEN} = 2.06 \quad \text{and} \quad Y_{CCP} = 2.12$$

Finally, by adopting the simple notch tip damage zone concept of macroscopic fracture mechanics introduced by Waddoups, Eisenmann, and Kaminski (2) and utilizing the preliminary notched, σ_n, and unnotched strength, σ_o, data from Table 1:

$$\sigma_o = \frac{K_C}{\sqrt{\pi a_o}} \quad , \quad \sigma_n = \frac{K_C}{Y_{DEN} \sqrt{a+a_o}} \qquad (2)$$

where a_o represents the dimension related to the size of the notch tip damage zone which for an unnotched sample can be thought of as equivalent to the "inherent" flaw size. Next, we obtain

$$\frac{\sigma_n}{\sigma_o} = \frac{\sqrt{\pi a_o}}{Y_{DEN} \sqrt{a+a_o}} = 0.86 \sqrt{\frac{a_o}{a+a_o}} \qquad (3)$$

Hence, a value of $a_o = 1.17$ mm is implied.

Using equation (2) a fracture toughness for the coated CMC material is estimated as:

$$\boxed{K_C = 8.66 \text{ MPa } \sqrt{m} \quad \text{COATED CMC}}$$

Following a similar procedure the fracture toughness for the uncoated CMC material is estimated as:

$$\boxed{K_C = 2.50 \text{ MPa } \sqrt{m} \quad \text{UNCOATED CMC}}$$

with a value of $a_o = 0.54$ mm.

Clearly, further research is required to substantiate the above and extend the correlations for a range of notch sizes and configurations. More extensive microscopy is required to establish reasons for the shortfall in strength of the "uncoated" fibre laminate which is suspected to be due to fibre damage incurred during processing. Thus, an opportunity exits here for material enhancements through improved processing techniques and to further investigate fibre/matrix interface phenomena.

Some of our exploratory fatigue testing on notched specimens with coated fibres at cyclic loads corresponding to 50% and 80% of static notched strength suggested that residual strengths after 10^6 cycles were negligibly different from the initial static strength. More extensive research is required before any general and definitive conclusions can be drawn on this subject however.

5. REFERENCES

(1) M.T. Kortschot and P.W.R. Beaumont, "Damage Mechanics of Composites", Parts 1-4, Composites Science and Technology, 39 pp. 289 (1990) and 40, pp. 147 (1991).

(2) M.E. Waddoups, J.R. Eisenmann and B.E. Kaminski, "Macroscopic Fracture Mechanics of Advanced Composite Materials", J. Comp. Materials, Vol 5, pp. 446-454 (1971).

THE INFLUENCE OF PROCESSING ON THE HIGH TEMPERATURE MECHANICAL PROPERTIES OF A WHISKER-REINFORCED ALUMINA COMPOSITE

Kenong Xia
Comalco Research Centre
Thomastown, Victoria, Australia

John R. Porter
Rockwell International Science Ctr.
Thousand Oaks, California

Terence G. Langdon
Departments of Materials Science and Mechanical Engineering
University of Southern California
Los Angeles, California

ABSTRACT

It is well known that the creep behavior of alumina at elevated temperatures may be markedly improved by incorporating SiC whiskers into the matrix to form a composite material. Recent experiments demonstrate that the creep properties of SiC whisker-reinforced alumina depend critically on the processing procedure used to fabricate the composite. In this paper, typical creep date for a conventionally processed commercial composite are described and the results are compared with data obtained under similar experimental conditions for a composite produced using dispersion processing.

INTRODUCTION

Many experiments have been conducted to investigate the creep behavior of unreinforced polycrystalline alumina: Cannon and Langdon (1983) summarized much of this work in tabular form. In general, alumina exhibits a stress exponent, n, close to 3 at high stress levels with a transition to $n = 1$ at low stresses (Cannon and Langdon, 1988), and this behavior is attributed to a transition from some form of dislocation creep at high stress levels to diffusion creep at the lower stresses. Chokshi and Langdon (1991) discussed the precise significance of a stress exponent of $n = 3$ in ceramic materials and also the methods for predicting the transitions in creep behavior between different stress exponents.

The creep resistance of polycrystalline alumina may be markedly improved by incorporating SiC into the matrix. Experimental results are available showing this improvement using SiC in the form of either whiskers (Chokshi and Porter, 1985; de Arellano-López et al., 1990; Lin and Becher, 1990, 1991; Lipetzky et al., 1988, 1991; Nutt et al., 1990) or particles (Niihara et al., 1986). This trend is consistent with the experimental observation that the incorporation of SiC whiskers into an alumina matrix leads to a three or four-fold increase in the fracture toughness by comparison with monolithic alumina (Becher and Wei, 1984).

It is well known from the models for the creep of fibrous composites that the creep properties of whisker-reinforced composites may depend on the average length of the whiskers (Lilholt, 1988). Nevertheless, this is an area which has received very little attention. Yasuda et al. (1991)

investigated the effect of whisker shape and size on the fracture toughness and four-point bending strength of SiC whisker-reinforced Al_2O_3 at room temperature, and this paper reports observations on the effect of processing and whisker morphology on the creep properties at high temperatures.

EXPERIMENTAL MATERIALS AND PROCEDURES

The experiments were conducted using two different sources of SiC whisker-reinforced polycrystalline alumina.

First, a series of creep tests was undertaken using a conventional alumina composite produced commercially by the Advanced Composite Materials Corporation. This material, commercially designated SA25, was prepared by hot pressing high purity Al_2O_3 with three different quantities of SiC whiskers: 9.3, 18 and 30 volume per cent of whiskers, respectively. The Silar whiskers used in the fabrication are α-SiC single crystals having a high aspect ratio with typical lengths of ~20-60 μm and diameters of ~0.5-1.0 μm. These whiskers are hexagonal in cross-section and heavily faulted on the basal plane. The grain size of each material was of the order of 1-2 μm.

Careful microscopic inspection prior to testing revealed differences in the distributions of the SiC whiskers in the three sets of commercial composites: these differences are described in more detail elsewhere (Xia and Langdon, 1988a, 1989). Briefly, the SiC whiskers become densely packed into agglomerates as the density of whiskers is increased, with the agglomerates separated by channels, about 2-10 μm wide, which are relatively free of whiskers. In addition, the agglomerates are elongated perpendicular to the hot pressing direction, as demonstrated in Fig. 1 for the composite containing 18 vol % SiC for (a) a section perpendicular to the hot pressing direction and (b) a section parallel to the direction of hot pressing and with the hot pressing direction lying vertical in the photomicrograph. For this material, the agglomerates have longitudinal dimensions less than ~50 μm. No agglomerates were visible in the material containing only 9.3 vol % SiC whiskers, although in all materials the whiskers were located preferentially in the plane perpendicular to the hot pressing direction.

Second, experiments were conducted on a dispersion processed (DP) alumina composite containing 15 vol % SiC whiskers in high purity Al_2O_3. This material was fabricated using a colloidal method described elsewhere (Porter et al., 1987; Porter, 1988) and the grain size was ~2 μm. The composite was produced by hot pressing using identical SiC whiskers as for the commercial SA25 material.

All specimens were creep tested in air under conditions of constant load using a simple four-point bending apparatus. A description of the testing assemby is given elsewhere (Xia and Langdon, 1989). Each specimen was ~26 mm long and with a cross-section of ~2.0 × 3.0 mm. The specimen was placed between four pivot rods, made of high purity sapphire, with the pivot points having separations of 6.4 and 19.0 mm above and below the specimen, respectively. The load was applied to each specimen in the direction of hot pressing using an alumina loading ram, and the loading assemby was contained within a vertical split furnace with the temperature controlled during each test to within ±2 K of the required testing temperature of 1673 K. Each specimen was held at the testing temperature for at least 30 minutes prior to application of the load. The displacement of the upper loading ram was continuously recorded and the standard procedure of Hollenberg et al. (1971) was used to calculate the outer fiber stresses and strains.

Following testing, some foils were prepared for transmission electron microscopy from the areas of tensile loading.

EXPERIMENTAL RESULTS AND DISCUSSION

It is convenient to describe the results on the conventionally-processed SA25 composite and then to compare these data with results from the dispersion processed samples. Some creep results were published earlier for the SA25 composite (Xia and Langdon, 1987, 1988a, 1988b, 1989) and for the dispersion processed composite (Porter et al., 1987, 1990).

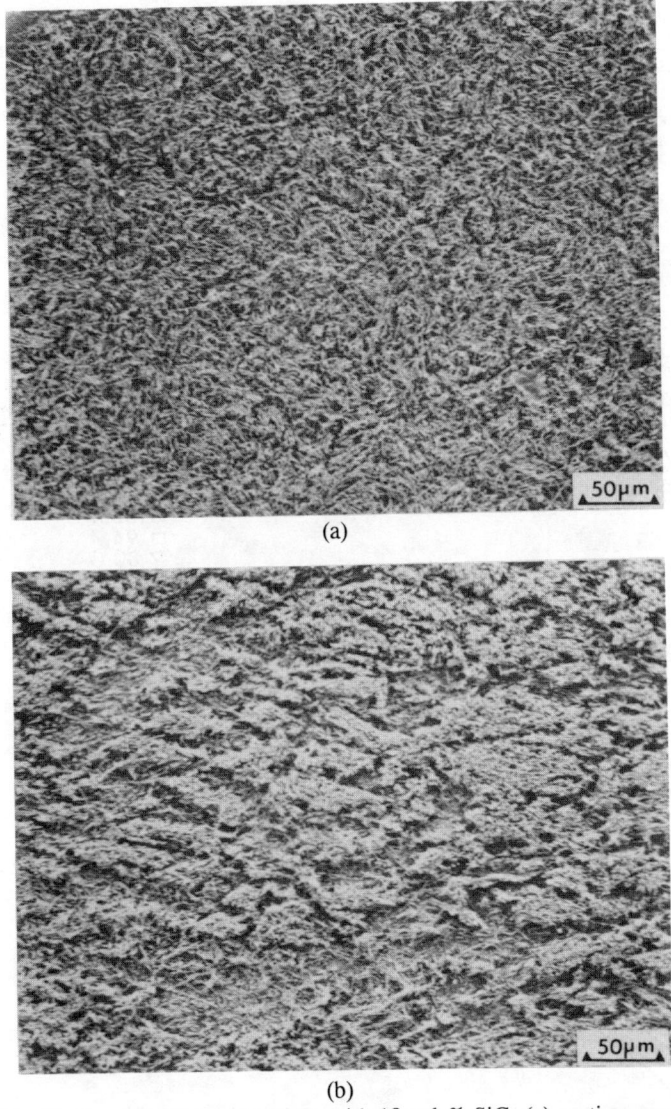

Figure 1   Distribution of whiskers in Al_2O_3 with 18 vol % SiC: (a) section perpendicular to hot pressing direction, (b) section with hot pressing direction lying vertical.

Figure 2 shows creep curves of strain, ε, versus time, t, for a series of samples containing 9.3 vol % of SiC whiskers tested under different levels of applied stress, σ, at an absolute temperature, T, of 1673 K: similar creep curves are given in Figs 3 and 4 for the composites containing 18 and 30 vol % of SiC whiskers, respectively. Most specimens were tested to failure and the points of failure are denoted by small crosses at the ends of the creep curves. Some of the slower tests, at the lower stress levels, were not taken to failure but, nevertheless, an attempt was made to continue testing until a minimum or essentially steady-state creep rate was reasonably established. The various creep curves in Figs 2-4 confirm the normal three stages of creep, with an initial primary region where the creep rate decreases, a region where the creep rate is reasonably constant, and a tertiary region where the creep rate accelerates to failure. In general, the tertiary stages were rather brief and failure occurred abruptly at low total strains.

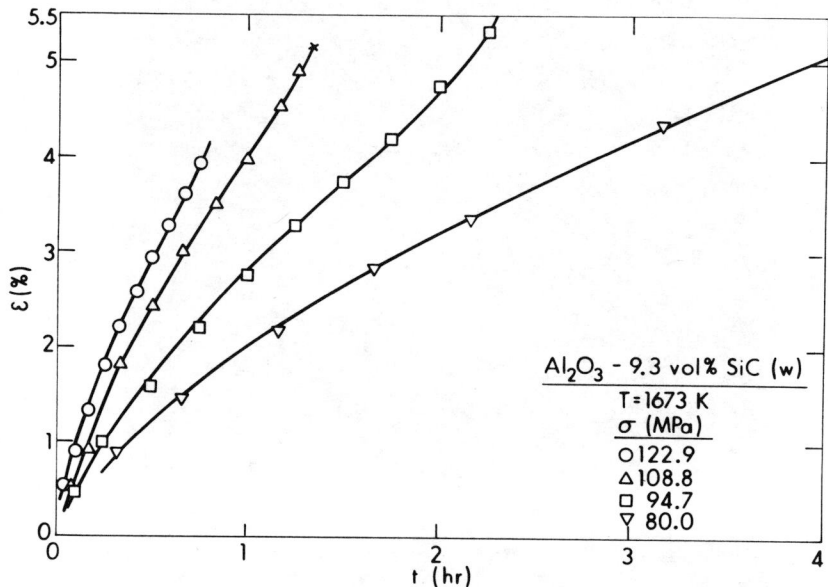

Figure 2　　Strain versus time for Al$_2$O$_3$ with 9.3 vol % SiC whiskers.

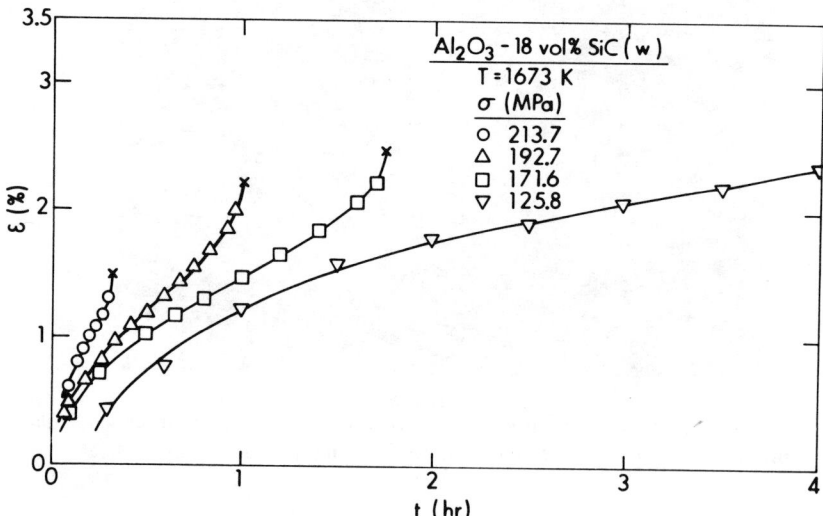

Figure 3　　Strain versus time for Al$_2$O$_3$ with 18 vol % SiC whiskers.

A comparison of Figs 2-4 shows that the overall strains to failure tend to decrease as the volume percentage of SiC whiskers increases, and this is especially marked for the composite containing 9.3 vol % SiC whiskers where failure strains of >5% were achieved.

Figures 5-7 show plots of the instantaneous creep rate, $\dot{\varepsilon}$, versus the strain, ε, using the creep curves from Figs 2-4 for the three different composites. Inspection shows that the primary region is not well defined in the material containing 9.3 vol % SiC but the total strains to failure in this material are significantly larger and even exceed 8% at the lowest applied stress of 40.0 MPa. The

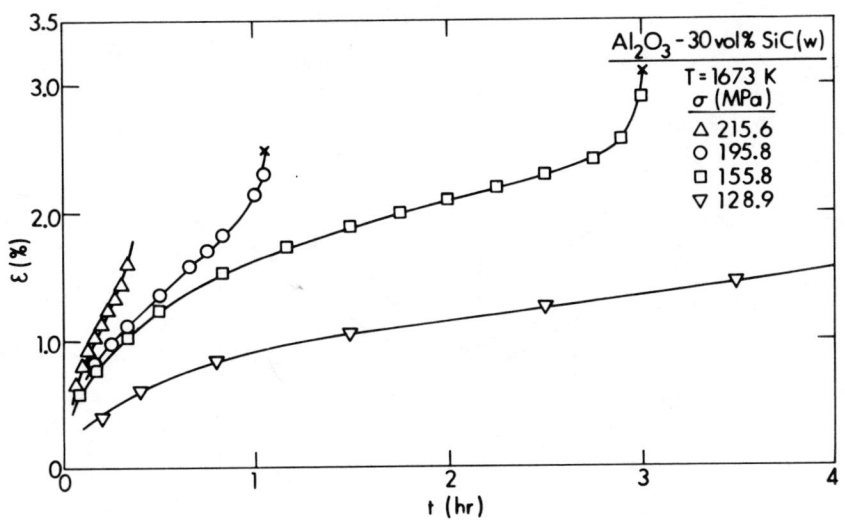

Figure 4   Strain versus time for Al_2O_3 with 30 vol % SiC whiskers.

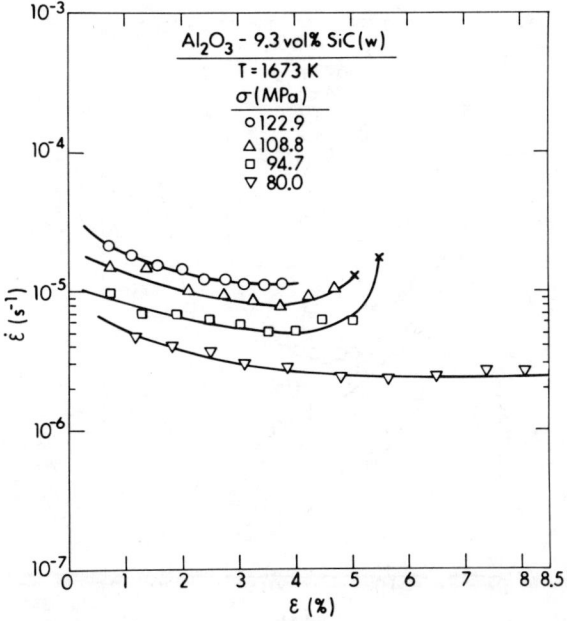

Figure 5   Instantaneous strain rate versus strain for Al_2O_3 with 9.3 vol % SiC whiskers.

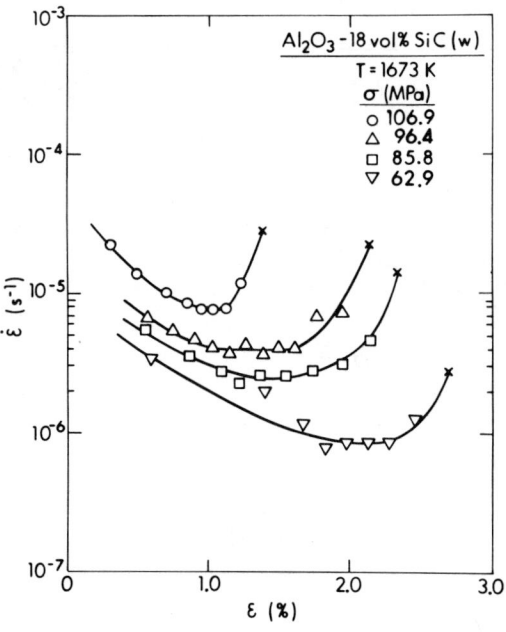

Figure 6   Instantaneous strain rate versus strain for Al_2O_3 with 18 vol % SiC whiskers.

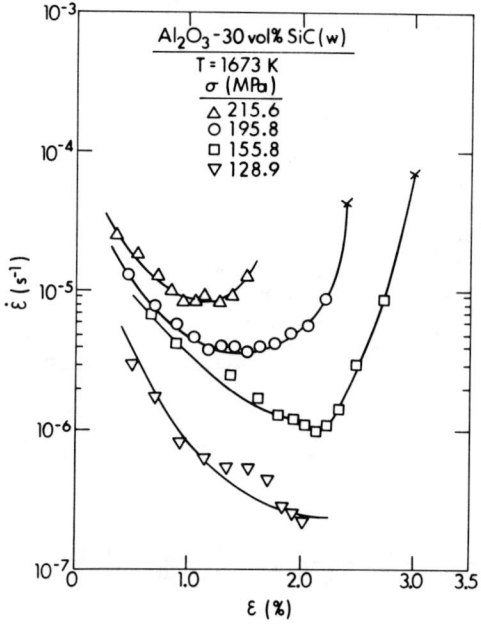

Figure 7   Instantaneous strain rate versus strain for Al_2O_3 with 30 vol % SiC whiskers.

steady-state or minimum creep rate is reasonably well defined for this material. For the composites containing 18 and 30 vol % SiC whiskers, the primary region is extensive, the period of minimum creep is extremely brief and the advent of the tertiary stage leads to a very rapid acceleration and failure.

The dispersion processed (DP) alumina was tested at the same temperature (1673 K) with 15 vol % of SiC whiskers. Figure 8 shows a typical creep curve for this material tested with an outer fiber stress of 81.0 MPa. This creep curve is unusual because it exhibits a primary region extending over more than 100 hours and then an essentially quasi-steady-state region where the creep rate is reasonably constant. Tertiary creep was never attained in the dispersion processed alumina even when the testing was continued for >500 hours. This is in marked contrast to the commercial material where, as shown in Figs 2-4, tertiary creep began after only a few hours of testing.

Figure 9 gives a typical variation of instantaneous strain rate, $\dot{\varepsilon}$, with strain, ε, for the dispersion processed material using the creep curve from Fig. 8. This plot is representative of all specimens tested at 1673 K and it demonstrates clearly that the total creep deformation consists of an extended primary stage so that a true steady-state condition is never attained. In addition, the very low creep rate ($< 10^{-8}$ s^{-1} for much of the test) leads to outer fiber strains which are extremely small.

It is possible to make a direct comparison between the dispersion processed and the conventional composite by logarithmically plotting the nominal steady-state creep rate against the applied stress as shown in Fig. 10. Each datum point refers to a test conducted on a different specimen at a constant absolute temperature of 1673 K, with the upper points (circles) taken from published data for monolithic hot pressed alumina produced commercially and with a grain size of 1.6 μm (Chokshi and Porter, 1986), the central points (squares) from this research program for the commercial material containing 18 vol % SiC whiskers (as documented in Figs 3 and 6) and the lower points (triangles) also from this research program for the dispersion processed (DP) alumina containing 15 vol % SiC whiskers.

Two conclusions arise from inspection of Fig. 10.

First, the creep rates in the alumina composites are significantly slower than in the unreinforced alumina. The decrease is slightly over one order of magnitude for the conventionally processed commercial composite but it is more than three orders of magnitude for the dispersion processed composite. Thus, dispersion processing of the alumina composite enhances the creep resistance and leads to a material where, for normal levels of the applied stress, the creep rates are remarkably low. Furthermore, well-defined steady-state creep rates were never attained in the dispersion processed samples so that the datum points in Fig. 6 may overestimate the true minimum creep rates at the lower stress levels.

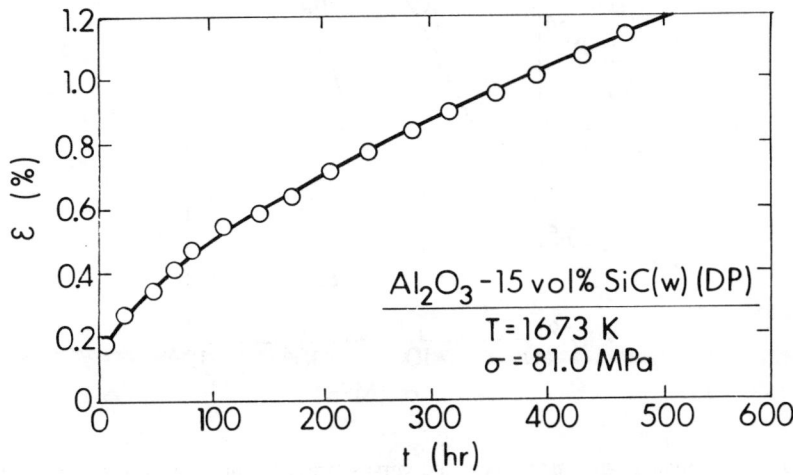

Figure 8   Strain versus time for the dispersion processed (DP) alumina.

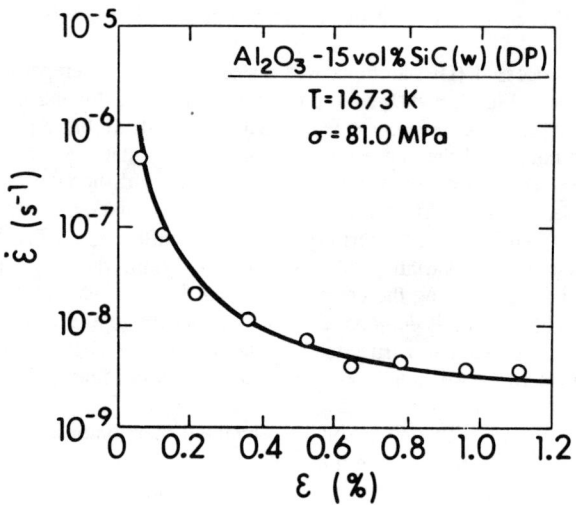

Figure 9 Instantaneous strain rate versus strain for the dispersion processed (DP) alumina.

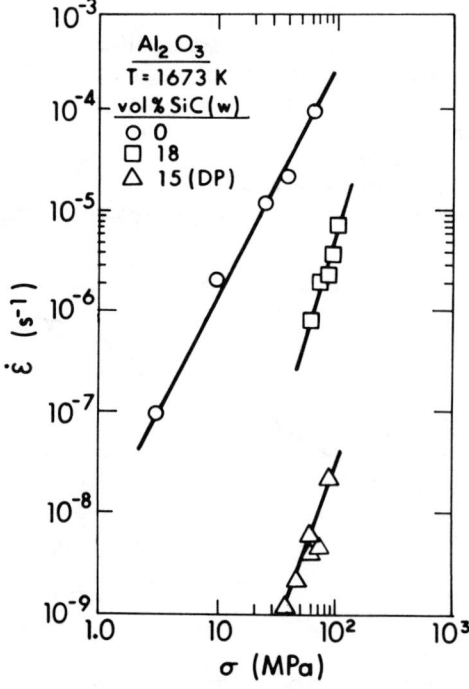

Figure 10 Nominal steady-state creep rate versus stress for unreinforced alumina and the two alumina composites.

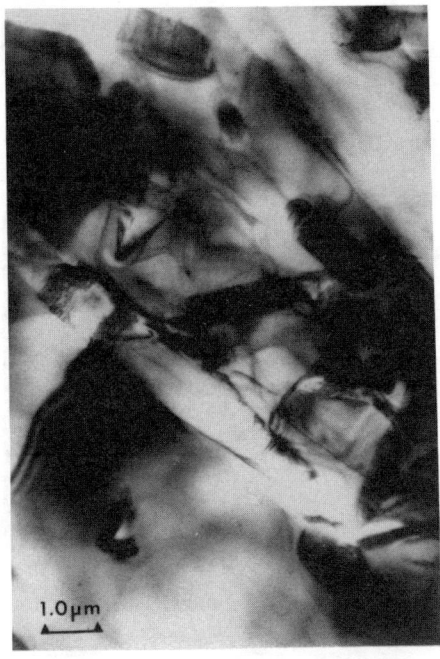 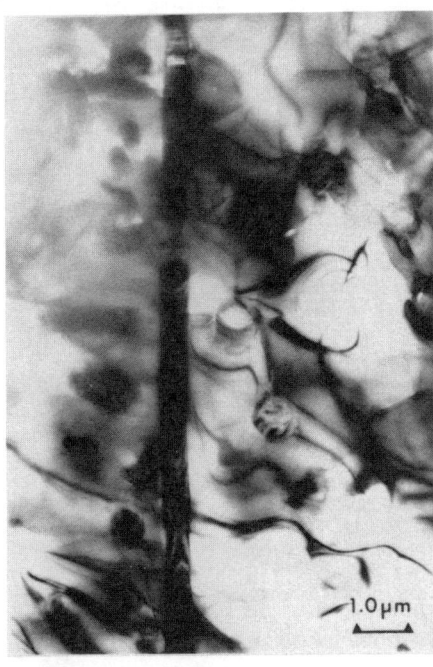

(a) (b)

Figure 11 Transmission electron micrographs of tensile portions of dispersion processed (DP) composite after deformation to a strain of 1.2% at 1673 K under a stress of 67 MPa: (a) tapered whisker terminations imply that the whiskers exit through the surface of the foil, (b) in-plane whiskers maintain a high aspect ratio.

Second, the datum points for the composite materials lie along lines of steeper slope than for the monolithic alumina. In unreinforced Al_2O_3, the stress exponent n ($= \partial \ln \dot{\epsilon}/\partial \ln \sigma$) is consistently very close to 3 in the dislocation creep region (Cannon and Langdon, 1988) but the commercial composite in Fig. 10 with 18 vol % SiC whiskers exhibits a stress exponent of $n \simeq 4$.

It is important to note that the small difference in the volume percentage of the SiC whiskers for the two composites shown in Fig. 10 should have only a very minor effect on the creep behavior and it cannot account for the marked difference in the overall creep rates. In fact, there appears to be a tendency for slightly higher creep resistance, and therefore lower creep rates, in alumina composites containing higher percentages of SiC whiskers (Porter et al., 1987; Porter and Chokshi, 1987) and this is opposite to the apparent trend in Fig. 10.

Transmission electron micrographs of the dispersion processed (DP) and commercial composites are shown in Figs 11 and 12, respectively. These photomicrographs were taken in the tensile regions of the specimens after termination of the test or, for the conventionally processed material, after failure. Figure 11 is for a dispersion processed sample tested at 1673 K to a strain of 1.2% and with an outer fiber stress of 67 MPa, and Fig. 12 is for a conventionally processed sample tested at 1673 K to a failure strain of 3.2% and with an outer fiber stress of 75 MPa.

In general, careful inspection revealed essentially similar microstructures in both the dispersion processed and commercial specimens. Very little cavitation was visible in either material after testing to failure over a range of stresses at a temperature of 1673 K, although the cavities appeared to be slightly larger in the commercial specimens. The dislocation configurations were also very similar in both types of material and there was no obvious difference in either the arrangements of dislocations within the grains or in the overall dislocation densities. Both composites were also

 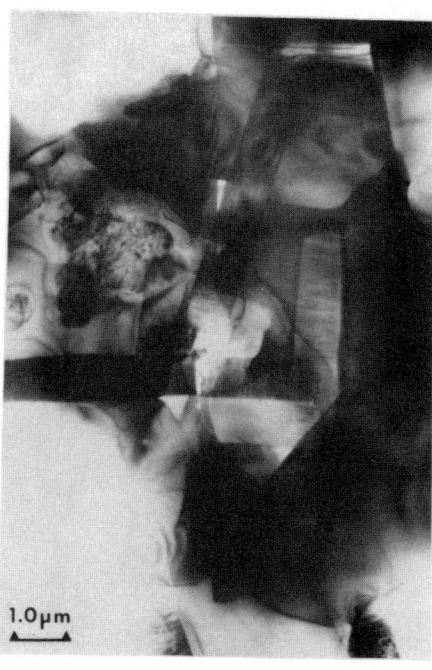

(a)   (b)

Figure 12   Transmission electron micrographs of tensile portions of conventionally processed composite after deformation to a strain of 3.2% at 1673 K under a stress of 75 MPa: (a) whisker terminations lying perpendicular to whisker axes in center of photomicrograph imply the termination of whiskers within the foil, (b) additional whisker terminations lying perpendicular to the fiber axes.

essentially free of any glassy phase at the grain boundaries. Therefore, the nature of the deformed microstructures within the crystalline matrix provides no clear explanation for the marked difference in creep behavior, as documented in Fig. 10.

Although the matrix microstructures were similar in both materials, there were significant differences associated with the size and distribution of the SiC whiskers. Dispersion processing leads to a relatively homogeneous distribution of whiskers whereas, as shown in Fig. 1(b), the conventionally processed material contains large agglomerates of whiskers separated by channels which are reasonably whisker-free. This difference in whisker distribution was confirmed using transmission electron microscopy but, in addition, there was evidence also that, despite using identical Silar whiskers from the same source for both types of composite, there was a difference in the average aspect ratios of the whiskers in the two types of material. As indicated in Fig. 11, the whiskers in the dispersion processed material tend to either terminate with a taper, indicative of an exit from the top or bottom surfaces of the foil [Fig. 11(a)], or they lie in the plane of the foil and show that the aspect ratio is very large [Fig. 11(b)]. By contrast, the conventionally processed composite has more whisker ends [Fig. 12 (a) and (b)] and these ends are generally perpendicular to the whisker axes so that the whiskers terminate within the foil.

The microstructural evidence implies that the average whisker aspect ratio plays an important, and perhaps dominant, role in determining the precise creep behavior of the composite. Indeed, the large differences in the experimental creep rates recorded in Fig. 10 demonstrate that the aspect ratio is significantly more important than the precise volume percentage of SiC whiskers. For the conventionally processed material, the average whisker aspect ratio is reduced during processing and the composite is then more creep resistant than monolithic alumina but significantly less creep resistant than a similar composite fabricated by dispersion processing where the whisker aspect ratio remains very large.

The present results are not complete. On the one hand, they demonstrate clearly the importance of processing in influencing the high temperature mechanical properties of whisker-reinforced ceramic composites; but on the other hand, the results are insufficient either to give the form of the dependence of creep rate on whisker aspect ratio or to permit the development of a detailed model to explain the precise nature of the differences in creep resistance. Experiments are currently in progress to provide a more detailed description of these factors in influencing the creep behavior of SiC whisker-reinforced alumina.

CONCLUSIONS

1. Creep tests were conducted at a temperature of 1673 K on two types of SiC whisker-reinforced alumina: a conventionally processed commercial alumina containing either 9.3, 18 or 30 vol % SiC whiskers and a dispersion processed alumina containing 15 vol % SiC whiskers.

2. The distributions of the SiC whiskers in the two types of composite were different. In the conventionally processed material, there was a tendency, with increasing SiC content, for the whiskers to become densely packed into agglomerates separated by essentially whisker-free channels. In the dispersion processed material, the distribution of SiC whiskers was relatively homogeneous.

3. The creep curves in the conventionally processed material exhibited the normal three stages with a tertiary region leading to abrupt failure, whereas the creep curves for the dispersion processed material exhibited an extended quasi-steady-state behavior and there was no tertiary stage.

4. The nominal steady-state creep rates for the conventionally processed composite containing 18 vol % SiC whiskers and the dispersion processed composite were markedly different. The results show that the conventionally processed and the dispersion processed composites exhibit creep rates which are slightly more than one order of magnitude and more than three orders of magnitude slower than for unreinforced alumina, respectively.

5. The differences in creep behavior between the two composites are attributed to a reduction during processing in the average whisker aspect ratio in the conventionally processed material.

ACKNOWLEDGMENT

This work was supported by the U.S. Army Research Office under Grant No. DAAL03-91-G-0230.

REFERENCES

Becher, P.F., and Wei, G.C., 1984, "Toughening Behavior in SiC Whisker-Reinforced Alumina," *Journal of the American Ceramic Society*, Vol. 67, pp. C267-C269.

Cannon, W.R., and Langdon, T.G., 1983, "Creep of Ceramics: Part 1 - Mechanical Characteristics," *Journal of Materials Science*, Vol. 18, pp. 1-50.

Cannon, W.R., and Langdon, T.G., 1988, "Creep of Ceramics: Part 2 - An Examination of Flow Mechanisms," *Journal of Materials Science*, Vol. 23, pp. 1-20.

Chokshi, A.H., and Langdon, T.G., 1991, "Characteristics of Creep Deformation in Ceramics," *Materials Science and Technology*, Vol. 7, pp. 577-584.

Chokshi, A.H., and Porter, J.R., 1985, "Creep Deformation of Alumina Matrix Composite Reinforced with SiC Whiskers," *Journal of the American Ceramic Society*, Vol. 68, pp. C144-C145.

Chokshi, A.H., and Porter, J.R., 1986, "High Temperature Mechanical Properties of Single Phase Alumina," *Journal of Materials Science*, Vol. 21, pp. 705-710.

de Arellano-López, A.R., Cumbrera, F.L., Domínguez-Rodríguez, A., Goretta, K.C., and Routbort, J.L., 1990, "Compressive Creep of SiC-Whisker-Reinforced Al_2O_3," *Journal of the American Ceramic Society*, Vol. 73, pp. 1297-1300.

Hollenberg, G.W., Terwilliger, G.R., and Gordon, R.S., 1971, "Calculations of Stresses and Strains in Four-Point Bending Creep Tests," *Journal of the American Ceramic Society*, Vol. 54, pp. 196-199.

Lilholt, H., 1988, "Models for Creep of Fibrous Composite Materials," *Materials Forum*, Vol. 11, pp. 133-139.

Lin, H.-T., and Becher, P.F., 1990, "Creep Behavior of a SiC Whisker Reinforced Alumina," *Journal of the American Ceramic Society*, Vol. 73, pp. 1378-1381.

Lin, H.-T., and Becher, P.F., 1991, "High-Temperature Creep Deformation of alumina-SiC-Whisker Composites," *Journal of the American Ceramic Society*, Vol. 74, pp. 1886-1893.

Lipetzky, P., Nutt, S.R., and Becher, P.F., 1988, "Creep Behavior of an Al_2O_3-SiC Composite," *Materials Research Society Symposium Proceedings*, F.D. Lemkey et al., ed., Materials Research Society, Pittsburgh, Pennsylvania, Vol. 120, pp. 271-277.

Lipetzky, P., Nutt, S.R., Koester, D.A., and Davis, R.F., 1991, "Atmospheric Effects on Compressive Creep of SiC-Whisker-Reinforced Alumina," *Journal of the American Ceramic Society*, Vol. 74, pp. 1240-1247.

Niihara, K., Nakahira, A., Uchiyama, T., and Hirai, T., 1986, "High Temperature Mechanical Properties of Al_2O_3 Composites," *Fracture Mechanics of Ceramics*, R.C. Bradt et al., ed., Plenum Press, New York, N.Y., Vol. 7, pp. 103-116.

Nutt, S.R., Lipetzky, P., and Becher, P.F., 1990, "Creep Deformation of Alumina-SiC Composites," *Materials Science and Engineering*, Vol. A126, pp. 165-172.

Porter, J.R., 1988, "Dispersion Processing of Creep Resistant Whisker-Reinforced Ceramic-Matrix Composites," *Materials Science and Engineering*, Vol. A107, pp. 127-132.

Porter, J.R., and Chokshi, A.K., 1987, "Creep Performance of Silicon Carbide Whisker-Reinforced Alumina," *Ceramic Microstructures '86*, J.A. Pask and A.G. Evans, ed., Plenum Press, New York, N.Y., pp. 919-928.

Porter, J.R., Lange, F.F., and Chokshi, A.H., 1987, "Processing and Creep Performance of SiC-Whisker-Reinforced Al_2O_3," *Ceramic Bulletin*, Vol. 66, pp. 343-347.

Porter, J.R., Xia, K., and Langdon, T.G., 1990, "Microstructural Aspects of Creep in SiC Whisker-Reinforced Al_2O_3," *Metal and Ceramic Matrix Composites: Processing, Modeling and Mechanical Behavior*, R.B. Bhagat et al., ed., The Minerals, Metals and Materials Society, Warrendale, Pennsylvania, pp. 381-389.

Xia, K., and Langdon, T.G., 1987, "An Investigation of the Creep Properties of an Al_2O_3 Composite Containing 25% SiC Whiskers," *Proceedings of the IX Inter-American Conference on Materials Technology*, Facultad de Ciencias, Físicas y Matemáticas, Universidad de Chile, Santiago, Chile, pp. 259-263.

Xia, K., and Langdon, T.G., 1988a, "Creep and Fracture in an Alumina Composite," *Microstructural Science*, H.J. Cialone et al., ed., ASM International, Metals Park, Ohio, Vol. 16, pp. 85-96.

Xia, K., and Langdon, T.G., 1988b, "The Mechanical Properties at High Temperatures of SiC Whisker-Reinforced Alumina," *Materials Research Society Symposium Proceedings*, F.D. Lemkey et al., ed., Materials Research Society, Pittsburgh, Pennsylvania, Vol. 120, pp. 265-270.

Xia, K., and Langdon, T.G., 1989, "High Temperature Creep of Alumina Composites Containing SiC Whiskers," *Proceedings of the MRS International Meeting on Advanced Materials*, M. Doyama et al., ed., Materials Research Society, Pittsburgh, Pennsylvania, Vol. 4, pp. 185-190.

Yasuda, E., Akatsu, T., and Tanabe, Y., 1991, "Influence of Whiskers' Shape and Size on Mechanical Properties of SiC Whisker-Reinforced Al_2O_3," *Journal of the Ceramic Society of Japan*, Vol. 99, pp. 52-58.

AUTHOR INDEX

Processing, Fabrication, and Manufacturing of Composite Materials — 1992

Abhiraman, A. S.	1
Adley, Mark D.	187
Beaumont, Peter W. R.	247
Biner, S. B.	131
Chen, J.	61
Colligan, K.	11
Conway, T. A.	27
Costello, G. A.	27
Desai, P.	1
Dharani, L. R.	209
Earthman, J. C.	221
Evans, W. J.	103
Feng, C. R.	231
Gentry, John	115
Gonzalez-Doncel, G.	201
Gupta, Manoj	149
Hong, Wei	209
Hoyle, Chris	115
Isaac, D. H.	75, 87, 103
Jain, Sanjeev	45
Jasiuk, I.	61
Kedward, Keith T.	247
Kim, H. K.	221
Langdon, Terence G.	253
Lavernia, Enrique J.	149, 177, 221
Liang, X.	221
Mohamed, Farghalli A.	149, 177
Ozbek, S.	75, 87
Park, Kyung-Tae	177
Porter, John R.	253
Qin, Jialing	187
Ramani, Karthik	115
Ramulu, M.	11
Sadananda, K.	231
Saib, K. S.	103
Sherby, O. D.	201
Srivatsan, T. S.	149
Taggart, David G.	187
Thitipoomdeja, S.	87
Thorpe, M. F.	61
Tryfonidis, Michail	115
Wade, B.	1
Wolfenstine, J.	201
Xia, Kenong	253
Yang, Daniel C. H.	45

Book Number: G00733